UNDERSTANDING BIOINSTRUMENTATION

UNDERSTANDING BIOINSTRUMENTATION

By

Dr. M. Prakash

Dept. of Zoology
M.M.H. Post Graduate College
Ghaziabad (U.P.)
(India)

DISCOVERY PUBLISHING HOUSE PVT. LTD.
NEW DELHI-110 002

First Published – 2009

Reprinted – 2026

ISBN: 978-81-8356-479-3

Understanding Bioinstrumentation

Published by:

DISCOVERY PUBLISHING HOUSE
4383/4B, Ansari Road, Darya Ganj
New Delhi-110 002 (India)
Phone: +91-11-23279245; 23253475; 43596065
Mobile: +91 9811179893 / +91 9871656464
E-mail: discoverybooksindia@gmail.com
orderdphbooks@gmail.com
namitwasan9@gmail.com
web: www.discoverypublishinggroup.com

Printed at:
Infinity Imaging Systems
Delhi

Preface

The present title "Understanding Bioinstrumentation" has been written for those students interested in careers in diverse fields of biological sciences. It provides a structured approach to learning by covering all the important topics in a uniform, systematic format. The book has been comprehensively designed incorporating recent advances in this fast moving field. It also provides accessible information on bio-instrumentation in compact form for undergraduate students in biology and related life sciences. It is intelligible to the educated layman, though it deals with some complex ideas. It is an adequate text for all the requirements of students in this area. In addition, busy lecturers who require a quick reference compendium will find it useful, particularly for tutional planning. Simple, yet hopefully clear figures and tables are provided throughout the book.

The over-riding goal of this book, and indeed of the whole *Understanding series*, is to present the essential information concering bioinstrumentation in a compact, readily accessible form which leads itself to student learning and revision. The convergence of various approaches has generated a rich panorama of detail, the significance of which we are still attempting to unraval. The present text has been written as an introduction to this rapidly growing field.

To make the work more comprehensive and informative, the author has consulted many authoritative books, research journals, abstracts, monographs etc., so there can be no claim to originality except in the manner of treatment.

The author expresses his thanks to his friends and colleagues whose continue inspirations have initiated him to bring out this book.

The author expresses his gratitude to Mr. Wasan and staff of M/s Discovery Publishing House Pvt. Ltd. for their whole hearted co-operation in the publication of this book.

In the mean time, the author will remain sincerely responsible for any shortcomings of the book and be grateful to the readers for their suggestions and constructive criticism for the continuous betterment of the book. He takes this opportunity to appeal to the readers to send their suggestions straightaway to his Publisher.

Author

CONTENTS

1

INTRODUCTION

Pharmacogenomics encompasses several major areas: the study of polymorphic variations in drug response and disease susceptibility, identification of the effects of drugs/xenobiotics at the genomic level, and genotype/phenotype associations. The most common type of human genetic variations is *single-nucleotide polymorphisms* (SNPs). Several novel approaches to detection of SNPs are currently available. The range of new methods includes modifications of several conventional techniques, such as PCR, mass spectrometry (ms), and sequencing, as well as more innovative technologies such as *fluorescence resonance energy transfer* (FRET) and microarrays. The application of each of these techniques is largely dependent on the number of SNPs to be screened and sample size. The current chapter presents an overview of the general concepts of a variety of genotyping technologies, with an emphasis on the recently developed methodologies, including a comparison of the advantages, applicability, cost efficiency, and limitations of these methods.

The human genome is made up of approx three billion nucleotides that code for all the macromolecules necessary for human life. The most common types of human genetic variations are single-nucleotide polymorphisms (SNPs), which are defined as DNA sequence variations that occur when a single nucleotide (A, T, C, or G) in the genome sequence is changed. It is estimated that only one in every thousand bases is different, or that the DNA code is approx 99.9% identical between human subjects. SNPs occur in both coding and noncoding regions and may or may not result in altered gene expression or gene products. Even SNPs that do not themselves change protein expression and cause disease may be in close physical proximity on the chromosome, or "linked" to deleterious mutations. Because of this linkage, SNPs may be shared among groups of people with harmful but unknown mutations and serve as markers for them. Such markers can help uncover the actual functional mutations and accelerate efforts to find therapeutic drugs.

One of the initial applications of the recent advances in the human genome sequencing project is the emerging field of pharmacogenomics. Pharmacogenomics encompasses several major areas: the study of polymorphic variations to drug response and disease susceptibility, identification of the effects of drugs/xenobiotics at the genomic level, and genotype/phenotype associations. The promise of pharmacogenomics is that studies using genome-based technology will lead to the identification of novel SNPs and the characterization of their impact on human health. The development and application of screening technologies is therefore of high priority.

The last decade or so has witnessed a veritable explosion in the design and development of molecular genetic technologies that can be used in pharmacogenomic and molecular toxicological studies to understand the biological basis of complex traits and diseases and their relationship to environmental exposures. It is well-recognized that characterization of DNA sequence variation will enable the identification of novel genetic risk factors for disease, novel targets for drug therapies, and avoidance of adverse drug reactions. The SNP consortium (a group of pharmaceutical and bioinformational companies, five academic centers, and a charitable trust) is currently producing an ordered high-density SNP map of the human genome.

For the last 25 yr, the most commonly used approach to identify genes that influence traits has been meiotic or linkage mapping. All linkage analysis methods involve the assessment of the transmission and cosegregation of alleles at regions on the genome known as marker loci, with disease alleles assumed to be carried by family members exhibiting the disease of interest. Unfortunately, linkage analysis has not proved powerful enough to detect genes influencing many common multifactorial diseases, primarily because the study of genes with a small to moderate effect on a trait or a disease requires the collection of hundreds if not thousands of families for reliable results.

There are a variety of reasons why SNPs have emerged as an alternative form of sequence variation for gene identification and mapping studies. Primary among them is the high frequency with which

SNPs are found in the genome, lending utility for the discovery of disease-related genes. SNPs are found throughout the genome, in exons, introns, intergenic regions, promoters, enhancers, and so on. Therefore, they are likely to be associated with a functional or physiologically relevant allele. Because SNPs occur in such great abundance over the genome, groups of neighboring SNPs may have alleles that show distinctive patterns of linkage disequilibrium and may create a haplotypic diversity that can be exploited in both genetic linkage and direct association epidemiologic studies. Another advantage to studying SNPs is that since they typically have only two variant alleles, SNPs will have allele frequencies that will drift as a function of the dynamics of different populations, creating allele frequency differences that can be exploited in population-based studies. Lastly and most importantly, owing to their simple structure, the development of technologies that enable rapid, efficient, and cost-effective genotyping of thousands of individuals for hundreds of SNPs has become possible.

A number of novel, high-throughput genotyping technologies have recently been developed, including various microarray formats, matrix assisted laser desorption/ionization time of flight (MALDI-TOF) mass spectrometry (MS), and TaqMan allele discrimination approaches. However, these current state-of-the-art approaches do not yet meet all of the requirements for maximum utilization of genotyping information. A major issue with each of these approaches is the cost per SNP detection. Currently, most procedures involve polymerase chain reaction (PCR) amplification of a target sequence, a somewhat costly and time-consuming method that limits possibilities for automation. As the scale of genotyping analyses increases, the cost per genotype will need to decrease from the current level of approx $1–3 to pennies or tenths of pennies. A second key requirement for any genotyping technology is flexibility. As new SNPs are identified, there will be a need for rapid inclusion of the novel SNP within the screening procedure. For several currently commercially available, preconfigured microarrays, this is a major problem. Although it is now possible to reconfigure an existing microarray or develop a new custom array more rapidly, these technologies are still associated with high costs for synthesis and further assay validation requirements. Additional requirements for an optimal genotyping approach include sensitivity (requiring less than 1 ng genomic DNA/genotype), scalability and automation compatibility, and efficient turnaround times. For most of these newer technologies, DNA template amount is not a problem, although the amount of input DNA for microarray analysis is relatively higher.

PCR-Based Techniques

SSCP-PCR

Single-strand conformational polymorphism (SSCP) analysis is one of the most widely used methods for mutation detection. DNA regions with potential polymorphisms are amplified by PCR, the products are denatured, and the single strands thus formed are electrophoresed on a polyacrylamide gel. A fragment with a single base modification migrates differently than wild-type DNA. Alternative conformation-based mutation screening methods include conformation-sensitive gel electrophoresis, chemical or enzymatic mismatch cleavage detection, *denaturing gradient gel electrophoresis* (DGGE), and denaturing *high-performance liquid chromatography* (HPLC). The underlying principle of these methods is that the melting characteristics of double-stranded DNA are defined by its sequence, and hence a single-base mismatch can produce conformational changes in the double helix that cause the differential migration of homoduplexes and heteroduplexes containing base mismatches during gel electrophoresis. This method is highly sensitive for identifying mutations in areas of highly GC-rich sequences.

PCR Mismatch Cleavage Detection

Mismatch cleavage detection takes advantage of the fact that mismatched bases are sensitive to cleavage by enzymes and chemicals. After PCR amplification, wildtype and variant alleles are subjected to denaturation/renaturation to create heteroduplex molecules. The products are electrophoresed side by side to detect the presence of mismatch cleaved molecules following incubation with resolvases.

Denaturing Gradient Gel Electrophoresis (DGGE)

In denaturing gradient gel electrophoresis, the PCR products are resolved on a denaturing gradient gel containing formamide and urea under temperature control. SNPs are revealed by their migrational differences from wild-type homoduplexes. The major advantage of this method is its accuracy; however, its disadvantages are low throughput and difficulty of optimization.

Denaturing HPLC

In this method, polymorphisms are detected by analyzing the mobility of DNA heteroduplexes using chromatography under denaturing conditions. The variant sample is first hybridized with wild-type DNA to form a mixture of homo- and heteroduplexes. The heteroduplexes can be separated from the homoduplexes by column chromatography at a temperature that partially denatures the mismatched DNA.

Restriction Fragment Length Polymorphism (RFLP)-PCR Analysis

For *restriction fragment length polymorphism* (RFLP)-PCR analysis, a specific target region of genomic DNA is amplified by PCR. The product is then digested with appropriate restriction enzyme(s) and visualized after being gel-electrophoresed. If the SNP produces a gain or loss of restriction site, the restriction pattern is altered, and homozygous wild-type, mutant, or heterozygote carriers are easily identified. A major limitation of this method is the requirement that the polymorphisms result in an altered restriction enzyme site.

Oligonucleotide Ligation Assay (OLA) Genotyping

The *oligonucleotide ligation assay* (OLA) approach is based on the premise that hybridization with specific oligonucleotide probes effectively discriminates between wild-type and variant sequences. Three probes are used in this assay, two allelespecific probes and a common fluorescent probe. The 5'-end of the common probe is immediately adjacent to the 3'-end of the allele-specific probe. The PCR product is incubated with the three probes in the presence of thermally stable DNA ligase. Ligation of the fluorescently labeled probe to the allele-specific probe occurs only when there is a perfect match between the probe and the template. The wild-type and the variant genotypes are differentiated following electrophoresis of the ligated products. The major disadvantage of this method is that highly GC-rich regions make the allelespecific ligation step difficult to optimize.

Branch Migration Inhibition (BMI)

The *branch migration inhibition* (BMI) technique is based on the fact that spontanecus strand exchange is inhibited by sequence differences between two DNA molecules. Genomic DNA is amplified using four primers. The two forward primers are 5'-labeled with either biotin or digoxigenin. The two reverse primers have similar priming sequences but different tail sequences, which consist of 20 nucleotides that are not complementary to the genomic target but are incorporated into the PCR products. The PCR products are then subjected to heat denaturation and reannealing of single strands to form, eventually, a doubly labeled, four-stranded cruciform DNA structure. When there is no mutation, the two arms of this structure are identical, and strand exchange via branch migration leads to its complete dissociation into two duplex molecules, producing no signal. In the presence of a sequence difference, as in mutation or polymorphism, branch migration in the presence of Mg^{2+} is inhibited, and the cruciform structure does not get resolved. Thus, the stable association of biotin and digoxigenin is detected by standard *enzyme-linked immunosorbent assay* (ELISA) techniques. One of the primary limitations of BMI is that it cannot distinguish between homozygotes for two alternative alleles. It only detects heterozygotes and therefore requires an additional step, in which a reference amplicon is added to each amplified sample corresponding to one of the two possible homozygotes. The denaturation and branch migration steps are then repeated.

Pyrosequencing

Pyrosequencing is a DNA sequencing technique based on the detection of released pyrophosphate (PPi) during DNA synthesis. In a cascade of enzymatic reactions, visible light is generated that is proportional to the number of incorporated nucleotides. The cascade starts with a nucleic acid polymerization reaction in which inorganic PPi is released as a result of polymerase-mediated incorporation of nucleotides. The released PPi is subsequently converted to ATP by ATP sulfurylase, which provides the energy to luciferase to oxidize luciferin and produce light. Since the added nucleotide is known, the sequence of the template can be determined. Pyrosequencing uses the Klenow fragment of *E. coli* DNA *Pol*I. The ATP sulfurylase used in pyrosequencing is a recombinant version from the yeast *Streptomyces cerevisiae*, and the luciferase is from the American firefly *Photinus pyralis*. One picomole of DNA yields 6×10^{11} ATP molecules, which generates more than 6×10^{9} photons at a wavelength of 560 nm. A charge-coupled device (CCD) camera easily detects this light. Two different pyrosequencing strategies are currently available: solid and liquid phase. Solid-phase pyrosequencing utilizes immobilized DNA, and the excess substrate is washed off after each nucleotide addition. In liquid-phase pyrosequencing, a pyrase, a nucleotide-degrading enzyme is introduced, thereby enabling the removal of the solid phase support and intermediate washing. For SNP analysis using pyrosequencing, the 3'-end of the primer is designed to hybridize one or a few bases before the polymorphic position. Each allele combination provides a distinct pattern on the program readout. These programs can be analyzed manually or by pattern recognition software.

Array Pyrosequencing

Pyrosequencing can be applied to both ordered and random arrays; for example, the PSQ 96 System employs a DNA array, a nucleotide delivery module, and a CCD camera. A sprayer is used to deliver all four different nucleotides. Current imaging technologies require a minimum of more than 5000 template molecules. Several optimizations are still under way to allow the use of this technology for reliable high-throughput DNA sequencing, but the range of applications is growing as more institutions acquire the technology.

Specialized software has been designed to automate the classification of genotypes for samples screened by pyrosequencing in a microtiter plate using a SNP genotyping algorithm. Based on pattern recognition, this algorithm both scores the genotype and provides a value for the quality of each SNP that is scored. The assignment of this value is based on a number of different parameters, including differences in expected and obtained sequences around the SNP, signal-to-noise ratio, and variance in peak height and peak width.

Dynamic Allele-specific Hybridization (DASH)

Dynamic allele-specific hybridization (DASH) is essentially an enhanced form of allele-specific hybridization that uses a convenient microtiter plate format, a simple duplex-DNA intercalation for signal production, and a dynamic low–high temperature sweep to capture all phases of probe–target-DNA melting. For the purpose of DASH assay design, one needs to anticipate target-DNA secondary structure problems, and a maximum negative threshold of -4.0 kcal/mol should be expected. Furthermore, probe target ratios of C+G percentages should be more than 1.0. Two probes are designed for each SNP, representing both allelic sequences complementary to the biotinylated strand of the PCR product. The plates containing the bound product, the probes, and a DNA intercalating dye are subjected to a range of different temperatures, to follow the decrease in fluorescence as the temperature increases. The assay is repeated by using alternative allele-specific probes, and genotypes are scored from the fluorescence curves obtained. Devices that support the DASH procedure have been used to analyze 89 intragenic SNPs.

Allele Discrimination Using Fluorescence Resonance Energy Transfer Detection (FRET)

Fluorescence resonance energy transfer (FRET) occurs when two fluorescent dyes are in close proximity to one another and the emission spectrum of one dye molecule overlaps the excitation spectrum of the other fluorophore. Commonly used FRETbased technologies include the TaqMan assay and Molecular Beacons.

TaqMan Genotyping

The basis for FRET allele discrimination and quantitation is to measure PCR product accumulation continuously using a dual-labeled fluorogenic oligonucleotide probe, called a TaqMan probe. This probe is composed of a short (~20–30-base) oligodeoxynucleotide labeled with two different fluorescent dyes. On the 5'-terminus is a reporter dye, and on the 3'-terminus is a quenching dye. This oligonucleotide probe sequence is homologous to an internal target sequence present in the PCR amplicon. When the probe is intact, energy transfer occurs between the two fluorophors, and emission from the reporter is quenched by the quencher. During the extension phase of PCR, the probe is cleaved by the 5'-nuclease activity of DNA polymerase, thereby releasing the reporter from the oligonucleotide quencher and producing an increase in reporter emission intensity. The Applied Biosystems Sequence Detection systems use fiberoptic systems that connect to each well in a 96-well PCR tray format. The laser light or tungsten-halogen lamp excitation source excites each well, and a CCD camera measures the fluorescence spectrum and intensity from each well to generate real-time data during PCR amplification. The system software examines the fluorescence intensity of reporter and quencher dyes and calculates the increase in normalized reporter emission intensity over the course of the amplification. The results are then plotted vs time, represented by cycle number, to produce a continuous measure of PCR amplification. Several other companies also market real-time PCR detection systems including Stratagene and Bio-Rad.

Lee et al. first demonstrated that the 5'-nuclease assay could be used for allelic discrimination. In the assay, two TaqMan probes as described above are included in the reaction, one specific for each allele. The probes are distinguished through the use of different fluorescent reporter dyes (usually 6-carboxy-fluorescein [FAM] and 6-carboxy-4,7,2',7'-tetrachlorofluorescein [TET]). A mismatch between probe and target greatly reduces the probe hybridization efficiency and specific cleavage. Following PCR, an increase in the level of a FAM fluorescent signal without an increase in the TET-specific signal indicates that only the FAM-specific sequence (allele) was present and that the sample is homozygous (and vice versa). An increase in both reporter signals indicates heterozygosity. The software makes three separate calculations to arrive at the result for allele discrimination. First, using multicomponent analysis, the software determines the contribution of each component dye to the observed fluorescence spectrum. Following this, these dye component results are normalized based on control reactions (which have no template), known allele 1 template, or known allele 2 template, which are run on the same plate. An allele 1 score (on a scale of 0–1) and an allele 2 score are calculated for each sample. Finally, the allele 1 and 2 scores are normalized for the extent of the reaction, based on the results of the no-template control. A number of factors contribute to allelic discrimination based on a single mismatch. First is the thermodynamic contribution owing to the disruptive effect of a mismatch on hybridization. A mismatched probe will have a lower melting temperature than a perfectly matched probe. Second, the assay is performed under competitive conditions; therefore the mismatch is prevented from binding because stable binding of an exact match probe blocks hybridization of the mismatch. Third, the 5'-end of the probe must start to be displaced before cleavage occurs. Once a probe starts to be displaced, complete dissociation occurs faster with a mismatch than with an exact match.

Molecular Beacons

Molecular beacons are oligonucleotide probes that have two complementary DNA sequences flanking the target DNA sequence and a donor/acceptor dye pair at opposite ends of each probe. The probe adopts a hairpin loop conformation with the reporter and the quencher dyes close together when it is not hybridized to the target, and therefore, no donor fluorescence is generated. When hybridized to the right target sequence, the two dyes are separated and the fluorescence increases. Thermal instability of the mismatched hybrids increases the specificity of molecular beacons. For SNP genotyping, two molecular beacons with exact sequence matches to the wild-type and variant alleles are used in the same PCR. The use of two differentially labeled molecular beacons in the same PCR reaction allows the simultaneous detection of three possible allelic combinations.

Multiplex Automated Primer Extension Analysis (MAPA)

Multiplex automated primer extension analysis (MAPA) is a semiautomated fluorescent method that can accurately and easily genotype multiple SNPs simultaneously. This technique is a modification of a commercially available protocol that uses the extension of a primer designed to end one nucleotide 5' of a given SNP with fluorescent ddNTPs, followed by automatic sequencing on an ABI PRISM 377 Sequencer. The MAPA modification includes the incorporation of several primers corresponding to several SNPs in the same reaction and loading the primer extension products on a single gel lane. There is a limit to the number of SNPs one can multiplex with this method, dictated by the range of primer lengths (16–50 nucleotides) and the minimum spacing in the primer length that allows for separation.

Therefore, the maximum number of SNPs this method can multiplex is approx 10–12 SNPs per sample. Another drawback is that primer orientation appears to affect the accuracy of genotyping heterozygotes, perhaps because of the formation of strandspecific secondary structures.

Capillary Electrophoresis

In 1981, Jorgenson and Lukacs were the first to demonstrate electrophoretic separation of samples inside narrow-bore capillaries filled with electrophoretic media. Capillary electrophoresis (CE) was found to separate small molecules with a very high resolution. In recent years this technique has been modified for the detection of point mutations and SNPs. The most widely employed of several modifications is a technique known as *constant denaturant capillary electrophoresis* (CDCE), coupled with high-fidelity PCR. This application has lent itself extremely well to high-throughput analysis of samples. CDCE combines the principles of CE and *denaturing gradient gel electrophoresis* (DGGE) in linear polyacrylamide matrices. The denaturing conditions in CDCE are achieved by heating a section of capillary in a temperature-controlled water jacket. CDCE offers high resolution and amenability to automation, and, coupled with high-fidelity PCR, it is possible to measure point mutations at frequencies as low as 10^{-6} in human genomic DNA. The CDCE instrument has been further improved by the addition of a two-wavelength detector. This allows the use of two sets of samples labeled with two different fluorescent dyes, thus permitting comparison of two separate channels. Separation of PCR products is generally conducted in capillaries with an internal diameter of 75 μm at a constant current of 9 μA. Future integration of multiple capillary arrays and automation systems should increase the speed and the scale of this technique.

MALDI-TOF Mass Spectrometry

Karas and Hillenkamp first introduced MALDI-TOF MS in 1988 as a revolutionary method for ionizing and mass-analyzing large biomolecules. They discovered that irradiation of crystals formed by suitable small organic molecules (called the *matrix*) with a short laser pulse at a wavelength close to a resonant absorption band of the matrix molecules caused an energy transfer and desorption process, producing gas phase matrix ions. They also found that when a low-concentration of a nonabsorbing analyte, such as a protein or a nucleic acid molecule, was added to the matrix in solution and embedded in the solid matrix crystals, the nonabsorbing, intact analyte molecules were also desorbed into the gas phase and ionized upon irradiation, allowing their mass analysis.

Originally, MALDI-TOF MS was proposed as an alternative high-throughput technology for DNA sequencing to replace the conventional method. Enzymatic DNA sequencing coupled with MALDI-TOF MS analysis has been shown to be effective at discovering previously unknown SNPs. However, there is a loss of signal intensity and mass resolution with increasing DNA size, owing to the size-dependent tendency of the phosphodiester backbone of DNA to fragment during the MALDI process. Consequently a robust MALDI-based approach to SNP discovery, which requires sequencing of PCR products up to 300 bp in length, has not been demonstrated. This limitation has also hampered attempts to analyze PCR amplicons containing SNPs directly. Additionally, during the MALDI process, double-stranded PCR products can dissociate into single strands of slightly different masses, which as a result are poorly resolved. Minisequencing has become the most widely used MALDI-TOF MS-based method for SNP analysis. It involves annealing of a primer to a template PCR amplicon downstream of a SNP. A mix of deoxynucleotide triphosphates and dideoxynucleotide triphosphates are added to a PCR template and primer, along with a DNA polymerase.

The polymerase extends the 3'-end of the primer by specifically incorporating nucleotides that are complementary to the sequence of the PCR product. Extension terminates at the first position in the template where a nucleotide complementary to one of the ddNTPs in the mix occurs. MALDI-TOF MS-based methods have been developed in which extended primers are solid phase-purified and detected by MS; the identity of the polymorphic nucleotide is determined by measuring the mass of the extended primer.

The greatest promise of MALDI-TOF MS for SNP analysis lies in its ability to genotype many SNPs rapidly, accurately, and simultaneously. Recently, another approach to MALDI-TOF MS has been developed that does not require a PCR amplification step. This direct approach involves the sequence-specific hybridization of two oligonucleotides to form an overlapping structure at the polymorphic position. Enzymatic cleavage and amplification of an allele-specific, short oligonucleotide signal molecule, which is derived from this overlap structure, follow this. The signal molecules produced in this reaction contain a biotin group, enabling solid-phase sample preparation by capturing these molecules on streptavidin-coated magnetic beads. They are then washed to remove contaminants, and the clean signal molecules are eluted for MALDI-TOF MS analysis.

Microarrays

The DNA microarray chip has revolutionized the application of high-throughput genotyping in the last few years. A DNA microarray is a small chip, generally about a square centimeter, most commonly made of glass, plastic, or silicon. SNP analysis with the DNA microarray chip is a hybridization-based genotyping technique that allows the simultaneous analysis of many polymorphisms. High-density microarrays are created by attaching hundreds of thousands of oligonucleotides to a solid surface in an ordered array. The DNA of interest is PCR-amplified to incorporate fluorescently labeled nucleotides and then hybridized to the chip. Each oligonucleotide in the array acts as an allele-specific probe. Well-matched sequences hybridize more efficiently than mismatched sequences and therefore give stronger fluorescent signals. The signals are quantitated by high-resolution fluorescent scanning and analyzed by sophisticated software programs. Many biotechnology companies have developed and are marketing DNA microarrays. The unique features for several different microarray approaches currently available are described in the following sections.

Affymetrix GeneChip Technology

Affymetrix uses light-directed synthesis for the construction of high-density DNA probe arrays using two methods: photolithography and solid-phase DNA synthesis. Synthetic linkers modified by photochemically removable groups are attached to a glass substrate, and light is directed through a photolithographic mask to specific surface areas to produce photodeprotection. This is followed by the chemical coupling of hydroxyl-protected deoxynucleosides at the illuminated sites. Next, light is directed to different regions of the substrate by a new mask, and the chemical cycle is repeated. Thus, for a given reference sequence, a DNA probe array can be designed that consists of a highly dense collection of complementary probes. The amount of nucleic acid information encoded on the array in the form of different probes is limited only by the physical size of the array and the achievable lithographic resolution. Because the arrays are constructed on a rigid material (glass), they can be inverted and mounted in a temperature-controlled hybridization chamber. A fluorescent-tagged nucleic acid sample injected into the chamber hybridizes to complementary oligonucleotides on the array. Laser excitation enters through the back of the glass support, focused at the interface of the array and the target solution, and the fluorescence emission is collected by a lens and passes to a sensitive detector through a series of optical filters. A quantitative 2D fluorescence image of hybridization intensity is obtained by simply scanning the laser beam or translating the array.

Different flow-through systems have been developed to allow a continuous measurement of real-time hybridization to an array, by adding cell lysis and amplification to a miniaturized fluidics system. This approach extends the two dimensions of the lateral microarray resolution by the time-resolved analysis of the binding process as a third dimension. Real-time hybridization also allows the calculation of binding kinetics for every spot under different temperatures and changed hybridization conditions.

Nanogen Biochip Cartridges

Nanogen has developed a microchip cartridge, called the *Nanochip*, to facilitate rapid identification and precise analyses of biological molecules using a process based on the electrical properties of biological molecules. This technology, termed *electronic addressing*, places DNA fragments at selective sites on a silicon microchip and involves active hybridization as opposed to the passive hybridization process described above. After the DNA is addressed, a test site or multiple test sites are electronically activated with a positive charge. The negatively charged probes move to the positively charged test sites, where they are concentrated and bound by a chemical process to that site. The microchip is then washed, another set of DNA probes is added on, and different sites are activated. Therefore, an array of specifically bound DNA probes can be assembled or addressed in a user-defined order. The highly advantageous feature of the Nanochip compared with other array approaches is its flexibility in experimental design. Because you can hybridize only the sites you want by specifying particular sites for electronic concentration and hybridization, you can run multiple experiments on the same chip. The current configuration of the Nanochip contains 100 sites on a single cartridge.

Single-Base Extension-Tag Array on Glass Slides

Also called minisequencing or template-directed incorporation, single-base extension-tag array on glass slides (SBE-TAGS) involves extension of a primer located adjacent to the position of the SNP, using DNA polymerase in the presence of fluorescently labeled ddNTP. The SBE-TAGS method marks each primer with distinct 5'-end sequence tags that allows separation of a multiplexed SBE reaction by hybridization to a microarray. Depositing unmodified nucleotides on a glass slide with routine spotting equipment can easily generate these arrays. Arrays are scanned using external argon lasers, and a matrix is applied to correct for the crosstalk between multiple overlapping fluorophores. This method has been used to genotype over 100 SNPs accurately.

Microsphere-Based Technology

Microsphere-based techniques have been described in the literature for a number of applications but have been further developed by Luminex. The Luminex technology couples existing flow cytometric technology with color-coded microspheres, each of which carries an individual assay. The approach is rapid and extremely flexible. The first use of flow cytometry for analysis of microsphere-based immunoassays was published in 1977 and was reviewed by McHugh in 1994. The flow cytometer is able to discriminate different particles based on size or color, providing the potential for multiplex analysis. The Luminex system is based on the principle that panels are created by combining up to 100 different microsphere-based assays into a single sample test. Multiplexed assays can be run on sample volumes as small as 5 μL. Each assay is individually constructed around a single microsphere set with its own identifying fluorescent color. Each set of microspheres is manufactured with unique relative proportions of red and orange fluorescent dyes. The system consists of 100 distinct sets of fluorescent microspheres and a standard benchtop flow cytometer interfaced with a personal computer containing a digital signal-processing board. Individual sets of microspheres can be modified with reactive components such as oligonucleotides, antigens, or antibodies and then mixed to form a multiplexed assay set. A further advantage is that the system is extremely flexible and permits easy incorporation of new endpoint measures. This contrasts with DNA microarray chip technology, which is not only more expensive but has less flexibility in making new probes available as new polymorphic alleles are identified. So instead of requiring the reconfiguration and synthesis of a new chip when a new allele is to be added to the screening panel, the Luminex system simply requires the addition of an additional oligonucleotide-hybridized microsphere.

The Luminex technology is very amenable to studies of SNP genotyping owing to its flexible format. One can visualize an assay in which a bank of prelabeled probes is held in reserve and an investigator or clinician can pick and choose the SNPs for which to screen. One drawback of the technology, however, is that it requires a considerable amount of time for assay optimization and validation, particularly in the multiplex format. For this reason, many investigators have decided to wait for the availability of commercial kits for use on the instrument.

Another method using the microsphere-based Luminex assays has also been developed for successful multiplexing of SNPs. The conventional Luminex assay using *single base chain extension* (SBCE) has been modified into an allele-specific primer extension reaction (ASPE). This method utilizes a pair of allele-specific primers that differ from each other at the 3'-end and encode different "ZipCode" sequences at the 5'-end, in the same reaction. The DNA polymerase extends only one primer if the template DNA sequence is homozygous, whereas both primers are extended in heterozygotes. The ASPE reaction eliminates the necessity of post-PCR cleanup and the addition of unlabeled nucleotides.

Genotyping Analysis of the CYP2D6 Gene

As mentioned above, the application of SNP genotyping analyses to pharmacogenetic endpoints is of growing clinical importance. The genetic polymorphisms associated with specific human CYP and phase II enzymes typically occur with variable frequency in different populations or ethnic groups and may result in *poor metabolizers* (PMs) or *extensive metabolizers* (EMs) for specific substrates. CYP2D6 metabolizes up to 20% of commonly prescribed medications, including antidepressant/psychotics, antiarrhythmics, and β-blockers, as well as many environmental agents. There is a wide intersubject variability in the pharmacokinetics of all CYP2D6-metabolized substrates, which exhibit variations in clearance over a 20–200-fold range depending on the agent under study. To date, over 74 allelic variants in the human CYP2D6 gene have been identified, many of which are associated with either decreased or enhanced metabolic activity. The CYP2D6 gene represents a challenge for genotyping: the numerous known polymorphisms are not caused only by single-nucleotide substitutions or deletions but also by gene deletions, duplications, and the presence of pseudogenes.

For CYP2D6, the 74 known variants are associated with 60 different polymorphic regions. The number of SNPs within a particular variant allele range from a single SNP (for example, CYP2D6*1B) to eight SNPs (for example, CYP2D6*4G). Four unique SNPs in exons 3, 4, 8, and 9 are observed among the *6 variant allele subfamily. Although the T1707Del SNP would identify a *6 variant, genotyping analysis would require the screening of all four of these SNPs to distinguish among the *6 A, B, C, and D variant alleles. Screening for only 4 of the 60 currently known possible SNPs results in redundant variant allele classifying information, i.e., screening for 56 SNPs is required for the accurate and complete elucidation of the CY2D6 genotype.

A number of novel, high-throughput genotyping approaches have recently been developed, yet these do not yet meet all the requirements for maximum utilization of genotyping information. For example, PCR-based techniques such as PCR-RFLP, PCR SSCP, OLA, and others are time-consuming and labor-intensive and do not provide large amounts of information quickly, i.e., it is not possible to multiplex using any of these techniques. Furthermore, gene duplications and gene deletions such as those present in CYP2D6 cannot be identified.

Sequencing-based techniques such as minisequencing and pyrosequencing are extremely accurate but, again, are laborious and not cost-effective. Furthermore, when multiple SNPs are necessary to define a variant allele and they lie in different regions of the gene, as they do in CYP2D6, numerous fragments would have to be sequenced in order to genotype an individual. FRET-based techniques do not provide a solution for multiplexing either, as the time and cost of optimizing assays far outweighs the ability to genotype for all the various SNPs. However, TaqMan-based assays are extremely useful in genotyping for the common variants of CYP2D6 using commercially available kits. Microsphere-based methods using both SBCE and ASPE techniques coupled with flow cytometry and microtiter-based assays are promising in their applications for multiplexing. Multiple PCR products can be screened simultaneously using specific probes, thus allowing detection of SNPs that are far apart. CYP2D6 genotyping, for example, therefore poses a great challenge, especially since none of the latest techniques allows for the detection of gene duplications and deletions in a speedy and cost-effective manner.

Measurement

The analysis of gene expression is an integral part of any research characterizing gene function. A wide variety of techniques have been developed for this purpose, each with their own advantages and limitations. This section seeks to provide an overview of some of the most recent as well as conventional methods to quantitate gene expression. These approaches include Northern blot analysis, *ribonuclease protection assay* (RPA), reverse transcription polymerase chain reaction, *expressed sequence tag* (EST) sequencing, differential display, cDNA arrays, and the *serial analysis of gene expression*

(SAGE). Current applications of the information derived from gene expression studies require assays to be adaptable for the quantitative analysis of a large number of samples and end points within a short period coupled with cost effectiveness. A comparison of some of these features of each analytical approach as well as their advantages and disadvantages has also been provided.

Gene expression analyses have long been used to provide insights into gene function. Many environmental pollutants, toxicants, and heavy metals affect cellular function by causing drastic changes in gene expression patterns. For both toxicological screening and chemical-specific mechanism of action studies, a wide range of approaches is available to evaluate changes in gene expression at the molecular level that may occur because of a toxic response. These approaches are not unique to the analysis of endpoints of interest in molecular toxicology, as for example, the expression of biotransformation enzymes, but are extremely valuable tools for all genomic studies. The recent rapid technological advances in this field were prompted by the ability to identify genes at the nucleic acid level, rather than proceeding from a known protein to its chromosomal counterpart. Expression studies have previously relied on techniques such as Northern blot analyses or the ribonuclease protection assay, each of which measures the expression of only a small set of genes at a given time. More recent technologies, including *serial analysis of gene expression* (SAGE), quantitative *reverse transcription polymerase chain reaction* (RT-PCR), cDNA microarrays, and high-resolution 2D gel electrophoresis, allow for the expression levels of tens to thousands of genes to be screened at once. The number of samples and number of genetic endpoints to be analyzed and taking into account both cost and throughput capability, one analytical approach for RNA expression analysis may be more appropriate for a particular application or research study.

Prior to any expression analyses, it is essential to verify the integrity of the RNA and to obtain accurate measurement of RNA concentration levels. The feasibility of obtaining meaningful quantitative gene expression data is dependent on the utilization of a validated approach with multiple quality control measures in place. These include the inclusion of either endogenous or exogenous standards or positive controls to assess reproducibility of all steps of the assay; verification of the absence of genomic DNA contamination by DNase treatment and/or in the case of PCR-based methods, the use of primers that span intron/exon junctions for amplification of cDNA only; quantitation analysis of samples collected during the exponential phase of PCR amplification; and negative controls to verify absence of contamination and specificity of the probe used for detection of target mRNAs.

Northern Blot Analysis

Northern blotting was developed as the RNA counterpart of Southern blotting. This technique mainly involves separation of RNA species on the basis of size by denaturing gel electrophoresis followed by transfer of the RNA onto a membrane by capillary, vacuum, or pressure blotting. The RNA is then permanently bound to the membrane either by heating at 80°C or by UV crosslinking. These membranes are then probed with partial or complete cDNA oligonucleotides that are labeled by radionucleotides or chemiluminescent moieties. Nonspecific hybridization is removed by washing, and then the blots are audioradiographed. The resulting visible band(s) indicates the size of the RNA, and the intensity corresponds to the relative amounts of the RNA. The band intensities are quantitated by densitometry using the appropriate image analysis software. Northern blotting is perhaps one of the few techniques that permits mRNA size determination and therefore is useful for the detection of alternatively spliced transcripts or mutations that result in modified mRNA sizes. One of the primary drawbacks of Northern blotting, however, is that the technique yields semiquantitative results. The other limitations involve the requirement of very high quality intact RNA concentrations, variability in transfer efficiencies, and high background levels on the audioradiograms. Typically, the expression of various housekeeping genes with similar copy numbers are used as external controls for sample loading variability and blot-to-blot comparisons. However, the expression of these housekeeping genes may vary with different stages in the cell cycle or among different cell, tissue, or disease types. Variations of the Northern blot such as dot, slot, and fast blots have been developed in an attempt to increase quantitation and simplify the assay. However, before any of these alternate procedures can be used, it is imperative to demonstrate, via the Northern blot, that the probe used in the application is specific to the target RNA, as there is no scope for size fractionation with these methods.

Ribonuclease Protection Assay (RPA)

The *ribonuclease protection assay* (RPA) is a variation of the Northern blot approach, except that it is performed in a solution containing a labeled antisense target RNA probe and the target mRNA without prior gel fractionation or blotting. The unhybridized probe and the sample RNA are degraded enzymatically following incubation for several hours. The remaining hybrids are electrophoresed on a denaturing polyacrylamide gel and visualized by autoradiography. Alternatively the RNase-resistant hybrids are precipitated and bound to filters for direct quantitation by scintillation counting. RPA is considered to be 10-fold more sensitive than Northern blot analysis.

Several issues need to be taken into account when one is designing RPA probes. If the RPA products are to be analyzed using gel electrophoresis, the RPA probe should contain some terminal sequences that will not hybridize with the target mRNA, so that undigested probe can be distinguished from probe-RNA hybrids on the basis of size. As is the case with Northern analysis, quantitation requires the concurrent hybridization of an invariant control mRNA. Probes may be multiplexed together in a single hybridization reaction if the sizes of the products do not overlap. This holds true if the products are analyzed by electrophoresis. However, if RPA products are going to be analyzed by the scintillation counter method, then the use of two different radionucleotides solves this problem. The two main advantages to RPA are sensitivity and the ability to determine absolute RNA levels. The disadvantages are difficulties encountered in designing adequately sensitive internal controls, and the high quantity/ quality of RNA required for the assay.

Expressed Sequence Tag (EST) Sequencing

The concept of *expressed sequence tag* (EST) sequencing was first described in 1991. The underlying goal was to create cDNA libraries, pick random clones, and then carry out a single sequencing reaction with a large number of clones. Each reaction generates approx 300 bp of sequences that represent a unique sequence tag for a particular transcript. EST sequencing can be carried out using both normalized (in which each transcript is represented in more or less equal numbers) and nonnormalized cDNA libraries. The advantage of using normalized libraries is that redundant sequencing of highly expressed genes is minimized. The advantage of nonnormalized libraries is that the abundance of the transcript in the original cell is accurately reflected in the frequency of clones in the library. Hence these libraries can be used to identify highly expressed but unknown genes as well as to compare the expression of highly expressed genes in different cells or tissues.

There are currently over 1.5 million human ESTs in the publicly available database of ESTs (dbEST) provided by the National Center for Biotechnology Information. These ESTs are derived from approx 1200 human cDNA libraries. In addition to public databases, several companies have generated larger collections of ESTs. These include Human Genome Sciences, Incyte Pharmaceuticals, and Celera Genomics Group.

Subtractive Cloning by Representational Difference Analysis (RDA)

Subtractive cloning methods have been in use for many years and offer an inexpensive and flexible alternative to EST sequencing and cDNA array hybridization. The PCR-based method commonly used is known as *representational difference analysis* (RDA). In this analysis, double stranded-cDNA is created from the two cell or tissue populations of interest (for example, tumor and normal tissue), linkers are ligated to the end of the cDNA fragments, and then the cDNA pools are amplified by PCR. The cDNA pool from which unique clones are desired is designated as the "tester," and the cDNA pool that is used to subtract shared sequences is designated as the "driver." Following PCR amplification, the linkers are removed from both cDNA pools, and unique linkers are ligated to the tester sample. The tester is then hybridized to an excess of driver DNA, and sequences that are unique to the tester cDNA pool are amplified by PCR. The primary limitation of this method is that subtle quantitative differences are missed because the cDNAs identified are usually those that differ significantly in expression level between the cell populations. In addition, because each experiment is a pairwise comparison, and the subtractions are based on a series of sensitive biochemical reactions, it then becomes difficult to compare a series of RNA samples directly.

Reverse Transcription-Polymerase Chain Reaction (RT-PCR)

RT-PCR is an in vitro method for amplifying defined target sequences of RNA. It is an extremely sensitive method and can be used to compare levels of mRNA in different sample populations and to characterize patterns of mRNA expression. RT-PCR analyses have been modified to increase its sensitivity and accuracy; some of the modifications include semi-nested, nested, and even three-step nested RT-PCR techniques. There are also a number of detection methods that can be used, yielding either semiquantitative or quantitative results. All of the components of an RT-PCR reaction and subsequent product detection are interdependent and require careful optimization to ensure specificity, sensitivity and reproducibility of the assay. Typically, all measurements are standardized to a calibrator sample so that data collected at different time points can be compared directly.

The first step in RT-PCR is the *reverse transcription* (RT) of the RNA template into cDNA, followed by its exponential amplification in a PCR reaction. Separation of the RT and the PCR steps is advantageous for long-term storage of the cDNA or analysis of multiple targets. The RT step can be primed using specific primers, random hexamers, or oligo-dT primers. Specific primers can sometimes cause marked variation in estimates of mRNA copy numbers; while random hexamers can overestimate mRNA copy numbers by about 19-fold. Numerous RT enzyme preparations are commercially available and vary in terms of efficiency and in range of primers that can be used for first-strand synthesis. In all RT-PCR applications, it is critical to include a no-RT template control to avoid the quantitation of false positives.

Like other methods for RNA quantitation, RT-PCR can be used for relative or absolute quantitation. Absolute quantitation, using competitive RT-PCR, measures the absolute amount or number of copies of a specific target mRNA sequence in a sample. In competitive RT-PCR, increasing amounts of DNA highly homologous to the target, but distinguishable by either size or restriction sites, are added to the PCR, and both target and competitive template are quantified. It is assumed that the amplification efficiency of both templates is identical; however, this may not be the case. Most gene expression analysis studies utilize a relative expression calculation similar to those used in standard assays such as the Northern blot. Expression of the gene of interest is reported relative to expression of an endogenous control gene, which is assumed to have equal expression in all tissues in the study. In this way, expression levels can be compared from tissue to tissue. The endogenous internal control in relative RT-PCR may be analyzed in a multiplexed reaction or in two separate reactions. Common internal controls include 18S rRNA, β-actin, β-glucuronidase, and GAPDH mRNAs.

Critical to either absolute or relative RT-PCR is quantitation of the product during the exponential phase of PCR. This represents a challenge because internal control RNAs are typically constitutively expressed housekeeping genes of high abundance, and their amplification reaches the plateau phase with very few PCR cycles. It is therefore difficult to identify comparable exponential phase conditions in which the PCR product from a rare target mRNA is detectable. Detection methods with low sensitivity, like ethidium bromide staining of agarose gels, are therefore not recommended. Detecting a rare message while staying in exponential phase with an abundant message can be achieved in several ways: (i) by improving the sensitivity of product detection; (ii) by decreasing the amount of input template in the RT or PCR reactions; and/or (iii) by decreasing the number of PCR cycles.

Modifications involving the application of fluorescence probes and instrumentation have led to the development of kinetic RT-PCR methodologies that facilitate the quantitation of nucleic acids with improved sensitivity and throughput and overcome many of the problems described above. There are currently at least three manufacturers of fluorescence resonance energy transfer detection (FRET)-based instrumentation systems.

The ABI PRISM 7700 contains a built-in thermal cycler with 96 wells and a fluorescence reader that can read wavelengths between 500 and 660 nm. The fluorescent light source in this case is a laser, and the emission is directed to a spectrograph with a *charge-coupled device* (CCD). The most recent commercially available model, ABI PRISM 7000, uses a tungsten–halogen lamp, and the fluorescence emission is directed through four optical filters to a CCD camera. The rest of the features are similar to the 7700. On the other hand, the ABI PRISM 7900HT has a 384-well capacity and allows the use of multiple fluorophores in a single reaction owing to the feature of continuous wavelength detection.

The LightCycler uses small-volume glass capillary tubes that are heated and cooled by an airstream. A blue light-emitting diode is the light source, and the fluorescence is read by three photodetection diodes with different filters. It can analyze up to 32 samples per run.

Bio-Rad has recently launched an optical module that fits into their conventional thermal cycler. This device can scan up to 96 samples simultaneously and at present can monitor four different fluorescent reporters.

To date, there are four different competing techniques available to detect the amplified product with the same sensitivity. The simplest method employs fluorescent dyes that bind specifically to double-stranded DNA. The other three utilize the hybridization of fluorescently labeled probes to specific amplicons. These four methods are molecular beacons, DNA binding dyes, hybridization probes, and hydrolysis probes.

Molecular beacons

Molecular beacons are probes that have a loop structure complementary to the target nucleic acid molecule and a stem structure that is formed by the annealing of complementary sequences on the ends of the probe sequence. A fluorescent marker is attached to one arm, and a quencher is attached to another. In solution, the free molecular beacons have a hairpin structure, with the stem keeping the arms in close proximity, thereby resulting in the efficient quenching of the probe. On encountering a complementary target, they undergo a conformational change that results in the formation of a probe-target hybrid. This hybrid forces the stem apart, leading to the separation of the fluorophore and the quencher and consequently the restoration of fluorescence, while the free molecular beacons remain nonfluorescent. The main drawback of molecular beacons is their ability to form alternate conformations that fail to place the fluorophore next to the quencher, resulting in large background signals.

DNA-binding dyes

DNA-binding dyes such as SYBR green, which exhibit no fluorescence alone in solution, can be incorporated into double-stranded DNA during the PCR elongation step. Detection of the fluorescence of the DNA-binding dyes therefore increases during the elongation step and decreases during denaturation. The specificity of target detection largely depends on the specificity of the PCR primers, and a separate probe is not added. An important failing of this method is that the number of dye molecules that are incorporated into the PCR product may vary with each PCR cycle and from sample to sample, and therefore the analysis is semiquantitative at best.

Hybridization probes

The LightCycler uses hybridization probes. One probe has at its 3' end a fluorescein donor, whose emission spectrum overlaps the excitation spectrum of an acceptor fluorophore, which is attached to the 5' end of the second probe. This acceptor labeled probe is blocked at its 3' end to prevent its extension during PCR. Fluorescent light is produced from FRET following excitation of the donor. The two dyes are apart when in solution; however, following hybridization of the probes to the target sequence, they are brought into close proximity, and FRET occurs. Therefore, the increasing intensity of wavelength of the second dye is directly proportional to the amount of DNA synthesized. Furthermore, a melting curve analysis can also be performed for multiplex analysis, as the probes are not hydrolyzed.

Hydrolysis probes

Hydrolysis probes are usually used in TaqMan assays; they use the 5' nuclease activity of the DNA polymerase to hydrolyze a hybridization probe after it has bound to its target. The PCR step, which follows the RT step, increases the specificity of the reaction by the use of three oligonucleotides complementary to the DNA. Two primers amplify a specific amplicon, followed by the use of a probe that hybridizes to the product during annealing/extension. The probe has a fluorescent dye at the 5' end and a quencher at the 3' end. If no complementary amplicon is generated, the probe remains intact. Conversely, if the probe binds to the complementary sequence as it is being amplified, it is eventually cleaved, thus separating the reporter and quencher dyes, causing emission of fluorescence. Because of the high Tms of the probe, the TaqMan system PCR annealing and extension steps can be combined and most reactions are carried out at 60–62°C. This also ensures maximum 5'-3' exonuclease activity of the Taq polymerase. The increase in the length of the annealing/extension step, coupled with increased Mg^{2+} or Mn^{2+} for longer amplicons, makes this system less efficient and flexible than others. Real-time RT-PCR assays are conclusively more reliable than conventional ones and can easily be adapted to a high-throughput setup.

Differential Display

Another widely used PCR-based method that is extremely popular is differential display or RNA fingerprinting. Differential display involves RT primed with either an oligo-dT or an arbitrary primer, in conjunction with the RT primer to amplify cDNA fragments that are then separated on a polyacrylamide gel. The presence or absence of bands on the gel visualizes differences in gene expression. Differential display has also been adapted for use in fluorescent DNA sequencing machines. It is efficient for analyzing as little as 5–10 ng of total RNA. A limitation of this method is the generation of false positives either during PCR or in the cloning of differentially expressed PCR products. Large amounts of RNA are required to discriminate true positives from false positives. A modification of the technique based on the analysis of 3'-end restriction fragments claims to result in fewer false-positive signals. In this method, double-stranded cDNA is prepared and digested with a restriction enzyme with a four-base recognition site. Linkers are then ligated to the restriction fragments, and the entire pool of transcripts is amplified by PCR. Gel electrophoresis of the 3'-end fragments reveals the differences in gene expression. The distinct advantage this modification offers is that every gene in the cell can be identified by the use of a series of restriction enzymes; furthermore, because the migration of the bands in the gel is determined by the restriction site at the 3'-end, known genes can simply be identified by measuring the size of the restriction fragment.

cDNA Microarrays

In a cDNA array, many gene-specific polynucleotides derived from the 3'-end of RNA transcripts are individually arrayed on a single matrix. This matrix is then simultaneously probed with fluorescently tagged cDNA representations of total RNA pools from test and reference samples, allowing one to determine the relative amount of transcript present in the pool by the type of fluorescent signal generated. An internal control is provided for each measurement. The adaptable nature of the fabrication of the array and hybridization methods allow the technique to be widely applied—the limitations being cost, the availability of clones for the solid phase, and the quality of the RNA extracted from cell lines or tissues. The targets for the arrays are labeled representations of cellular mRNA pools. A labeled product from the 3'-end of the gene is produced by RT with an oligo-dT primer. The purity of the RNA is critical, particularly when using fluorescence, as cellular proteins, lipids, and carbohydrates can mediate significant nonspecific binding of fluorescently labeled cDNAs to slide surfaces. For adequate fluorescence, the total RNA required per target, per array, is 50–200 ng. For mRNA present as a single transcript per cell, application of target derived from 100 ng of total RNA over an 800-mm^2 hybridization area containing 200-μm-diameter probes will result in approx 300 transcripts being sufficiently close to the target to have a chance to hybridize. Therefore, if the fluorescently tagged transcripts are 600 bp, have an average of 2 fluor tags per 100 bp, and hybridize to their probe, approx 12 fluors will be present in a 100-μm^2 scanned pixel. Such low levels of signal are at the lower limit of fluorescence detection and can easily be rendered undetectable by assay noise. A variety of means by which to improve signal from limited RNA have been proposed. For example, efficient mixing of the hybridization fluid should bring more molecules into contact with their cognate probe, increasing the number of productive events. Posthybridization amplification methods have also been reported in which detectable molecules are precipitated at the target by the action of enzymes "sandwiched" to the cDNA target.

A critical challenge of the high-throughput technologies available to measure gene expression is the accurate and adequate analysis of the vast amounts of data generated. At present the most widely used computational approach for analyzing microarray data is cluster analysis.

This analysis groups genes based on similar expression profiles and compares them with other clustered genes, providing clues to the function or regulation of the genes. The three broad categories of cluster analysis include a tree-based approach that uses a measure of the distance between genes such as a correlation coefficient to group genes into hierarchical trees. The second category minimizes variation within clusters so that between-cluster variation is maximized. The third category groups genes into two basic blocks, one in which the correlation is maximized and one in which the correlation is minimized. All these categories basically utilize the intensity differences between the mean intensity for each of the groups. However, relative mean comparisons ignore the premise that differences in expression level of less than 100% may exert meaningful biological effects. Various statistical models have been designed to approach the problem of gene expression data analysis, and several problems still remain associated with each of the strategies. Although technological advances have simplified the ability to study thousands of genes at once, the interpretation of this data and its subsequent analysis continue to pose a challenge.

Serial Analysis of Gene Expression (SAGE)

Serial analysis of gene expression (SAGE) utilizes isolated sequence tags from individual mRNAs that are concatenated serially into long DNA molecules that are then sequenced. Initially double-stranded cDNA is synthesized from mRNA using a biotinylated oligo-dT primer. The cDNA pool is then cleaved by a restriction enzyme, also known as an anchoring enzyme, and is then separated on a polyacrylamide gel. The total number of tags identified by this method to date number close to 5 million. SAGE requires relatively higher concentrations of RNA compared with RT-PCR or microarray analyses, and it is relatively technically difficult to create tag libraries. There are two major concerns when using SAGE. One concern is identifying sequencing errors, and the second is making valid tag to gene assignments. Several modifications have been developed to increase the utility of SAGE in terms of both methodology and data interpretation.

2

BIONANOTECHNOLOGY

Nanotechnology is not a set of particular techniques, devices, or products. It is, rather, the set of capabilities that we will have when our technology gets near the limits set by atomic physics. We can make predictions for such a technology without knowing the specifics of how it will be achieved. We can, for example, know the strength of a substance with a given pattern of atoms and covalent bonds without knowing the process by which it was formed. We do have to know however, the pattern of atoms and bonds. The laws of physics don't tell us directly how strong a material can be; they tell us how strong a particular one will be. It is similar with other things of interest to a technologist. Physical law does not tell us how powerful a motor can be; it lets us say how powerful a specific motor will be. So we can only get a grasp of the outlines of the capabilities of nanotechnology by analyzing a set of designs. So we have to come up with a set of designs.

These designs have to be simple enough so that we can analyze them now. The more they look and act like machines we already know about, the less predicting we have to do and the more likely we are to get it right. Note that this is not necessarily the way many actual *nanomachines* will be built, and almost certainly not the most efficient, powerful, or economical way to build them. But if we can design nanomachines like current-day macroscopic ones, we know all the things that need to be analyzed, from structural strength to frictional heating, and we will have good reason to believe that they will work. That means that the real nanomachines of future designs will almost certainly be better than our designs here. As we know for sure that they will be at least as good.

Another property that we would like to have in our future technology is *autogeny*. An autogenous technology is one whose manufacturing base—all the machines that make machines—is capable of producing any piece of machinery in the manufacturing base. In our context, it means designs that would, if they existed, be able to build more machines like themselves. An example is today's machine tool industry: it consists of machines of steel cut into shapes with a ten-thousandth of an inch accuracy, and it can make machines of steel cut into shapes with a ten-thousandth of an inch accuracy. It's a stable point in the panorama of machine-building possibilities.

Scaling Laws

An ant does not look like an elephant. An ant can lift hundreds of times its own weight, but an ant the size of an elephant could not even stand up. The reason is *scaling laws*. The one directly applicable to the ant/elephant question is called the *square/cube law*. The weight of an object varies with its volume, that is, with the cube of its length, where the strength of its legs varies with its cross-sectional area, and thus the square of the length. In other words, an ant the size of an elephant, one thousand times as long, would have legs a million times stronger, but it would weigh a billion times as much.

If we run in the opposite direction, down to *nanomachines*, it turns out we can just about ignore weight entirely.

Do not try to bend a windowpane of glass in your hands. It would not bend enough for you to notice, and if it breaks you will cut yourself. But an optic fiber and the fibers in the fiberglass fabric of an auto/boat patch kit, bend easily. You could tie knots too small to see in the ultrafine fibers of fiberglass insulation. They are all the same stuff: glass. Being thinner means being more flexible—another scaling law. Parts in nanomachines are almost inconceivably thin, and so we will have to use the stiffest materials we can, like diamond, to compensate.

Swing your arm in a full circle, taking one second to do so. Your hand will move at between one and two meters per second. A robot arm a million times shorter than yours would make a circle a million times shorter, so if it moved at the same speed, it'd go around a million times per second.

Of course, if you tried to swing your arm around a million times per second, it would fly right off. Your bones and tendons are way too weak to support that kind of force. But the scaling laws say that the tiny robot arm will experience the same stress per unit area (in general) as yours when its hand moves at the same speed. So its frequency can go up as its size goes down. You can see the same phenomenon at work in the animal kingdom. Pelicans and geese flap their wings no faster than you can wave your arms. Sparrows and wrens flap much faster, and a hummingbird's wings are a blur.

You could pick up and place, with reasonable accuracy, small items from one spot on your desk to another at about one per second, but a nanoscale robot arm should be able to do the same at about a million second. While car engines run at thousands of *revolutions per minute* (RPM) and small electric motors at tens of thousands, nanoengines and motors can turn at billions or tens of billions of RPM, well within conservative design parameters.

Structures

Today's macroscopic machinery is mostly made of metal, although plastic of various kinds has played an increasing role for the past half-century. *Steel* is king, with *aluminum* alloys playing the role of prince. But we cannot make *nanomachines* from steel, or aluminum or plastic either. It is those *scaling laws*.

The same mathematics of form that make fiberglass flexible means that a beam of steel at the nanoscale would be about as useful as one of Play-Doh in the everyday world. To make useful beams, frameworks, casings, and shafts, we need to use the stiffest, hardest stuff we have: diamond. (Some theorists have claimed that a structure of carbon nitride is harder, but no one's managed to make it yet.) Even diamond is a bit rubbery at the nanoscale. It would be nice to have something harder, but diamond will do

As I write, it is diamond's golden anniversary: synthetic diamond, that is. Just fifty years ago, diamond, real honest-to-goodness diamond, of the same molecular structure as natural gems, was first made in the laboratory. Today, it's made by the ton in factories: about a billion dollars' worth of synthetic industrial diamond is sold each year.

Nature produces diamond by compressing carbon at extreme temperatures and pressures. The first industrial method for *diamond synthesis* worked much the same way, producing spoonfuls of diamond grit in gargantuan one-hundred-ton hydraulic presses. But the way most diamond is made today is by chemical vapour deposition, or CVD. This is a method whereby, at relatively low temperatures and pressures, one can grow diamond crystals from carbon-bearing gas such as methane (natural gas) or acetylene. Even alcohol has been used.

Some reactions are very finicky. CVD is not. Robert deVries, retired General Electric diamond researcher, *quipped*, "You can almost make this at home with a *microwave oven.*" The process has been duplicated by amateurs in their garages. In any chemical process, you can control the conditions only on the average. The actual reactions involve random collisions between the gas molecules and the diamond being formed. But in a nanomachine, everything is precisely controlled. You can specify just which molecule touches which, when, and how hard. Robert Freitas and Ralph Merkle have made an extensive study of the process. Given how easy diamond making is with random gas collisions, it is unimaginable that it would not be quite straight forward in a nanomachine.

Diamond is a wonderful building material. It's five times as hard and twenty times as strong as steel. (That's high-alloy steel. It's one hundred times as strong as the "*mild steel*" that coat hangers are made of.) It is as slippery as Teflon. It is a good conductor of heat and a good insulator of electricity, both excellent qualities in a framework material. And its thermal expansion is tiny. So we'll be using a lot of diamond (and other gemstone or gemlike materials: sapphire, tungsten carbide, boron nitride, etc.) in our *nanomachines*. Cylindrical chunks of diamond work fine for shafts, and slabs of it work fine for walls and supports.

BEARINGS

The most important element of today machine technology is the bearing. The simplest thing a part can do repeatedly is turn, and *bearings* allow it to do this while supporting a force. In your car, the crank shaft, driveshaft, axles, and wheels all turn, supported by bearings. If the bearings for the power of the engine is converted by friction into heat instead of motion, and the car seizes up or catches fire. The simplest form of a bearing is a *sleeve bearing*, simply a shaft going through a hole. The harder the material of the shaft and hole, the better; that why jeweled bearings are used in, fine watches. A door hinge is a common example of a sleeve bearing. Like any macroscopic bearing, a hinge turns ignore easily with a few drops of oil on it.

At the *nanoscale*, a molecule of oil is a lot more like a loose part than a lubricant. On the other hand, we can do something unavailable at the macroscale, which is to make tile bearing surfaces atomically precise. That as smooth as the laws of physics allow. They are still "*bumpy*" with individual atoms, but atoms are soft and slippery. A common lubricant at the macroscale is graphite. Graphite is a form of carbon that comes in stacks of sheets, where each sheet is a single molecule, atomically smooth. The sheets slide over each other easily. At the *nanoscale*, we can roll the sheets up into seamless tubes. Two such tubes, nested inside each other, should you lake a very nice bearing.

They do such tubes, called *buckytubes* or *fullerene nanotubes,* can be made now, on a hit-or miss basis, in enough quantity to be experimented on in the laboratory. Two of them nested one inside the other are sometimes called a *double-walled nanotube* (DWNT). Shell nanotube bearings demonstrated in the lab, with the outer tube rotating on the inner and carrying a load. I am sure that lots of graphite hearings will be used in nanomachines. Forever, they have a couple of drawbacks. First, they have no resistance to an axial force: the tubes are just as happy to slide in a linear fashion as to rotate. For some applications, this is just what you want. Consider the brake cables on a bicycle. For others, it is not. You do not want the axles on your car sliding back and forth.

The other problem is that graphite is soft. In many applications in a machine, you want the bearings, and everything else, to be as hard and stiff as possible, If your door hinges were made of rubber, the door would sag, scrape the floor, and hit the jamb. The stiffer the bearings (and the parts), the more likely the parts will go where they were planned to. We can design (but not yet build at the nanoscale) bearings of diamond and similar hard crystals, though jeweled bearings have been used in fine watches for centuries. With careful attention to the configuration of surface atoms, these can be quite slippery to rotation, but resist axial and radial forces letter than graphite ones.

Macroscopic sleeve bearings are cheap and ubiquitous, but for more demanding applications, more sophisticated forms of bearings are used. One major reason is friction. There is no friction at the atomic scale, but there are effects that cause some of the same problems. At the macroscale, friction causes wear. A properly designed nanomachine will not wear. A wear occurs because the surface imperfections of macroscopic parts, or grit, dislodge material when surfaces move in contact. A *nanomachine*, if properly designed and built, has no surface imperfections and no grit. (This does not mean that a nanomachine will last forever. One good hit from a cosmic ray and it's a pile of junk.)

Heating, the other effect of friction, is something we do have to worry about. *Atomically* smooth doesn't mean mathematically smooth.

So when two surfaces made of atoms move past each other, the atoms of one will press alternately against the atoms of the other and the gaps in between. Now, there is no dragging force as one atom slides on another; atoms themselves are perfectly smooth. But an atom moving across a bumpy surface gets bounced up and down, and that bouncing is heat. Rub your fists together, with your knuckles in contact. You'll feel the vibration all the way up your arms. That vibration represents lost energy in a nanomachine.

You can cut clown this vibration a lot by giving the two surfaces in contact a different pitch, so that one has a different number of atoms per nanometer than the other. You can get an intuitive feeling for the difference by sliding a couple of machine screws along each other. Two screws of the same pitch will lock up as the grooves match; screws of different pitch will slide. Even so, there is still some energy dissipation from sliding surfaces.

At the macroscale, more sophisticated bearings use balls or rollers. Surfaces do not slide, they roll. The same techniques work fine at the nanoscale. You have to remember that even diamond is rubbery at the nanoscale; rolling across it is like rolling across a box spring. You can lose energy by making the whole thing jiggle. But that can be minimized as well, especially if the bearing is designed to run at one specific speed. The natural vibration of the part can be tuned to feed the energy back into the mechanical motion rather than dissipating it as heat.

The next machine element we need is the gear. Here the bumpiness of atoms is an advantage. It is quite possible to have gears whose teeth are atoms or rows of them. In this application, as with *macroscopic gears*, the important thing is to get; the pitch to match so they do not slide easily.

Since atoms are soft and diamond is rubbery, you can—in fact, you must—press gears together, where in a *macroscopic machine* this would cause jamming. In a *nanomachine*, however, this means that the gears are supported and fewer hearings are necessary. In fact, for some designs, there is no distinction. A planetary gear and a roller bearing can be the same thing.

Motors

The subject of fueled engines deserves its own chapter, but we'll consider electric *motors* here. *Electric motors* at the macroscale work by magnetism. Current flowing through wires creates magnetic fields that exert physical forces between the wires and other current-car crying wires or permanent magnets. Unfortunately this won't work at the *nanoscale*. It is those scaling laws again. The magnetic effect gets less power fail as the scale goes down. Luckily, the electric effect gets more powerful, so well use it instead. The electric effect is powerful enough to cause motion even at the macroscale—it's what causes static cling in clothing. At the nanoscale it is almighty powerful. In fact, it is what holds atoms and molecules themselves together.

The simplest way to design an *electrostatic motor* is much like a waterwheel: electrons are pumped in at the top, they fill buckets on the wheel, and they are sucked off at the bottom. Instead of gravity pulling them down, the electric force pushes them from a low voltage area to a high voltage one (low to high because electrons are negatively charged—blame Benjamin Franklin).

The power of a motor is its torque (rotary force) times its speed. For this kind of electric motor, ignoring friction and the like, power is also the voltage dines the current. In fact, the torque is proportional to the voltage and the speed determines the current: the faster those little buckets go, the more electrons are being carried across per second.

Drexler, in *Nanosystems*, analyzes a motor of this type. Its wheel is 390 nanometers in diameter and 25 nanometers thick. It rotates at 800 megahertz (48 billion RPM). The buckets of electrons are moving at about Mach 3 (faster than 2000 mph—the interior of the motor must maintain a vacuum). It draws 110 nanoamps of current at 10 volts, and produces 1.1 microwatts of power.

This may not seem like much, because the motor is tiny. A billion of them, producing 1.1 kilowatts, would be about the size of the ball at the point of a ballpoint pen. Enough of them to match the 100,000 horsepower of a large commercial jet engine would fit in the palm of your band.

Oh, and by the way, Drexler's design is overly bulky, since he was designing a combined motor and flywheel. The motor could be designed to work fine in at least fifteen times less volume. The bottom line is that for any macroscopic device using nanotechnology, motors will never be seen. Microscopic motors can be strewn throughout the fabric of whatever you are designing, capable of providing more physical force or power than you could ever need.

For applications where a precise position is needed instead of top speed and power, a stepping motor, similar to a *clock motor*, is indicated. Like other motors, macroscopic steppers are electromagnetic but the nano-sized ones would be electrostatic. One way to build such a motor would be to separate electric charges in the rotor (this is easy—even so common a molecule as water has a built-in charge separation) so that its has alternating areas of positive and negative charge. Then surround it with electrodes that you charge and discharge with alternating currents. Driving this kind of motor is a lot more complex than the other kind, but you can make it go for backward, stop, turn a specific angle, and so forth. Stepping motors are commonly used today in robotics.

From Here to Autogeny

Now that we have motors, shafts, gears, and bearings, we can build a robot arm. An arm is basically just a series of boxes on hinges with a motor to power each joint. Typically the motors are geared down to provide more force and higher precision (and stiffness) than the motor would have alone. Robot arms with grippers (a finger is just a miniature arm) can pick up things and move them, and in particular can build more robot arms. Macroscopic parts are made in many different ways, ranging from molding, to forging (beat it into shape), to machining. Machining is the most expensive, often the slowest method, but also the most accurate and least restricted in what it can make. A milling machine and a lathe can make just about any machine part (you need special attachments for screws and gears).

Milling machines and *lathes* can he thought of as very specialized robot arms. What they do is hold a knife and carve the shape you want out of the chunk of material you give them. To get the shape just right, they must hold the knife very firmly. This is technically known as *stiffness*, and it tells you how far the knife (or whatever) will move given a certain force against it. The higher the stiffness, the better.

A milling machine doesn't look like an arm. All the joints are big, heavy, precision-machined hunks of metal that don't wobble at all. If we build a nanomachine on the plan of a milling machine, the position of the tool could be have exactly controlled with respect to the position of the work piece. And this is the kind of machine that we'll make to build the parts that the slender robot arms will put together. There is just one big difference. A classic milling machine (or lathe) operates by taking a chunk of material and cutting off every thing that does not look like the part you need. At the nanoscale, a better strategy would be the opposite: to build up the part you want, an atom or molecule at a time. This is what allows you to have complete control over where every atom is in the finished product.

This method of controlling the reaction precisely is called *positional chemistry* or *mechanosynthesis*. Mechanosynthesis is what will allow us to make million—atom parts and get every atom exactly right. This, in turn, will allow us to build molecular milling machines and do mechanosynthesis. And that is our *autogenous technology*.

Pumps

Unless you live on an old-fashioned farm and milk your cows by hand, the milk you drink came through a milking machine and subsequent processing and handling stages, courtesy of a number of *pumps*. Unless you own an artesian well, the water you drink was pumped, either from the ground or by way of water mains from a reservoir. Commonly used fuels are pumped, many times, through processing plants, from one tank to another along pipelines, into your car at the service station, and again from your gas tank to your car's engine.

Pumps for air in low-pressure applications are called *fans*. There are probably fans in your computer to cool it, fans in your home heating system to heat and cool you, and more fans in your car. A typical hydronic home heating system has both fans and pumps.

Moving fluids around and managing their pressure is a substantial application of, and a necessity for, modern technology. Nanomachines will need to do it, too. If anything, most of the designs for dumps at the macroscale will work better at the *nanoscale*. The simplest pump is fan. Fans are noisy and inefficient because they cause turbulence. At the nanoscale they won't, because the smaller you get, the more syrupy fluids act. For gases, fans can be made that interact more or less individually with the molecules. Molecules of a gas bounce off one another like billiard balls, and fly some distance before they hit another molecule. The average distance for a given gas at a given pressure is called the mean free path. If the length of a molecule's path through the fan is smaller than the mean free path, molecules going through the fan tend to interact with the fan and not one another. If the fan blades can move for than the molecules tend to, the fan can attain something like a compression ratio of ten with one rotor and one stator (or pair of counter-rotating fans). Ordinary air has a mean free path of about 60 nanometers and an average speed in the neighbourhood of the speed of sound—figures easy for a *nanofan* to match.

One common pump for liquids, the *centrifugal pump*, is not recommended at the nanoscale for scaling-law reasons. But the kind you probably think of when someone says *pump*, the piston or positive displacement pump, should work fine. The positive displacement pump usually consists of two one-way valves and a cylinder in which a piston goes up and down. At the macroscale, the valves are often spring valves, but at the nanoscale it will probably be preferable to make them mechanically actuated in a timed sequence with the piston, like the valves in a car engine.

Can we make a *positive displacement pump* that interacts with the fluid one molecule at a time? Yes, and in a fairly elegant design. Consider a cylinder with a set of molecule-shaped dents in it. It is embedded in a wall like a revolving door. The molecules go through as the cylinder turns, just as people go through the door. But the trick is that a little piston at the bottom of each dent pushes the molecule out when it on the output side, and retracts to let a new molecule in on the input side.

Doing it this way allows you to make the dents much more attractive to one kind of molecule than to another, by making them fit snugly. This is how a large proportion of the molecular machinery in your body works—it is how *hemoglobin* grabs oxygen molecules to carry them in your red blood cells, for example. With a set of dents that strongly favours one kind of molecule, we can pump that fluid preferentially out of a solution—for example, oxygen out of air, fresh water out of seawater. This kind of selective pump is called a *molecular sorting rotor*.

Nanocircuitry

Light and radio are waves in the same stuff: electromagnetic fields. The only difference is the length of the waves, and thus the frequency. With fast enough *nanoswitches* and appropriated sized antennas, we should be able to handle light with the same facility that we now handle radio waves. Today, for example, we have phased-array antennas for radar. A flat array of antennas can produce a beam going in any desired direction by having each little antenna match the phase properly. What is more, it can detect the direction of an incoming beam without a lens or mirror. In essence, it a controllable hologram at microwave frequencies.

With *nanoelectronics nanoantennas*, we should be able to do the same thing with light—phased-array optics. Cover the surface of an object with optical antennas and it should be able to take on any properties you like that involve the reflection and absorption of light, or generate a completely synthetic pattern, making it appear to be anything you want. You could even have it present the image that would be there even in its absence, making it effectively invisible.

In order to make antennas and to connect switches, you need wires. Macroscopic wires are generally made of metal. Metal conducts electricity because the electron (some of them anyway) are not nailed down to particular nuclear or in particular bonds, as they are in diamond, for example, but are free to roam around at the behest of any passing electric field. Graphite conducts electricity because electrons can roam, but only within a given sheet. Rolled-up graphite, or buckytubes, can conduct or not depending on their particular structure. Various molecules have areas in which electrons can move around. These areas act as optical antennas at various frequencies, which is why different substances have different colours and albedos.

Conductive paths at the *nanoscale* can be made in many ways besides laying down wires of metal, but wires of metal do work. The other ways just: give us other options. Some arrangements of atoms form *superconductors—substances* with no electrical resistance at all. Since superconductors can be made with the statistical mixing of current-day chemistry it seems likely that we could make better ones with a technology that allowed us to place atoms in a specific, desired pattern. Superconducting wires would be useful at the nanoscale, since ordinary wires gain more resistance as they get thinner. It remains to be seen whether superconducting wires can be made thin enough to be useful, however.

Most of the elements—parts and mechanisms—of macroscopic machines can be scaled to work at molecular size, like sleeve bearings, or redesigned to, like electric motors. The few that cannot, like *centrifugal pumps*, have substitutes, plus there are many new designs, like sorting rotors, that don't have any macroscopic parallel at all. At we can confidently expect to be able to design and build machine systems at the *nanoscale* with a wide variety of capabilities and applications, and in particular, to design and build manufacturing systems.

Nanobiotechnology

In this section practical uses for *current* biotechnology, will be dealt and about what tools biotechnology can present toward making structures for molecular nanotechnology. Protein structures will be primarily dealt, although there are many other structural components of cells that might be useful for nanotechnology. In particular, the physical-plant of a cell in terms of the design of the structural motifs of proteins, which could provide useful clues for designing nano sized structures are to be discussed including the tools (found in cells) that are used for receiving molecules from other sources, and holding molecules in storage until they are needed. How manufacturing control in a cell is enabled, and how inventory control is done in the cell will also be discussed touching about the management information systems in a cell, and manufacturing methods in a cell.

It is well known that DNA controls the metabolic processes of the cells by means of transcription to form RNA molecules, which are then translated to make protein molecules that embody the information previously stored in the DNA. The protein molecules, among other functions performed in the cell, work through a feedback loop to control the expression a replication of the DNA.

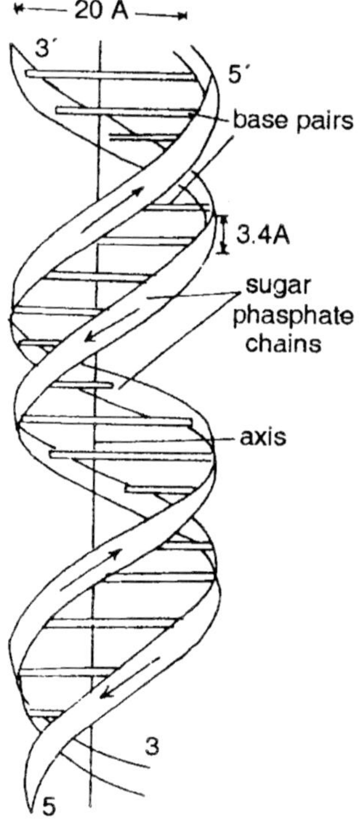

Fig. 2.1. Structure of DNA.

Structural Proteins

The physical-plant of the cell, and the components that surround the cell (termed the *extracellular matrix*), consist of structural proteins. Examples are the proteins collagen and keratin. *Collagen*, which is found in skin, adopts a triple helical structure. *Keratin* also has a triple helical structure. Collagen is, in fact, a repeating polymer of just three amino acids, with a few special amino acids inserted to provide for cross-linking and stability. This type of cross-linking is found throughout the animal and plant kingdoms to provide structural stability. Potentially, this type of motif could provide a basis (or platform) for building more complex components in a molecular manufacturing system.

Another macromolecular component of the physical-plant are the *carbohydrates*. Carbohydrate structure is not as well-understood as protein structure. The most abundant carbohydrates are cellulose and chitin. In fact, cellulose is a structural component that makes up more than half of all the carbon structures found in the world. As time progresses and carbohydrate analysis improves (especially the sequencing of carbohydrates), carbohydrate probably will be given serious consideration as an element in the design of structural components.

Fig. 2.2. *Collagen is a structural protein found in cells.*

Molecular Manufacturing within the Cell

DNA is used as the information source for transcribing messenger RNA, which in turn provides the information to produce proteins. One way to produce novel proteins is to modify the DNA, and thus, modify the messenger RNA. Novel DNAs can be created by chemical synthesis of DNA, or by selection and isolation of natural mutations.

Once the message has been created for the manufacture of a protein, a protein of nearly any imaginable amino acid sequence can be manufactured. The *messenger RNA* binds to a complementary section of the two *transfer RNA* molecules bound to the ribosome. One transfer RNA

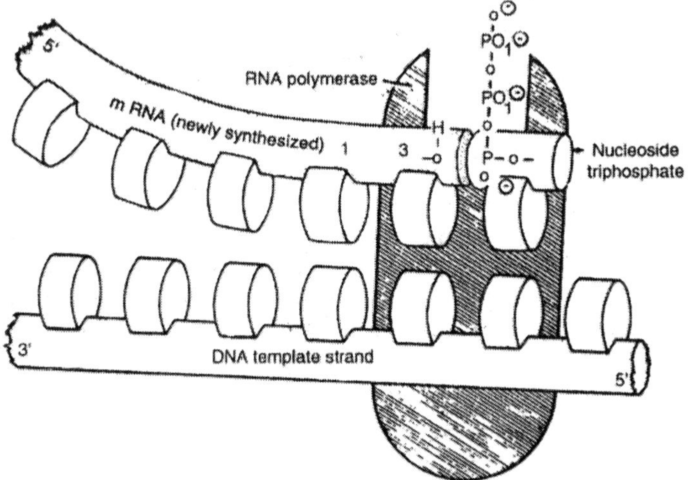

Fig. 2.3. Role of RNA polymerase in translation.

contains the growing protein chain, which is then joined to the next amino acid to be added to the protein, which is attached to the other transfer RNA. The messenger RNA then shifts (termed *translocation*) to allow the binding of a new transfer RNA with the next amino acid to be added. People have explored making unusual proteins by artificially loading transfer RNA with amino acids other than the usual 20 amino acids used for protein synthesis in nature. However, that is probably not a very practical protein production technique.

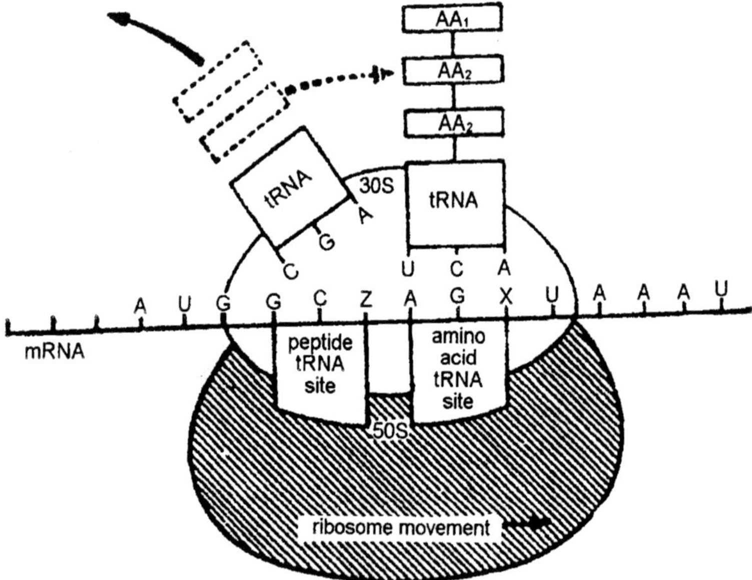

Fig. 2.4. Formation of polypeptide chain.

Inventory Control in the Cell

The cell has a large number of *enzymes*. Many types of feedback control mechanisms are used to keep an appropriate number of molecules of each enzyme, and to maintain a homeostatic balance in the cell. A substrate molecule binds to an enzyme and is cleaved to give two products. However, the smaller product also can bind to the enzyme and, thereby, inhibit the large substrate molecule from binding to the enzyme. The extent of this inhibition increases with the amount of the product that has already been produced. Thus, product binding to the enzyme effectively limits the concentration of the product that can be produced. This is a very common feedback mechanism in biosynthesis. The presence of many thousands of enzymes, each having

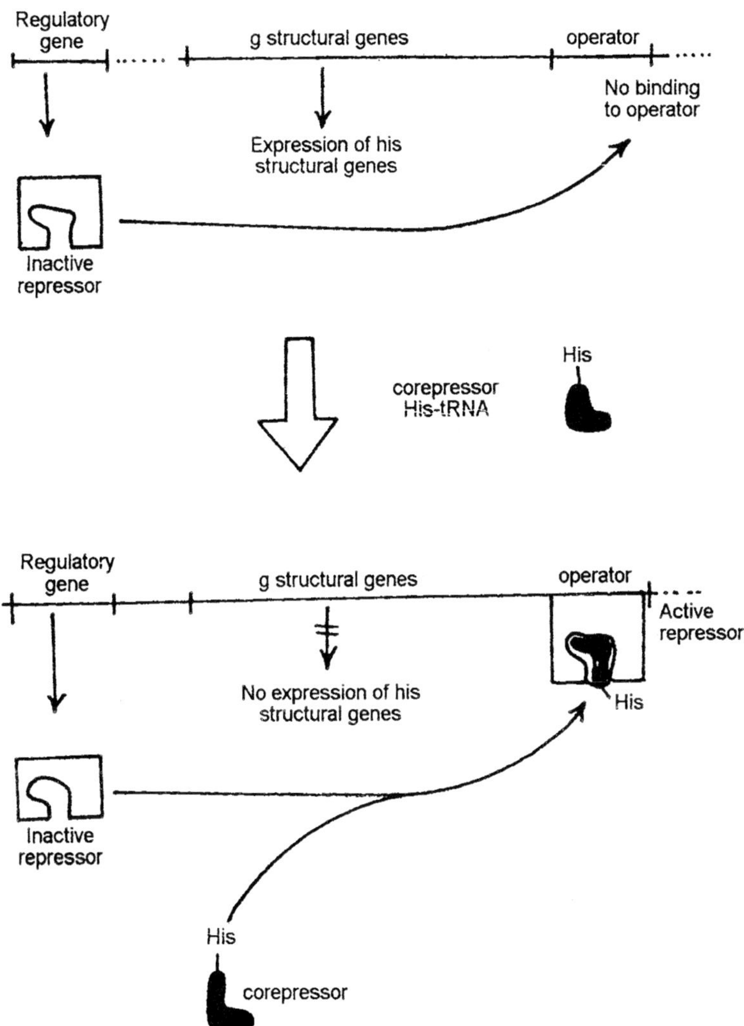

Fig. 2.5. Corepressor-repressor complex.

a variety of feedback loops, leads to a very complicated process of regulation. Considering these feedback loops leads to questions such as how to design a protein appropriately so that the binding of some reaction product gives the desired level of feedback.

Recent Technology for Producing Proteins

Proteins and peptides can be obtained in a variety of ways. One of these is *classical solution-phase chemical synthesis*. In this case, amino acids are modified with appropriate protecting groups to protect reactive portions of the amino acid, which if allowed to react, would create unwanted products, and then dissolved in solution. A reagent is added that causes two amino acids to combine, forming a peptide bond between the amino acids, and to exclude a water molecule. This is the classical approach for *small peptides*.

To make *larger peptides* (up to approximately 40 amino acid residues long), the only practical synthetic approach is *solid-phase synthesis*. Here, the first amino acid is attached (*covalently bonded*) to a solid support, and one amino acid (again with appropriate chemically reactive groups protected) is added at a time until the desired chain has been built, still attached to the solid support. The synthetic product is then cleaved from the solid support, stripped of the protecting groups, and extensively purified to give the final product.

Fragment condensation is a variation of these two techniques. Fragments of the desired peptide are made on a solid support, and then later combined in solution. This has some of the advantages of both methods, although it also has disadvantages (principally that the fragments often are not very soluble so that they cannot be coupled in solution).

In addition to these chemical approaches, several biological approaches can also be used. Proteins of interest can be purified from natural sources. Or, the gene for a protein of interest can be inserted into cells, the cells can be grown in culture to express large quantities of that protein, and the protein can be purified from the cultured cells. Some proteins can be purified from fermentation broths of cells.

Semisynthesis is a combination of the previously discussed methods. For example, a protein of interest could be obtained by biological methods, and the split into fragments by using chemical or biochemical methods. Changes could be made to one or more fragments by using the methods of organic chemistry, and then the fragments could be recombined to form a partially synthetic molecule. The advantages of using the synthetic approaches include the ability to introduce (into the molecule) segments that are not normally found in nature. This gives the molecule conformational (or physical) properties that could not be achieved with a molecule taken directly from a cell.

In general, to make *small peptides* (on the order of two to eight amino acid residues—often the size that is used in the pharmaceutical industry) the most economical approach is to use solution methods. To make peptides of 20 to 30 residues, solid-phase is almost the only way to do rapid development work. Beyond this size, it is generally necessary to use *biological synthesis* or *semisynthesis*.

Protein Structures Begin with Linear Chains of Amino Acids

Protein molecules are sometimes depicted as *amorphous blobs*. Peptides and proteins are chains of amino acids. One of the main characteristics of amino acids is that they have *handedness*. The chemical term for that is *chirality*. The side chain (referred to as an *R group*) is shown extending to the left of the central (*alpha*) carbon atom. The 20 amino acids used in biological protein synthesis have 20 different R groups, which give the amino acids their different properties. The amino group containing a nitrogen atom is on the far side. Linking these amino acids together to form a chain, excluding a molecule of water for each peptide bond formed, gives a polypeptide. Conventionally, peptides of 50 or more amino acid residues (about the size of *insulin*) are called *proteins*.

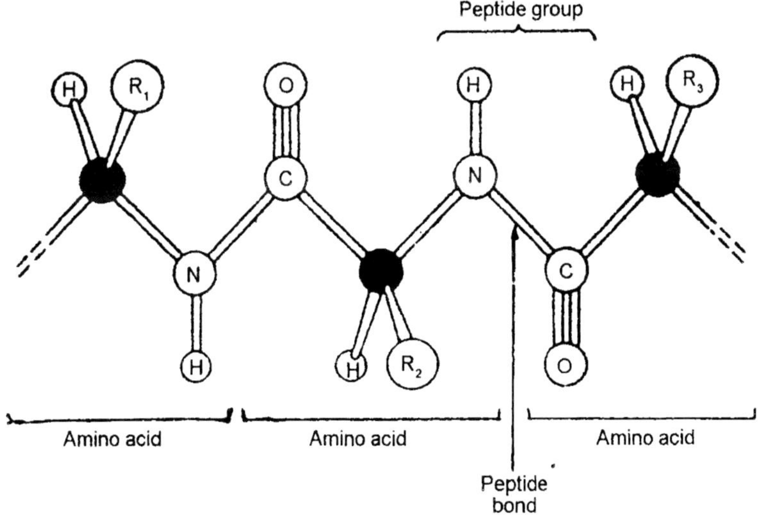

Fig. 2.6. The arrangement of atoms in amino acids joined by peptide bonds.

Complex Proteins

In addition to the *peptide bonds* linking amino acids into a linear array, other structural elements are important features of protein structures. These elements contribute to folding the linear chain in three dimensions. One of these is *hydrogen bonding*. Here, a relatively weak bond is formed between certain hydrogen atoms and an oxygen that is distant along the linear sequence, but near in three-dimensions as the chain folds. A large number of these weak bonds can lend significant stability to a protein structure.

A more stabilizing element that is not found in all proteins is a *covalent disulfide bridge* providing a cross-link between two sections of the molecule that are again distant along the linear sequence, but spatially near as the chain folds.

There are 20 amino acids which are used in proteins, and they have very different properties. One group is *hydrophobic* (that is, they do not like water and can be described as "oily"). Another set is very bulky, and also hydrophobic. Another set is electrically charged and *hydrophilic* (that is, they like water and they make very good hydrogen bonds). Another set is *polar* (which means that they like water, but they are not charged). A few amino acids also make unique conformational contributions to the structure of the protein molecule. For example, proline is a cyclic amino acid often found in turns (where the protein chain moves out from the interior of the globular structure, makes a turn, and moves back inside).

Secondary Structure in Proteins

As polypeptides become larger, the protein structure becomes more complete and distinct *secondary* and *tertiary* structures are found. One common secondary structure feature is for the chain to form a *helix*. Helices of different types can be described by the pitch of the helix, which is related to the number of amino acid residues per turn of the helix, and to the distance along the helix axis covered by each residue.

As illustrated several helices with different pitches. The example of the left is a case where the helical pitch is *zero* and the peptide is a covalent closed circle (in this case, of five residues). The next helix has four residues turn, and is similar to the *alpha helix* found in many proteins. The alpha helix a stable, rod like structure in which the CO of one amino acid residue is hydrogen bonded to the NH four residues ahead in the linear sequence. This forms tight coil with 3.6 residues per turn.

If the helix is stretched further, it will form 3_{10} helix, with three residues per turn. If the helix is less stretched, it will form a pi helix, with 4.4 residues per turn. These helices are less favourable energetically than the alpha helix, and are seldom seen in protein structures. The potential to form hydrogen bonds is the major reason that polypeptides form helices. The strength of the hydrogen bonds made possible by the different helix geometries determines why some potential helix structures are more stable than others.

If the helix is stretched further so that there are only two residues per turn, the result is not a helix, but an extended structure termed the *beta strand*, which is as extended as a peptide chain can get.

If two beta strands are placed together, a *beta sheet* is formed. One of the interesting properties of the beta sheet is that it has two sides. The side chains of consecutive amino acids alternate in the direction to which they are presented. Thus, the side chains of alternate amino acid residues point either down or up. This allows the presentation of different faces of this structure to different environments within the protein. For example, if the upper side of the beta sheet were exposed to water, the side chains on the upper side of the beta sheet would probably be largely *hydrophilic* (charged or polar) while those on the bottom (pointing to the interior of the protein) would likely be *hydrophobic*. Having the hydrophilic side of the beta sheet in contact with water and the hydrophobic side in contact with a hydrophobic environment stabilizes the bet structure.

Many *hydrogen bonds* are formed between adjacent strands of a beta sheet and these help to stabilize the structure. In general, beta structures are found the interior of proteins. They tend to have more hydrophobic residues than do alpha helices, although there are some alpha helices that are fully buried wit the interiors of proteins.

An example of a beta structure is *silk*. Silk fibers are rich in alanine glycine residues. These fibers are made of ***beta strands*** in which alanine residence are on one side of the beta sheet and glycine residues are on the other. The beta sheets are packed on top of one another such that alanine residues on sheet face alanine residues on the next sheet, and glycine residues face glycine residues. The opposing alanine residues keep their respective beta sheets from moving with respect to each other, because moving the sheets across one another would cause the alanine side chains to collide. That is why silk does not stretch, although it is flexible perpendicular to the beta sheets.

One structure that allows the protein chain to abruptly change direction is called the *beta turn* structure. Here, the protein chain enters the turn from the top right, swings around, and continues to the bottom right. These structures are seen very frequently (in about 20 to 25 percent of the residues) in proteins and allow the chain to change direction 180 degrees.

Structural Motifs in Proteins

These elements of secondary structure (*alpha helices*, *beta strands*, and *beta turns*) combine to form a number of *structural motifs*. These same structural motifs are seen in very diverse proteins. The proteins may often have very different functions, yet they are very similar in terms of the organization of the elements of secondary structure into structural motifs. The most prominent feature consists of the four alpha helices, which are oriented in a parallel fashion with adjacent pairs of helices antiparallel (the helices point in opposite directions) to each other. The interior of the four-helix bundle is hydrophobic and the exterior is hydrophilic.

This protein is very similar to structures that have been designed *de novo*. The approach that people have taken for *de novo* design was to look first at the proportion of the various amino acids found in helices in proteins of known three-dimensional structures. They found that certain amino acids are more likely than others to occur in helices. They have used those statistical properties to design molecules of about this size that fold into a four-helix bundle. These molecules were designed without using any direct sequence homology to known proteins. They found that they could design proteins that folded according to design to make compact structures.

Going beyond the design of a structure without function, John Stewart and colleagues of the University of Colorado placed certain catalytic residues the alpha helix bundle to *mimic* proteolytic enzymes. They placed the Residues dues at specific sites m the molecule chosen to be close enough spatially to each other to catalyze a hydrolysis reaction, and they did, in fact, obtain enzymatic activity.

Thus, some design work has been done and successful predictions have been made. Several key features have been essential to these successes. The researchers had a detailed knowledge of the structures that they were trying to mimic, and had identified the key features of these structures. They had a clear idea of the specific modifications that they wanted to make. The structures were small enough for current methodologies to enable the researchers to predict what the effects would be of the modifications that they intended to introduce.

Life is not very simple. Not all *alpha helices* in proteins are organized into a four-helix bundle. One of the other common structural motifs for alpha helices in proteins is seen in *hemoglobin*. Instead of the helices being oriented parallel to each other, they cross each other. The result is a structural motif with a very different shape from the four-helix bundle, even though the number of helices involved in each motif is similar. Such differences are one reason why it is very difficult to predict from the amino acid sequence of a protein what will be the folding pattern of that protein. You can predict a high probability that a given sequence will contain a substantial proportion of alpha helices, but the way in which the helices associate to form a structural motif will depend both upon the specific properties of the helices (determined by the specific amino acid sequence of each helix) and upon the way in which the helices interact with other portions of the protein molecule. Structural motifs like the four-helix bundle and the *hemoglobin helix motif* can be termed a *structural domain*. A domain is a group of interacting secondary structure elements that can be viewed as one structural unit.

A number of protein structural domains have been identified. The beta strands form a barrel-like structure, with the beta strands rep resenting the staves of the barrel. Accordingly, these structures are called *beta barrels*. A larger beta barrel structure is found in *superoxide dismutase*. This barrel is a less twisted barrel than the one in the previous example. Here, the beta strands are more parallel to each other. Like the previous examples of alpha helical domains, this protein also has metal atoms associated with it, but it has very little helical structure. Again, life is not so simple as to have all domains consist of either all alpha helices or all beta strands. The beta structure tends to be *hydrophobic* (not liking water) and inside the molecule. The helices tend to be hydrophobic along one surface and *hydrophilic* on the other surface, and are oriented with the hydrophobic side of the helices, essentially protecting the beta strands from water, and thus stabilizing the structure. As seen from the end view, the beta strands form a twisted sheet, which is typical of beta strands because of steric effects of the amino acids usually found in beta strands. It can also be seen that both sides of the beta sheet are covered by helices.

An illustration of the fact that proteins with entirely different functions can have similar structures is provided by carboxypeptidase. *Carboxypeptidase* is an enzyme that cleaves residues from the carboxy terminus of proteins. *Lactate dehydrogenase* is a *redox protein* (an enzyme that catalyzes the simultaneous oxidation of one substrate and the reduction of another substrate). Both enzymes have a similar supersecondary structural motif formed from a twisted beta sheet covered by alpha helices. It can be seen that proteins can be organized into families according to the presence of well-characterized, super secondary structural motifs.

Bacteriochlorophyll protein is an example of a *membrane protein*. It is almost all beta structure, with the beta structure being in contact with the membrane. This protein harvests light. Someone asked if photons could be used to transfer information to control *nano machines*. The chromophores of this protein absorb light, so this protein could be used to receive information by way of photons.

An example of a more complicated protein is *triose phosphate isomerase*. It is similar to the previous examples in that it consists of beta structure surrounded by alpha helices to protect the barrel from solvent. Panel B shows this beta barrel (which is more hydrophobic on both sides than the previous examples) to be different in shape from those examples.

Lessons for Molecular Nanotechnology

These many examples demonstrate that design criteria can be applied to building fairly large molecules. Biological molecules afford examples of many defined structural types that could be incorporated into designs. It is probably not possible to perform a *de novo* prediction of the complete structure of large, complicated molecules. However, such approaches are feasible with smaller molecules, given the help provided by the knowledge base of known protein structures.

Practical Considerations for Biological Production of Proteins

In general, using mammalian cells to produce proteins means a substantial time lag between completing the design and producing a molecule of known purity. A number of problems are associated with using mammalian cells. One problem is that most proteins produced by mammalian cells have some carbohydrate side chains attached to the protein. Carbohydrate side chains are often variable in structure, so that the protein obtained may not be of defined structure (that is, not all molecules of the protein will have identical carbohydrate side chains attached). This is a significant problem for precise characterization of the protein produced by mammalian cells.

If bacterial cells are used to produce protein, the protein will not have any carbohydrate attached. However, some carbohydrate side chains are often necessary for the protein to be soluble. Therefore, the proteins are often produced by bacteria as insoluble "*inclusion bodies*" so that purification can be difficult. Sometimes it is not possible to produce a particular protein in bacteria.

In general, it takes two to three months from the time you decide to produce a specific mutation in a protein to the time that you can produce very small amounts of a fairly pure, mutated protein. That sounds relatively gloomy. However, you can make many proteins in parallel. As an example, a group of five or six people was able to produce about 250 mutant proteins, obtain information about their reaction kinetics and other structural information, on relatively small amounts of material, over a period of about a year to a year and a half. That means that a fair number of possibilities can be tried without enormous resources.

These capabilities mean that, at least for small proteins, it is possible to design a protein and to make a number of structures related to the design to test the design principles. What turns out to be the limiting factor in this whole process is the time that is required to determine the three-dimensional structures of the proteins that are produced. Often it is very difficult to crystallize proteins. Even if a protein can be crystallized, it still takes a significant amount of time to determine the structure by *X-ray crystallography*.

What seems to be indicated to accelerate progress with protein molecules are methods for solving structures (at least at modest resolution) in days. What we need for proteins is comparable to the capabilities that exist with small molecules to run a synthetic reaction—place a sample into a *nuclear magnetic resonance* (NMR) spectrometer, determine whether the desired synthesis was achieved, and if not, to walk back into the laboratory for another attempt the same day. The development of analytical techniques to do something similar for protein structures will be a key event in pushing forward the use of biotechnology to develop *molecular manufacturing*.

3

Atomic Force Microscopy

The *atomic force microscope* (*AFM*) is one of the many recently developed types of nanometer-resolution microscopes or *scanning probe microscopes* (*SPM*). The AFM has a much wider range of applicability than the other SPMs: it can be used to observe and manipulate nanometer-sized objects of both conductive and insulating nature in both vacuum, air, gasseous, and liquid environments. The following sections provide a more detailed treatment of the physical principles behind the AFM.

Basic Principles of the AFM

A sketch of the layout of a typical AFM setup is shown in Fig. 3.1. The central component is the nanometer-sized tip mounted on the elastic cantilever. By use of the piezo-electric *xyz* scan driver the tip

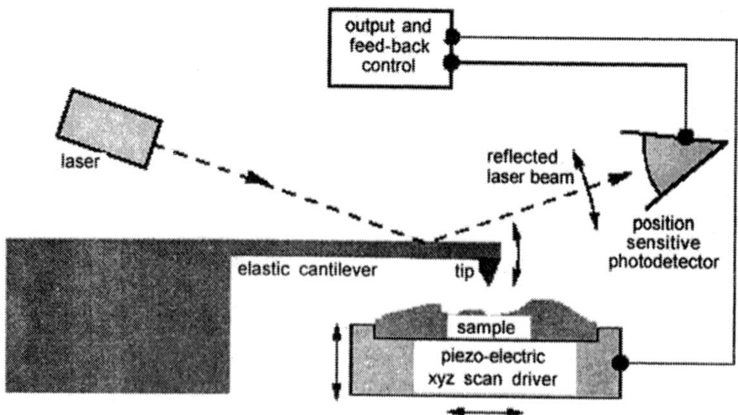

Fig. 3.1. A sketch of the layout of a typical AFM set-up showing the key components.

is scanned across the sample. The distance between the tip and the sample can be anywhere between 0 and 100 nm. Dependent on the topography and contents of the surface of the sample the force acting on the tip changes during the scan resulting in a position dependent deflection of the elastic cantilever. The deflection is monitored by reflecting a laser beam on the cantilever and recording the movement of the reflected beam with a position sensitive *photodetector*. The signal from the photodetector can be used in a feedback loop with the *xyz* scan driver to control the motion of the tip.

To build a well-functioning AFM several requirements has to be fulfilled:

1. The spring constant of the cantilever should be small enough to allow detection of minute atomic forces.
2. The resonance frequency of the cantilever should be as high as possible to minimize sensitivity to external mechanical vibrations.
3. The tip should be as sharp as possible to allow atomic resolution.
4. The tip should be as narrow as possible to allow penetration into deep troughs on the surface.

We shall look into these requirements in the following sections, where the different modes of operations are described.

Cantilever: Spring Constant and Resonance Frequency

The actual design geometry of an AFM cantilever can vary. For simplicity we shall in the following analysis just consider a beam of length L and rectangular cross section with height h and width w. Due to interactions with the surface atoms of the sample an external force F_{tip} acts on the tip and bends the cantilever.

It is a simple exercise of applied elasticity theory to find the bending Δz of the tip due to the force F_{tip}. Hooke's law for elastic solids relates the force per area (the stress) σ_{xx} with the relative change in length (the strain) Δ/L_0,

$$\sigma_{xx} = Y\frac{\Delta L}{L_0}, \qquad ...(1)$$

where Y is Young's modulus, a material parameter we already encountered and calculated. Due to the bending the cantilever is compressed on the lower surface, $L < L_0$ at $z = -h/2$, elongated on the upper surface, $L > L_0$ at $z = +h/2$, and unchanged at the center plane, $L = L_0$ at $z = 0$. Thus the length $L(z)$ of the cantilever is a function of z. We assume that the bended cantilever has the circular geometry depicted in Fig. 3.2(b). This is an approximation that will

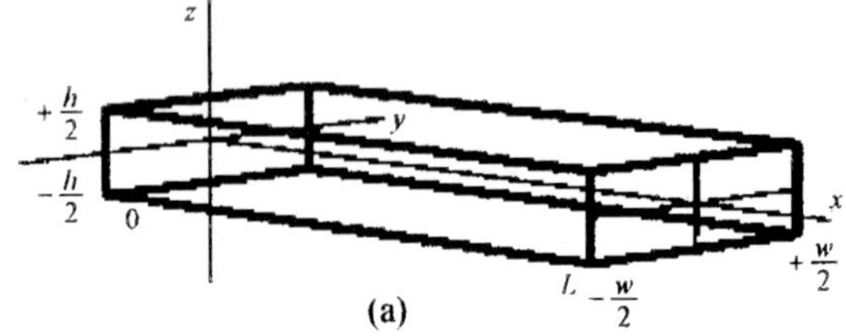

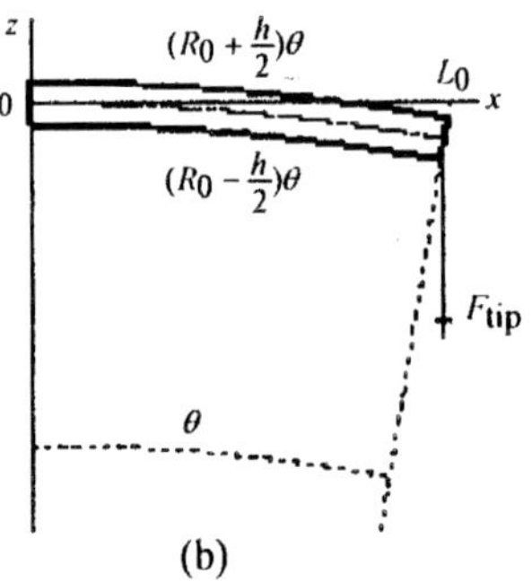

Fig. 3.2. (a) An AFM cantilever of dimensions L×w×h. (b) The approximately circular bending of the cantilever due to the external force F_{tip} acting on the tip.

allow us to obtain decent results without too much math. The math is simplified due to the fact that a circle has a constant radius of curvature. The exact results are quoted along the way.

At Fig. 3.2(b) we see that there is an angle θ between the end planes at the two ends at $x = 0$ and $x = L$. If we denote the radius of the bended but unstretched center plane $z = 0$ by R_0 we have $L_0 = R_0\theta$ and deduce that

$$L(z) = (R_0 + z)\theta = \left(1 - \frac{z}{R_0}\right)L \Rightarrow \frac{\Delta L}{L_0} = \frac{L - L_0}{L_0} = \frac{z}{R_0}. \quad \text{...(2)}$$

The *external force* F_{tip} acts on the tip and bends the cantilever. However, even in this situation the cantilever is in equilibrium, and consequently the external *force torque* L_0F_{tip} around the line $(x, z) = (0, 0)$ at the left end of the cantilever must balance the internal torque set up by the internal stress forces σ_{xx} times the arm z:

$$L_0 F_{tip} = \int_{-\frac{w}{2}}^{+\frac{w}{2}} dy \int_{-\frac{h}{2}}^{+\frac{h}{2}} dz\, z\, \sigma_{xx} = w \int_{-\frac{h}{2}}^{+\frac{h}{2}} dz\, z\, Y \frac{z}{R_0} = \frac{1}{12} Y \frac{wh^3}{R_0}$$

$$= \frac{1}{6} Y \frac{wh^3}{L_0^2} \Delta z, \quad \text{...(3)}$$

where Δz is the vertical displacement of the cantilever at $x = L$. It is given by $\Delta z = R_0(1 - \cos\theta) \approx \frac{1}{2}R_0\theta^2 = \frac{L_0^2}{2R_0}$ or $1/R_0 = 2\Delta/L_0^2$. So the force F_{tip} can be expressed by the displacement Δz of the tip and a spring constant K:

$$F_{tip} = K\Delta z, \quad \text{...(4a)}$$

$$K = \frac{1}{6}\left(\frac{h}{L}\right)^3 wY \text{ (the circular shape approximation),} \quad \text{...(4b)}$$

$$K = \frac{1}{4}\left(\frac{h}{L}\right)^3 wY \text{ (the exact result).} \quad \text{...(4c)}$$

From the spring constant we can estimate the (smallest) resonance frequency ω_0 of the cantilever (corresponding to a quarter wavelength oscillation) when it is oscillating freely. It is given by $\omega_0 = \sqrt{K / m_{eff}}$. Note that it is not the cantilever mass m that appears in the formula but an effective mass m_{eff}. This is due to the fact that the various parts of the cantilever do not oscillate with the same amplitude. For $x = 0$ the amplitude is zero, while for $x = L$ it is the maximal Δz. It has already been shown how the circular shape approximation leads to $m_{eff} = m/5$ and thus

$$\omega_0^2 = \frac{K}{m_{eff}} = \frac{5}{6}\frac{Yw}{m}\left(\frac{h}{L}\right)^3 = \frac{5}{6}\frac{Yw}{\rho whL}\left(\frac{h}{L}\right)^3 = \frac{5}{6}\frac{Y}{\rho}\frac{h^2}{L^4}, \quad \text{...(5)}$$

where we have introduced the mass density ρ. The approximate and exact results are

$$\omega_0 = 0.91\sqrt{\frac{Y}{\rho}}\frac{h}{L^2} \text{ (the circular shape approximation),} \quad \text{...(6a)}$$

$$\omega_0 = 1.02\sqrt{\frac{Y}{\rho}}\frac{h}{L^2} \text{ (the exact result).} \quad \text{...(6b)}$$

Eqs. (4c) and (6b) are central to the design of an AFM. The ratio h/L should be small to give a small spring constant and hence a good force sensitivity, but h/L^2 should be large enough to ensure a high resonance frequency that minimizes the sensitivity to external mechanical vibrations. The solution is to make small cantilevers. One good method is to make them in silicon since one then can take advantage of standard silicon processing techniques perfected by the microelectronics industry.

A typical choise is $L = 100\ \mu m$, $w = 10\ \mu m$, and $h = 1\ \mu m$. A deflection $\Delta z \approx 1 Å$ is easily detected and since for silicon $Y = 1.6 \times 10^{11}$ Pa and $\rho = 2.33 \times 10^3$ kg m^{-3}, we obtain the following spring constant K, tip-force F_{tip}, and resonance frequency ω_0 from Eqs. (4a), (4b), and (6a):

$$K = 0.40 \text{ N m}^{-1} \qquad \text{...(7a)}$$

$$F_{tip} = 40 \text{ pN} \qquad \text{...(7b)}$$

$$\omega_0 = 927 \text{ kHz}. \qquad \text{...(7c)}$$

These numbers are good news. The typical atomic forces acting on an AFM-tip is of the order 1 nN, and mechanical noise above 1 kHz is strongly suppressed.

Contact Mode

When operating in the so-called *contact mode* the AFM tip is forced down into the surface until repulsive contact with the *core electrons* of the surface atoms is obtained. During a scan the AFM tip simply follows the topographic features of the sample with atomic resolution (if the tip is sharp enough).

To illustrate the physics of the contact mode it is sketched in Fig. 3.3(a) how the electron orbitals get deformed as two atoms approach each other. At long range the weak attractive *van der Waals force* is the dominant force. This force will be treated in the next section. At intermediate range the cloud of valence electrons surrounding the atom get in contact and become deformed as a diatomic molecule is formed. At the closest range also the core electron clouds of the two atoms touch each other and get deformed. At this point the energy increases rapidly as a function of decreasing inter-atomic distance. It is extremely costly energy wise to compress the tightly bound core electrons further.

In general it is very difficult to calculate the exact form of the energy versus distance. There are no small parameters in the problem and many atomic orbitals must be taken into account in a serious calculation. It is customary, however, to parametrize the potential by the Lennard-Jones model potential

$$E(r) = B\left[\left(\frac{A}{r}\right)^{12} - \left(\frac{A}{r}\right)^{6}\right]. \qquad \text{...(8)}$$

The r^{-6}-term is well understood. The r^{-12}-term on the other hand is just an ansatz. There is no particular reason why the power of 12 has been chosen except that it is bigger than 6 and mimmicks the

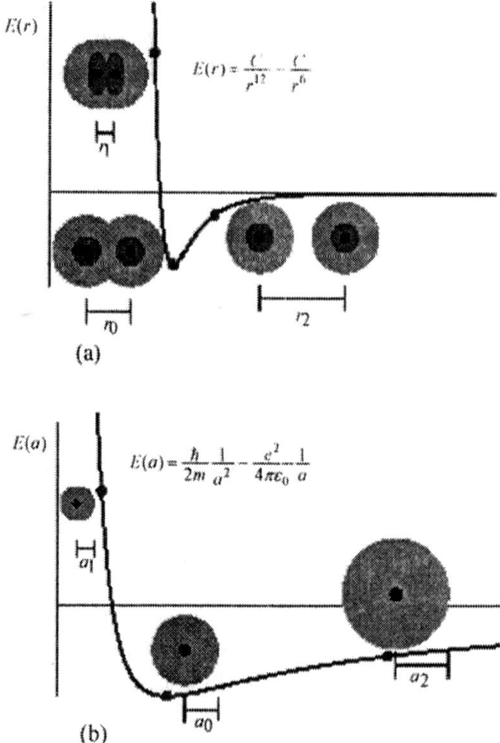

Fig. 3.3. Similarities and differences in the energy of diatomic molecules and single atoms. (b) The simple model of the ground state energy E(a) of the hydrogen atom as a function of the average distance a between the electron and the proton.

very strong repulsion observed in many experiments. The energy versus distance curve $E(r)$ ressembles the simple energy versus size-of-atom curve $E(a)$. For comparison $E(a)$ is depicted in Fig. 3.3(b).

Non-contact Mode

When operating in the *non-contact mode* the AFM-tip is not touching the surface. Instead it is influenced by the long range van der Waals forces. The advantage of the non-contact mode is that it does not destroy even the softest samples. But what is the origin of the van der Waals forces? This question will be answered now.

Atomic Polarization

The first step in understanding the van der Waals forces is a study of *atomic polarization*. When an *electric field* ε is applied to a charge neutral atom the positive and negative charges are displaced slightly from their equilibrium positions: the positive charges in the direction of the ε-field and the negative charge the opposite way. But how big a displacement d are we talking about? Well, an estimate can be obtained from the simple model of the hydrogen atom. We make a Taylor-expansion of $E(a)$ around the minimum $a = a_0$ to second order in $(a - a_0)$ to find the spring constant of the electron-proton system:

$$E(a) \approx E(a_0) + \frac{1}{2}E''(a_0)(a - a_0)^2. \qquad ...(9)$$

The spring constant in this expression is easily found to be

$$K = E''(a_0) = 3\frac{\hbar^2}{ma_0^2}\frac{1}{a_0^2} - 2\frac{e^2}{4\pi\varepsilon_0 a_0}\frac{1}{a_0^2} = \frac{2E(a_0)}{a_0^2} = 1.55\times 10^3\,\mathrm{m}^{-1} ....(10)$$

The displacement d between the electron and proton is therefore found by simple force balance between the spring force Kd and the electric force $e\varepsilon$:

$$e\varepsilon = kd \;\Rightarrow\; ed = \frac{e^2}{K}\varepsilon. \qquad ...(11)$$

The quantity ed is called the *dipole moment* of the atom. An *electric dipole* consists of an equal amount of positive and negative charge q that has been displaced the distance d (in general a vector pointing from $-q$ to $+q$) from one another. The dipole moment of the dipole is just the product qd.

What is then the *electrostatic energy* $U(r)$ of the dipole qd in the *electric field* $\varepsilon(r)$? The electric field is due to some potential $V_\varepsilon(r_0)$, that is $\varepsilon(r_0) = -V_\varepsilon'(r_0)$. We just study the case where the dipole is parallel to the electric field, so we place $+q$ at $r_0 + d/2$ and $-q$ at $r_0 - d/2$. We then obtain the energy $U(r)$ as

$$U(r_0) = +qV_\varepsilon\left(r_0 + \frac{d}{2}\right) - qV_\varepsilon\left(r_0 - \frac{d}{2}\right)$$

$$\approx q\left(+\frac{d}{2}\right)V_\varepsilon'(r_0) - q\left(-\frac{d}{2}\right)V_\varepsilon'(r_0) = -qd\varepsilon(r_0). \qquad ...(12)$$

We have seen that an electric field can induce a dipole moment, but the converse is also true: a dipole moment induces an electric field. Consider in 1D a dipole at the center of the coordinate system, $+q$ at $d/2$ and $-q$ at $-d/2$. What is then the *electric potential* $V(r)$ and the corresponding electric field $\varepsilon(r)$? Due to the opposite charges we can write their respective potentials as $+V_0(r - \frac{d}{2})$ and $-V_0(r - \frac{d}{2})$, where $V_0(x) = q/(4\pi\varepsilon_0 x)$ is the usual *Coulomb potential*. For $V(r)$ we thus get

$$V(r) = V_0\left(r - \frac{d}{2}\right) - V_0\left(r + \frac{d}{2}\right)$$

$$\approx -V_0'(r)\frac{d}{2} - V_0'(r)\frac{d}{2} = -dV_0'(r) = \frac{dq}{4\pi\varepsilon_0}\frac{1}{r^2}. \qquad ...(13)$$

The electric field is just the derivative of the potential, so

$$\varepsilon(r) = -V'(r) = \frac{2dq}{4\pi\varepsilon_0}\frac{1}{r^3}. \qquad ...(14)$$

We see here the characteristic r^{-3}-dependence of the electric field from a dipole.

van der Waals Forces

The *van der Waals forces* are now deduced as follows. If two charge neutral hydrogen atoms are placed at a distance r from one another they can gain energy by spontaneously generate dipole moments. Consider for example what happens if atom no. 1 by quantum fluctuations generates a dipole moment d_1. According to Eq. (14) this results in an electric field ε_1 at the site of atom no. 2. But Eq. (11) tells us that atom no. 2 as a consequence will generate a dipole moment ed_2,

$$ed_2 = \frac{e^2}{K}\varepsilon_1, \qquad ...(15)$$

and thus, according to Eqs. (12) and (14), gain the energy U

$$U(r) = -ed_2\varepsilon_1(r) = -\frac{e^2}{K}(\varepsilon_1(r))^2 = -\frac{e^2}{K}\left(\frac{2d_1 e}{4\pi\varepsilon_0}\frac{1}{r^3}\right)^2$$

$$= -\left(\frac{e^2}{4\pi\varepsilon_0}\right)^2 \frac{(2d_2)^2}{K}\frac{1}{r^6}. \qquad ...(16)$$

We can make the following simple estimate for $U(r)$. First we note that the only length scale relevant for the charge displacement $2d_1$ is the size of the atom, a_0. Thus we take $2d_1 \approx a_0$. With the usual abbreviation $E_0 = E(a_0)$ we have $2E_0 = e^2/(4\pi\varepsilon_0 a_0)$, and finally use Eq. (10) and write $K = 2E_0/(a_0^2)$. Thus we arrive at

$$U(r) \approx -(2E_0 a_0)^2 \frac{(a_0)^2}{2E_0/(a_0^2)}\frac{1}{r^6} = -2E_0\left(\frac{a_0}{r}\right)^6. \qquad ...(17)$$

Since force is the derivate of potential energy we obtain the following expression for the van der Waals force $F_{vdW}(r)$:

$$F_{vdW}(r) = -\frac{dU(r)}{dr} = 12\frac{E_0}{a_0}\left(\frac{a_0}{r}\right)^7 . \qquad ...(18)$$

To see a typical number we insert an atomic diameter $r = 2a_0$ and find

$$F_{vdW}(2a_0) = 5.8 \text{ nN}. \qquad ...(19)$$

Tapping Mode

The *tapping mode* combines the good sides of the contact and the *non-contact mode* of AFM-operation. The *AFM-cantilever* is forced to oscillate with an amplitude of the order 50 nm in such a way that the tip gently touches the sample once during each cycle. The benefits are threefold. First, the tapping mode has nearly the same resolution as the contact mode. Second, the tapping mode is almost as gentle, so it can be used for soft samples. Third, the tapping mode is in fact a simple and roboust way of operating the AFM.

4

Radiation Technique

While the field of forensic medicine is said to have begun at some indefinite time five or six centuries ago, the origins of forensic radiology can be described more precisely. Wilhem Conrad Rontgen—professor of physics, director of the Physics Institute, and Rector of the University of Wurzburg—observed an unusual phenomenon while experimenting with cathode ray tubes on November 8, 1895. After 50 d of intensive investigation, he determined that he had discovered a new kind of ray ("eine neue Arte von Strahlen"), one that could penetrate solid, opaque materials and produce photographic representations of their contents. He called them "X-rays" because "x" was the symbol of the unknown. A manuscript was produced and immediately accepted for presentation at the January 23, 1896, meeting of the Wurzburg Physical Medicine Society.

As sometimes happens today, word of his findings were "*leaked*" to the popular press and flashed by telegraph and cable throughout the electrified world, reaching New York on January 8, 1896. The potential for applying this new ray to the task of forensic problem solving was recognized almost immediately.

Professor Arthur William Wright, director of the Sloan Physics Laboratory at Yale University, is accorded primacy in the production of X-ray images (of inanimate objects) in the United States on January 28, 1896. A few days later, he bought a rabbit at a market and exposed the carcass to an X-ray beam for an hour. Examination of the photographic plate revealed lead shot within the body. Professor Wright had, for the first time, established a cause of death through radiography—put another way, by forensic radiology! In consonance with the purpose of this book, we will attempt a chronology of early forensic applications of radiology limited to the lower extremity.

Chronology of Early Forensic Radiology of the Lower Extremity

On a cold Christmas Eve in Montreal, George Holden shot Tolson Cunning in the leg. The wound healed, but the injured limb remained symptomatic. His surgeon, Dr. R. C. Kirkpatrick, requested an X-ray photograph of the area to be taken by Professor John Cox of the Physics Department at McGill University. This was accomplished on February 7, 1896 and showed the bullet lodged between the tibia and fibula. This image was submitted to the court during Mr. Holden's trial for attempted murder. Successful prosecution resulted in a sentence of 14 yr in the penitentiary.

In England, during September 1895, a burlesque and comedic actress known to us now only as Miss Folliott fell on the steps leading to her dressing room in the Nottingham Theater. Her foot injury kept her bedfast for a month, after which she was still unable to tread the boards. Dr. Frankish sent her to University College Hospital, where both feet were "*photographed*" by X-rays (perhaps the first comparison of films?) and the films demonstrated that the left cuboid clearly was displaced. This finding could be appreciated by both judge and jury when the negatives were displayed in court. The conviction of the theater owners for maintaining an unsafe workplace was somewhat mitigated by the charge that Miss Folliott was guilty of contributory carelessness. Because this case was commented on in British publications as early as April 1896, the actual radiography must be almost contemporaneous with the Holden case.

On September 2, 1895, Frank B. Bolling was thrown from his buggy "while driving a fractious horse" on the streets of Chicago and sustained a fracture of his right ankle. The fracture was set by two surgeons, and Mr. Bolling was able to return to work by May 1, 1896, but his ankle still hurt. To evaluate this persistent pain, Dr. Otto L. Smith and Professor W. C. Fuchs examined the offending ankle with 35 to 40 min of exposure to an X-ray tube placed just 5 in. from the ankle. Subsequent radiation damage and pain led to an amputation in November 1896 and two more amputations later because of pain and recurrent infections. Bolling filed the first malpractice suit for radiation damage and was awarded $10,000.

The first trial in which X-ray evidence was accepted by a United States Court took place in Denver in the waning months of 1896. The case began on June 15, 1895, when James Smith fell from a ladder while pruning a tree and injured his hip. Perhaps for pecuniary reasons, he waited almost a month before soliciting the professional services of Dr. W. W. Grant, who was widely known and well respected and a founder of the American College of Surgeons. He is credited with performing the first appendectomy in the United States, in 22-yr-old Mary Gartside.

Dr. Grant found no evidence of a fracture and did not restrict Mr. Smith's activity, but requested that he return in 1 wk. The diagnosis of "no fracture" was again asserted. Dr. Grant heard no more about Mr. Smith until April 1896, when the poorly paid law clerk engaged two of the best young lawyers in Colorado to file a $10,000 civil action against Dr. Grant, claiming limb shortening and disability as a result of his failure to diagnose a femoral fracture. Quick to take advantage of new technology, these bright young attorneys had engaged the services of Dr. Chauncey Tennent Jr, MD, of Denver Homeopathic College and a local photojournalist, Harry H. Buchwalter, to examine their client with X-rays. On four occasions between November 7 and 29, 1896, several attempts to obtain an image of Mr. Smith's hip (with exposures ranging up to 80 min) were finally successful in showing the outlines of an impacted fracture of the proximal femur. (It is not clear whether this was a femoral neck fracture or an intertrochanteric fracture.)

The fundamental legal issue was admission of a radiograph as evidence. Photographs were already recognized for their ability to show something a witness could testify to as an accurate representation of what had actually been seen. The X-ray image revealed structures or objects hidden from the eye and had been refused admission in some jurisdictions, which felt that it was "like offering the photograph of a ghost." The argument raged all day before District Judge Owen E LeFuvre who, after sleeping on the matter, handed down his decision, eloquently phrased in the elegant language of those days:

We... have been presented with a photograph taken by means of a new scientific discovery... it knocks for admission at the temple of learning. What shall be do or say? Close fast the door or open wide the portals? These photographs are offered in evidence to show the present condition of the head and neck of the femur bone, which is entirely hidden from the eye of the surgeon... Modern science has made it possible to look beneath the tissues of the human body, and has aided the surgeon in telling of the hidden mysteries. We believe it is our duty to be the first... in admitting in evidence a process known and acknowledged as a determinate science. The exhibits will be admitted in evidence.

The first appellate decision regarding the admission of radiographs as evidence in a US courtroom was rendered in 1897. It involved the case of Mr. Beall, who was in an elevator in the warehouse of WS Bruce and Company when it fell five stories. He sustained injuries to his leg and sued for compensation, claiming negligence in the construction and maintenance of the elevator. A witness, Dr. Saltman, testified that "overlapping bones of one of the plaintiff's legs, at the point where it was broken by this fall" were shown on a radiograph he was permitted to submit to the jury. The defense objected to admission of the radiograph and appealed. The Supreme Court of Tennessee ruled "...no sound reason was assigned at the bar why a civil court should not avail itself of this invention, when it is apparent that it would serve to throw light on the matter in controversy."

Dr. H. Graeme Anderson was one of the pioneer physicians assigned to the British Royal Flying Corps during World War I. He also became a licensed pilot, hence, one of the earliest flight surgeons. In his 1919 book, The Medical and Surgical Aspects of Aviation, Dr. Anderson devoted several pages to 17 cases of injuries to the talus that had been sustained during aircraft accidents. He termed this generic injury "*Aviator's Astralagus*" and illustrated it with radiographs. This probably is the first recognition of "*pattern injuries*," of which more will be said later. (It surely is worth mentioning that Dr. Anderson also reported that "Nemirovsky and Tilmant have lately organized an aeroplane...to carry a pilot, a surgeon, and a radiographer who can act as an assistant surgeon.... The elective current from the aeroplane can be used...to work the X-ray apparatus." Surely this is a first—and maybe the last— example of in-flight radiography. The author will welcome information about other examples).

Feet played an important role in solving the infamous Ruxton murder. On September 15, 1935, the wife of a Dr. Ruxton and her nursemaid disappeared from the family home in Lancaster and were never again seen alive. Two weeks later, a discovery of human remains triggered a search that began in the surrounding area and continued for another month until most of two female bodies could be reassembled. However, the faces had been mutilated, the teeth extracted, the terminal digits of the hands amputated, and other distinguishing topographical features had been excised from soft tissues, all to preclude identification.

Three feet were recovered. Casts were made and fitted into the shoes of the missing women. The presumptive left foot of Mrs. Ruxton was mutilated where she was known to have had a bunion and elsewhere (perhaps as a distraction), but a radiograph showed an exostosis of the first metatarsal head consistent with the bunion deformity. Other identifying anatomic features were found, and this case featured an early use of the photographic superimposition of facial features on a skull. Incriminating evidence also was found in the Ruxton home. The doctor was convicted and hanged for the murders.

In 1946, Dr. John Caffey MD—self-taught radiologist at New York's Babies Hospital (and the father of Pediatric Radiology)—published the first of several papers alerting the profession (and the public) to the peculiar concatenation of unusual skeletal injuries in the extremities of children that had often been associated with skull injuries, subdural hematomas, or both in the absence of any history of trauma. This led to the now widespread recognition of the intentional physical abuse of children—perhaps radiology's greatest contribution to forensic medicine.

In 1949, the Great Lakes liner Noronic caught fire and burned in Toronto, with many fatalities. Dr. Arthur C Singleton, a professor and head of the Radiology Department at the University of Toronto, was asked by the Attorney General of Ontario to assist in the identification of the bodies, thus becoming the father of mass casualty radiology. He was able to positively identify 24 of 119 fatalities by radiologic comparison alone. One of the illustrations in his paper on the matter showed comparison radiographs of a dismembered foot, noting similar features.

In 1981, Evans and Knight's book, *Forensic Radiology*, became available to the English-speaking world. It described applications of radiology for the purpose of identification and to identify evidence of abuse, mishaps, and malpractice, as well as for age determination and other anthropological conditions and correlation with forensic pathology, gunshot wounds, and other inflicted trauma. Unfortunately, of the more than 100 radiographs reproduced in this small but fairly comprehensive volume, only six were of the lower extremity.

In the ensuing quarter century, the utilization of radiology in the forensic sciences has increased and now includes some applications of newer modalities—nuclear radiology, ultrasonography, magnetic resonance imaging, and *computed tomography* (CT). Still, the great potential for radiology in forensics that was predicted by enthusiasts in the first couple of years after Rontgen's discovery remains largely unrealized.

Anthropological Considerations

It was recognized early after Röntgen's discovery that the X-ray could greatly assist in the identification of human remains, especially those in which superficial identifying features were distorted or destroyed by fire, immersion, decomposition, mutilation, or fragmentation. The first step in the procedure is the establishment of certain anthropological parameters, sometimes called the "*biologic profile*": age, sex, stature, and race or population ancestry. When the body or its parts are skeletonized, or when there are time and available facilities to remove flesh from the remains, the physical anthropologist can do the job with equal or sometimes superior accuracy. Otherwise, the radiologic method must suffice. In either event, it is useful to radiograph the remains for purposes of subsequent comparison with antemortem examinations.

First, one must determine whether the remains are human or animal. Skeletal similarities can be confusing to the inexperienced viewer.

Next, it must be determined whether the remains are those of a single human or commingled with other humans or animals—a sometimes difficult and time-consuming task. Only then can a biologic profile be established. For the purposes of this discussion, it is assumed that human body parts from the lower extremities are being analyzed.

Age Determination

The radiologic determination of maturity or prematurity at birth is based on the ossification of secondary centers at the knee. The distal femoral epiphysis will be partially ossified in 90% or more of full-term fetuses. The proximal tibial epiphysis will be similarly visible in 80% or more of mature neonates. Thereafter, chronological age is estimated according to skeletal maturation, as indicated by the appearance, growth, and ultimate fusion of epiphyses and apophyses (non-articulating secondary ossification centers).

Between mid-adolescence and middle age, the fusion line of the physis gradually disappears, tendinous attachments may become more prominent, and degenerative change insidiously commences. Estimation of skeletal age by radiologic evaluation of the lower extremities during these decades is fraught with difficulty and inaccuracy. The effects of advancing age become more apparent with advancing degenerative

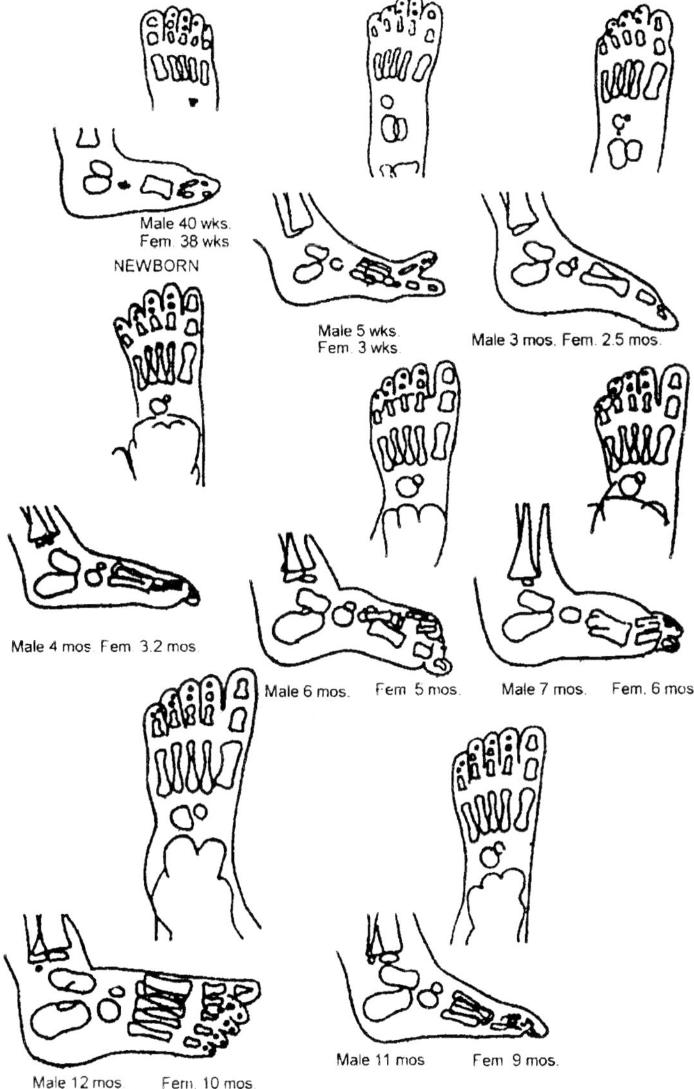

Fig. 4.1. Chronological development of the foot and ankle.

change and skeletal demineralization, but even then the use of radiology to estimate age may result of an error of a decade or so.

Sex Determination

Skeletal maturation accelerates in females at a greater rate than in males after the third or fourth year of life. However, this difference is not a useful determinant. In general, the male skeleton becomes more robust and heavier with aging and develops more prominent attachments for muscles and tendons. With further aging, there is a tendency for more degenerative hyperostotic changes in the male. Male long bones are approx 110% longer than their female counterparts. The male femoral head is larger. All of these findings are helpful but not definitive in establishing the sex of unidentified remains.

Other parts of the skeleton are far more helpful than those of the extremities. Bipartite patella, an anatomical variant occurring in approx 2% of the adolescent population, is nine times more common in boys than in girls.

In individuals of African ancestry, the tibia is long relative to the femur, but the ratios vary and overlap in US populations, probably due to racial mixing. The femoral shaft is bowed anteriorly in white and Asian populations compared with black populations; however, there is still considerable variability. However, a markedly bowed femur is unlikely to belong to a black decedent.

Craig developed a method of determining race based on the angle of the intercondylar shelf on the femur that can be used with either skeletal or fleshed remains. It requires radiography with true lateral positioning of the distal femur. The angle between the roof of the

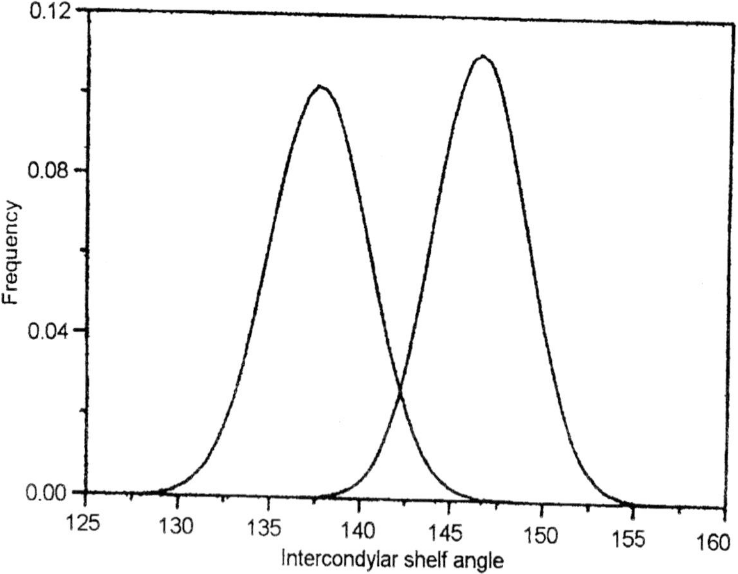

Fig. 4.2. Graphic representation of racial distribution of intercondylar shelf angles.

intercondylar notch (or intercondylar shelf) and the long axis of the femoral shaft are measured. Thus, this may serve as a fairly useful determinant for assigning race or population ancestry.

Steinbach and Russell suggested that a measurement of the soft tissue in the heel pad that exceeds 21 mm is a reasonably accurate indication of acromegaly. However, the fallibility of this diagnostic indicator has been described by Puckett and Seymour, who demonstrated a greater-than-"average" heel pad measurement in Blacks, 40% of whom had heel pads exceeding 21 mm compared with only 9% of Caucasians. Local soft-tissue swelling can skew this measurement even further. Hence, heel pad thickness is of doubtful value in the process of determining race or population ancestry in unidentified body parts.

Determination of Stature

Extensive research on World War II and Korean War casualties has enabled investigators to develop methods of estimating stature based on measurements of long bones. The length of the femur is the most reliable basis for calculating stature. The tibia is also useful, but there has been some controversy over the accuracy of tibial measurements, particularly the most appropriate location of the more distal measuring point. Apparently, the plafond of the tibia is the preferred site of measurement rather than the tip of the medial malleolus. The tables and equations furnished to estimate stature from long-bone measurements are based on direct measurements of defleshed or skeletonized specimens. However, measurements from radiographs can be equally useful if corrections for magnification are applied. The effect of magnification can be nullified with an extremely long distance between the X-ray tube and the object–film combination (≥6 ft), but this is impractical in most laboratory settings. An old but useful technique for measuring long bones involves the use of X-ray beams from conventional tube–object distances (36–40 in.) that are carefully collimated so that only nondivergent central rays are used. A partially opaque ruler can be used for direct measurements. Finally, if CT equipment is available, direct measures can be easily obtained from the image with a cursor.

Identification

The accuracy of radiology when used to identify human remains depends on the ability to find identical points of comparison between antemortem diagnostic radiological studies and postmortem images obtained with similar techniques. Hence, postmortem studies must be obtained with the parts positioned to simulate the routine radiologic positions and projections used in medical diagnostic studies. Equipment and technical factors employed in the production of such postmortem radiologic images will be discussed later in this chapter. Unless identity is readily established, it is wise to obtain radiographic studies of all available body parts prior to their release, because it is not easy to predict which body regions will be represented on the antemortem radiographs subsequently acquired.

Statistical Considerations Related to Body Parts

The abdominal trunk—particularly the soft tissue and bony structures of the back, including the spine and pelvis—are most likely to survive the rigors of separation, incineration, decomposition, and the activities of carnivores. The extremities are susceptible to damage by all of these factors. Further, the odds of finding antemortem studies of the extremities are less than those of some other body parts. In a survey of films in a large university hospital radiology department, investigators found that the lower extremity accounted for only 11% of all roentgenograms and an even smaller percentage of studies using other radiologic modalities. In the Bass and Driscoll study of incomplete skeleton retrieval in Tennessee, the femur was the second most common skeletal element found in 58 fragmented skeletons; the tibia ranked fifth, the fibula eighth, and the patella 13th. In a series of 30 cases in which the identification of the victim was established by comparing antemortem and postmortem radiologic findings, Murphy and coworkers found that the extremities contributed to 20% of cases.

Identification of Individual Skeletal Elements

Individual bones can be matched on antemortem and postmortem radiographs by comparing overall configurations or comparable abnormalities associated with (1) anomalous, congenital, or developmental lesions; (2) disease or tissue degeneration; (3) tumors or tumor-like conditions; (4) trauma; (5) iatrogenic lesions; and (6) trabecular patterns and vascular grooves.

Again, the position of the postmortem specimen must duplicate that of the antemortem radiograph. This may require trial and error or the "*shadow positioning technique*" of Fitzpatrick and Mascaluso.

Anomalous, Developmental, and Congenital Variations

Riddick et al. were able to identify the charred remains of a kidnapped murder victim on the basis of a developmental anomaly in a patella—a "*dorsal defect*" found in only 1% of the population. Clubfoot deformities were discovered in one examination of decomposed human remains. Finding custom-made orthopedic shoes in the room, investigators were able to locate the vendor who, in turn, helped them antemortem radiographs of the feet. These were successfully compared with postmortem studies.

Disease or Degenerative Change

Degenerative changes are frequently helpful or even definitive in the identification of unknown remains. We have already seen examples in the Ruxton case and in the case of heel spurs of the calcaneus. Judging from the literature, it is rare to be able to match skeletal remains by lesions that arise secondary to disease processes. However, certain diseases have such distinctive features that they could be used for identification purposes.

Tumors and tumor-like lesions

Malignant tumors may change so rapidly that they are of little use for comparison studies, especially when the films were obtained over a long interval of time. Other such lesions, however, are relatively stable and may be useful on occasion.

Trauma

Dr. Fovau d'Courmelles accurately predicted in the October 1898 American X- Ray Journal that "knowing the existence of a fracture in a person, who has been burned or mutilated beyond recognition, we can hope to identify him by the X-ray".

Residual deformities from fractures, retained bullets, embedded knife blades, or posttraumatic calcifications can all provide identification if appropriate comparative radiographs become available.

Iatrogenic alterations

Residual findings from previous treatments or manipulations can be helpful in using radiology for identification purposes. Orthopedic hardware, prostheses, drill holes and drill bits, osteotomies, fusions, and manipulations may all leave recognizable traces that can be seen on the radiograph.

Trabecular Patterns and Vascular Grooves

The trabecular pattern of some bones can be quite distinctive and can be matched on antemortem and postmortem films. Weight-bearing or stress-bearing large trabeculae in the femur, tibia, and os calcis are particularly useful. Kahana and Hiss have described a computerized system of matching trabecular patterns by superimposing densitographs and reported a case identified by that system using only the bones of a thumb. However, we find that in most cases, one can satisfactorily match the pattern by sight alone while manipulating the specimen.

Identification in Mass Casualty Situations

The same techniques and variables that are taken into consideration when identifying individual skeletal elements apply to mass casualty situations, except for the logistical problems that are complicated by the size of the matrix (i.e., the number of remains involved and the fact that they are sometimes commingled and often fragmented). Most mass casualty operations take on the characteristics of a field exercise, because a fear of contaminated blood and body fluids has closed the doors of most hospitals to mass casualty remains. The organization of a mass casualty operation is beyond the scope of this chapter, but there is extensive literature on the subject. Apart from obtaining radiographs of mass casualty victims for identification purposes, it is also useful to look for foreign bodies that may provide valuable information, such as serial or part numbers, sources of bombs or weapons, trace evidence of explosives or chemicals, or even an unexploded bomb.

Pattern Injuries

Some traumatic episodes result in radiologically demonstrable lesions that are relatively consistent. These are called pattern injuries, and they can frequently be used to predict or deduce the mechanism of injury, the offending agent, or both.

Abuse

Child abuse

In Caffey's first paper on the subject, he remarked on a pattern of skeletal lesions found in children who also had a subdural hematoma. The pattern was one of metaphyseal fragmentation, another finding he called "*involucrum*," and fractures in different stages of healing. Later, Caffey added other components to the pattern: "*bowing fractures*", *metaphyseal cupping*, and "*ectopic ossification centers*," which nowadays translates to epiphyseal separation. Caffey and others added still more observations to the total pattern, including transverse and spiral fractures of long bones that are inappropriate to the age and activity of the child.

The interpretation of pediatric radiographs requires considerable training and experience. Radiographically, it is quite true that "children are not just little adults." Other conditions, both normal and abnormal, may be confused with child abuse. One of the best examples of this is the so-called "*toddler's fracture*"—an undisplaced spiral fracture of the tibia that is the common result of the somewhat uncoordinated, often pigeon-toed effort at locomotion by children in this age group—which is not an indication of abuse. By contrast, a spiral fracture of the tibia in a nonambulatory child is almost invariably an indication of child abuse. A single thin line of periosteal calcification may exist in normal healthy infants up to age 4 mo. However, it is bilateral, symmetrical, unilamellar, and asymptomatic and is not indicative of child abuse. A number of other conditions may be confusing, including that darling of defense attorneys: osteogenesis imperfecta.

It must be remembered that the most important feature of radiologic findings in child abuse is that the injury is inconsistent with the age, development, and activity of the child.

Physical abuse of adults

Spousal abuse or abuse of intimate partners does not usually involve the lower extremity. Abusive blows to women are almost exclusively directed to the head, neck, and face. Domestic violence and automobile accidents are the most common causes of facial injury in women. Patterns emerge here: automobile accident victims tend to show massive and multiple injuries to the facial bones and mandible, whereas battered women show mostly fractures of the mandibular body or angle and the contralateral mandibular ramus. Of course, nasal, orbital, zygomatico-facial, and dental fractures are seen along with dislocations of the mandible. Defensive injuries of the upper extremity (*fending fractures*) are seen, but the lower extremities are rarely involved. Blows to the body are uncommon except in pregnant women, where the breast and abdomen may be pummeled.

Abuse of the aged

Here, the patterns of injury are somewhat like those seen in child abuse, i.e., twisting, pulling, and squeezing injuries of the extremities as a result of punishment or attempts at restraint. However, blows to the head may simulate those seen in spousal abuse. Senile osteoporosis complicates the diagnostic issue in the elderly—particularly in the extremities, where routine handling, lifting, turning, and restraint may produce fractures.

Pedal Injury Patterns

This topic is something of a double entendre, because it refers to injuries to the pedal extremity—which are caused mostly by pedals, either brake or rudder!

Dr. Andrew's collection of "*aviator's astragalus*" cases was amassed from air crashes, and he believed that the mechanism was one of dorsiflexion against the rudder pedal during impact. More contemporary studies have suggested a spectrum of injuries caused by this mechanism. In Group 1 injuries of this type, there is a relatively undisplaced fracture of the neck of the astragalus or talus. In Group 2, there is fracture of the talar neck with displacement of the body of the talus into equinous deformity. In Group 3, there is a posterior dislocation of the talus, which may involve the entire bone or only a major portion of the body of the talus, with the anterior fragment remaining roughly in place between the tibial plafond and the navicular. True posterior extrusion of the talus is quite rare, because it requires rupture of the Achilles tendon. Talar dislocations and extrusions also occur in plantar flexion in association with either pronation or supination, in which case the soft tissue injury and extrusion will be either anteromedial or anterolateral, respectively.

According to Simson, flight surgeons during World War II recognized a lesion they called "*Aviator's fracture*"—a pattern of dorsally displaced mid-foot metatarsal fractures that occur as a result of impact against a rudder pedal. It is impossible to tell from the rather poorly reproduced roentgenogram the extent of additional injuries to the tarsal bones. It appears that there are fractures of not only the metatarsals but also fracture/dislocations in the hind foot.

Tarsometatarsal fracture/dislocation (the so-called *Lisfranc lesion*) can be acquired through a variety of mechanisms, with dislocations occurring in both the dorsal and plantar direction. It is not uncommon for the history to include the fact that the foot was forced against or trapped beneath a brake or gas pedal. However, such injuries can also occur as a result of a fall, the foot being run over by a vehicle, or a heavy weight being dropped on the foot. Thus, the pattern is not specific. The key injury is disruption of the locking mechanism of the base of the second metatarsal between the first, second, and third cuneiforms. Ordinarily the lateral metatarsals are displaced both laterally and dorsally. There is a divergent Lisfranc deformity involving a fracture of the base of the second metatarsal and lateral subluxation of the four lateral metatarsals.

Bumper fracture

The bumper fracture has been recognized for years as a well-established pattern suggesting the injury, the object, the direction in which it was traveling, and the position of the victim.

Patterns of Torture and Terrorism

Physical beating is a widespread form of torture. Palmatoria is a form of localized torture virtually unique to the small West African country Guinea-Bissau. Palmatoria involve repetitive blows by a slender rod to the shin, where the tibia lies closest beneath the skin. Ordinary radiographs (of somewhat limited quality) have shown a periosteal reaction, presumably as a result of subperiosteal hemorrhage and hematoma. Somewhat peculiar endosteal and medullary changes have also been seen. Recent case reports in the United States and France have shown that this specific injury can produce a hidden endosteal fracture that is likely to be undetected on plain films but becomes obvious on a CT scan. It seems possible (perhaps likely) that some of the cases that took place in Guinea-Bissau would show similar findings when more sophisticated imaging modalities were used.

Falaca is a form of torture characterized by beating the foot, primarily the sole of the foot. This is a widespread form of torture in the Middle East, especially in Turkey and Iraq, as well as in Asia and in some Spanish-speaking areas (where it is called *bastinado*). Substantial injuries can be produced by this form of torture, including edema, hematoma, fractures, and injuries to ligaments, tendons, fascia, and aponeuroses. The clinical findings are usually diagnostic if they are made immediately after such torture. Radiography can confirm or exclude fractures and allows the investigator to estimate the extent of soft tissue injury and the time interval since torture. Early nuclear medicine studies show a massive increased uptake in the affected areas. Later radiographic studies demonstrate bone and soft tissue injury and evidence of subsequent healing.

Knee capping

The punitive destruction of a major joint by gunshot wound was originally thought to be the signature injury of certain criminal elements in the United States. More recently, it has been seen as an act of terrorism in other parts of the world, especially Northern Ireland. The punishment is not limited to the knee; other major joints have also been targeted.

Gunshot Wounds

From the day that Professor Wright examined his rabbit, radiology has played an important role in the evaluation of gunshot wounds. The number, location, and caliber of bullets; the angle and direction of fire; the discovery of concealed gunshot wounds; weapon ballistics;

types of bullets; and unconventional loads all fall within the purview of modern radiologic investigation.

Number of Bullets

Since more than one bullet can enter through a single entrance wound, it is important to determine the number of bullets and correlate them with entrance and exit wounds. Any discrepancy requires a search for spent bullet casings at the scene, additional entrance wounds, or radiographic investigation.

Caliber of Bullets

The use of radiographs to estimate the caliber of a bullet can be fraught with difficulty, particularly because the range of bullet sizes among calibers that are commonly used in handguns, for instance, is quite minimal. The magnification factor further confuses the picture and increases the risk of comprising the accuracy of the findings.

Location

Locating the bullet would seem to be a rather simple task. However, bullets may end up at sites quite distant from the entrance wound. They may be diverted from their expected path by striking bone or tissue. They may migrate great distances through blood vessels, the alimentary tract, the genitourinary tract, the spinal canal, the tracheobronchial tree, or within pleural and peritoneal spaces.

Types of Bullets

The ordinary load for handguns and rifles is a bullet made of lead. The lead may be fully or partially covered by another metal, called a jacket, which has several purposes: it protects the barrel of the weapon from leading; it hardens and lubricates the bullet, thereby somewhat diminishing the possibility of bullet deformity when it strikes tissue, particularly bone; and it may be shed as the bullet traverses tissues in the body. Each weapon has grooves inside the barrel that leave identifiable marks on the bullet as it passes through. This is important for ballistics identification. In the unjacketed bullet, the pattern will be on the lead itself; in semi- or fully-jacketed bullets, the ballistic markings will be on the jacket. Hence, it is important to identify and retrieve the separated jacket for ballistic purposes. Copper jacketing is usually quite easily identified. The so-called *silvertip load*, which has become quite popular in the United States, is jacketed with aluminum and can be quite difficult or impossible to locate with conventional radiography.

Military weapons tend to have high-velocity characteristics and, according to the Hague Piece Conference of 1899, must be fully jacketed. "Civilian" weapons and loads are mostly of a lower velocity. Some of the bullets used in civilian handguns and rifles have identifying characteristics, including the following:

1. The unjacketed bullet commonly used in high-velocity hunting ammunition fragments so extensively that the resulting pattern is called a "*lead snowstorm.*"
2. The Black Talon mushrooms into a characteristic six-petal flower or star configuration.
3. The Glaser safety slug carries multiple small lead pellets in a copper cup, producing an unusual mixed pattern of densities.
4. The Winchester Western .25 caliber cartridge contains a copper-coated lead hollowpoint bullet filled with a single no. 4 steel pellet and, thus, produces the unusual finding of a single small bullet accompanied by a single small shot.

Finally, unconventional loads—plastic bullets, rubber bullets, and ceramic bullets—are being seen with increasing frequency because they are used in crowd control.

Production of the Radiograph

As previously mentioned, it has become increasingly difficult to use hospital or clinical radiologic installations for forensic investigation. The medical examiner's office, whether a one-room morgue or a separate building, needs its own in-house radiologic facility. The basic requirements are functional radiographic equipment and space to store and use it; X-ray films, cassettes, film storage space; a dark room and film processing facility, a film identification system; and envelopes and storage space for archival radiographs.

X-Ray Equipment

Either mobile or fixed X-ray equipment will suffice. The mobile X-ray unit is entirely self-contained and mounted on wheels for easy maneuverability. Stand-alone units may also be used; these are mounted on the floor, wall, or ceiling. A disadvantage lies in the fact that the staff must bring the body or body parts to the machine. An advantage is that high-performance X-ray equipment is not required, because there is no need to overcome the problem of physical or physiologic motion in the body or body part being examined. Thus, adequate equipment can be obtained at fairly reasonable cost, particularly because second-hand equipment has been made available as a result of recent technological advances and the introduction of new modalities to the field of clinical radiology.

Film

X-ray film is sensitive not only to X-rays but almost every color of light. It is packaged in light-proof cardboard boxes and must be handled in a darkroom equipped with an approved X-ray safety light. X-ray film is also sensitive to heat, low humidity, and stray radiation; hence, it must be carefully stored. X-ray film comes in a variety of sizes. It is easier for the forensic facility to employ one standard size; 14- × 17-inch film is recommended.

Film cassettes

Unexposed film is loaded in the darkroom onto rigid holders called cassettes, which are made of plastic and metal. The cassettes are flat, rectangular structures that are hinged at one end. The unexposed film is sandwiched between two fluorescent intensifying screens, which help reduce the amount of X-ray exposure required. Some cassettes come with a built-in grid of parallel lead lines, which improves the quality of images made of thick body parts. Both grid and non-grid cassettes are required for different examinations.

Film processing

Exposed films must be unloaded from a cassette in the dark and processed. There are many small automatic processors on the market, and these are highly recommended. If available, a nearby hospital or clinic processing facility may be a satisfactory alternative to in-house processing.

Film Identification

If the radiographic output is quite low, one can successfully write the name or other identifier of the body or body parts directly on the film with an indelible marker after the film has been processed. A better and more dependable method involves affixing readily available lead letters and numbers on the exposable surface of the cassette with adhesive tape, thus putting identification on the film at the same time as the exposure.

Exposure Factors

All radiographic equipment will have a control panel allowing the technician to select exposure factors—milliamperes (ma), exposure time(s), and kilovoltage (kVp)— appropriate for the body part and its thickness. Milliamperage and time may be combined as milliampere-seconds (mAs).

Simplified exposure chart is suggested as a starting point for the production of useable images. Suggested milliampere-seconds, kilovoltage settings, and the selection of a grid or non-grid cassette are supplied for different body parts and tissue thicknesses. It is recommended that the X-ray tube be 40 in. from the film or cassette holder. A ruler or simple aluminum caliper should be used to measure the thickness of the part to be examined.

Positioning

Standard radiographic positions for the lower extremity are required in order to match features on antemortem radiographs. Figures provide information that can be used to determine how to position the body part and central beam of the X-ray for each standard position. Each positioning photo is accompanied by a sample of the image to be expected from it.

5

BIONANOELECTRONICS

Instrumentation and measurement technologies are some of the most promising areas for near-term advancements and enabling effects. As optical, fluidic, chemical, and biological components can be integrated with electronic logic and memory components on the same chip at marginal... to perform basic diagnostic functions in a minimally invasive way, providing new abilities to remedy health problems.

Other possible applications include: pervasive, self-moving sensor systems; nanoscrubbers and nanocatalysts; even inexpensive, networked "nanosatellites." For example, so-called "nanosatellites" are targeting order-of-magnitude reductions in... proach by Prentiss uses lasers to place molecules in a desired location. Chemical assembly techniques are being addressed by a number of groups, including Whiteside's approach to building structures one molecular layer at a time.

A third major area for development within molecular manufacturing is systems design and engineering....understanding of fundamental properties of structures at the nanoscale has been greatly enhanced by the ability to fabricate very small test objects, analyze them experimentally using new capabilities.

MICRO/NANOSCALE INSTRUMENTATION

Coal and diamonds, sand and computer chips, cancer and healthy tissue: throughout history, variations in the arrangement of atoms have distinguished the cheap from the cherished, the diseased from the healthy. Arranged one way, atoms make up soil, air, and water; arranged another, they make up ripe strawberries. Arranged one way, they make up homes and fresh air; arranged another, they make up ash and smoke. Our ability to arrange atoms lies at the foundation of technology. We have come far in our atom arranging, from chipping flint for arrowheads to machining aluminum for spaceships. We take pride in our technology, with our lifesaving drugs and desktop computers. Yet our spacecraft are still crude, our computers are still stupid, and the molecules in our tissues still slide into disorder, first destroying health, then life itself. For all our advances in arranging atoms, we still use primitive methods. With our present technology, we are still forced to handle atoms in unruly herds.

But the laws of nature leave plenty of room for progress, and the pressures of world competition are even now pushing us forward. For better or for worse, the greatest technological breakthrough in history is still to come.

Styles of Technology

Our modern technology builds on an ancient tradition. Thirty thousand years ago, chipping flint was the high technology of the day. Our ancestors grasped stones containing trillions of trillions of atoms and removed chips containing billions of trillions of atoms to make their axheads; they made fine work with skills difficult to imitate today. They also made patterns on cave walls in France with sprayed paint, using their hands as stencils. Later they made pots by baking clay, then bronze by cooking rocks. They shaped bronze by pounding it. They made iron, then steel, and shaped it by heating, pounding, and removing chips.

We now cook up pure ceramics and stronger steels, but we still shape them by pounding, chipping, and so forth. We cook up pure silicon, saw it into slices, and make patterns on its surface using tiny stencils and sprays of light. We call the products "chips" and we consider them exquisitely small, at least in comparison to axheads.

Our microelectronic technology has managed to stuff machines as powerful as the room-sized computers of the early 1950s onto a few silicon chips in a pocket-sized computer. Engineers are now making ever smaller devices, slinging herds of atoms at a crystal surface to build up wires and components one tenth the width of a fine hair.

These microcircuits may be small by the standards of flint chippers, but each transistor still holds trillions of atoms, and so-called "microcomputers" are still visible to the naked eye. By the standards of a newer, more powerful technology they will seem gargantuan. The ancient style of technology that led from flint chips to silicon chips handles atoms and molecules in bulk; call it bulk technology. The new technology will handle individual atoms and molecules with control and precision; call it *molecular technology*. It will change our world in more ways than we can imagine. *Microcircuits* have parts measured in *micrometers*—that is, in millionths of a meter—but molecules are measured in *nanometers* (a thousand times smaller). We can use the terms "nanotechnology" and molecular technology" interchangeably to describe the new style of technology. The engineers of the new technology will build both nanocircuits and nanomachines.

Molecular Technology Today

One dictionary definition of a machine is "any system, usually of rigid bodies, formed and connected to alter, transmit, and direct applied forces in a predetermined manner to accomplish a specific objective, such as the performance of useful work." Molecular machines fit this definition quite well. To imagine these machines, one must first picture molecules. We can picture atoms as beads and molecules as clumps of beads, like a child's beads linked by snaps. In fact, chemists do sometimes visualize molecules by building models from plastic beads (some of which link in several directions, like the hubs in a Tinkertoy set). Atoms are rounded like beads, and although molecular bonds are not snaps, our picture at least captures the essential notion that bonds can be broken and reformed. If an atom were the size of a small marble, a fairly complex molecule would be the size of your fist. This makes a useful mental image, but atoms are really about 1/10,000 the size of bacteria, and bacteria are about 1/10,000 the size of mosquitoes. (An atomic nucleus, however, is about 1/100,000 the size of the atom itself; the difference between an atom and its nucleus is the difference between a fire and a nuclear reaction.)

The things around us act as they do because of the way their molecules behave. Air holds neither its shape nor its volume because its molecules move freely, bumping and ricocheting through open space. Water molecules stick together as they move about, so water holds a constant volume as it changes shape. Copper holds its shape because its atoms stick together in regular patterns; we can bend it and hammer it because its atoms can slip over one another while remaining bound together. Glass shatters when we hammer it because its atoms separate before they slip. Rubber consists of networks of kinked molecules, like a tangle of springs. When stretched and released, its molecules straighten and then coil again. These simple molecular patterns make up passive substances. More complex patterns make up the active nanomachines of living cells. Biochemists already work with these machines, which are chiefly made of protein, the main engineering material of living cells. These molecular machines have relatively few atoms, and so they have lumpy surfaces, like objects made by gluing together a handful of small marbles. Also, many pairs of atoms are linked by bonds that can bend or rotate, and so protein machines are unusually flexible. But like all machines, they have parts of different shapes and sizes that do useful work. All machines use clumps of atoms as parts. Protein machines simply use very small clumps. Biochemists dream of designing and building such devices, but there are difficulties to be overcome. Engineers use beams of light to project patterns onto silicon chips, but chemists must build much more indirectly than that. When they combine molecules in various sequences, they have only limited control over how the molecules join. When biochemists need complex molecular machines, they still have to borrow them from cells. Nevertheless, advanced molecular machines will eventually let them build nanocircuits and nanomachines as easily and directly as engineers now build microcircuits or washing machines. Then progress will become swift and dramatic. Genetic engineers are already showing the way.

Ordinarily, when chemists make molecular chains—called "*polymers*" —they dump molecules into a vessel where they bump and snap together haphazardly in a liquid. The resulting chains have varying lengths, and the molecules are strung together in no particular order. But in modern gene synthesis machines, genetic engineers build more orderly polymers—specific DNA molecules—by combining molecules in a particular order. These molecules are the nucleotides of DNA (the letters of the genetic alphabet) and genetic engineers don't dump them all in together. Instead, they direct the machine to add different nucleotides in a particular sequence to spell out a particular message. They first bond one kind of nucleotide to the chain ends, then wash away the leftover material and add chemicals to prepare the chain ends to bond the next nucleotide. They grow chains as they bond on nucleotides, one at a time, in a programmed sequence. They anchor the very first nucleotide in each chain to a solid surface to keep the chain from washing away with its chemical bathwater. In this way, they have a big clumsy machine in a cabinet assemble specific molecular structures from parts a hundred million times smaller than itself.

But this blind assembly process accidentally omits nucleotides from some chains. The likelihood of mistakes grows as chains grow longer.

Like workers discarding bad parts before assembling a car, genetic engineers reduce errors by discarding bad chains. Then, to join these short chains into working genes (typically thousands of nucleotides long), they turn to molecular machines found in bacteria. These protein machines, called *restriction enzymes*, "read" certain DNA sequences as "cut here." They read these genetic patterns by touch, by sticking to them, and they cut the chain by rearranging a few atoms. Other enzymes splice pieces together, reading matching parts as "glue here"—likewise "reading" chains by selective stickiness and splicing chains by rearranging a few atoms.

By using gene machines to write, and restriction enzymes to cut and paste, genetic engineers can write and edit whatever DNA messages they choose. But by itself, DNA is a fairly worthless molecule. It is neither strong like Kevlar, nor colourful like a dye, nor active like an enzyme, yet it has something that industry is prepared to spend millions of dollars to use: the ability to direct molecular machines called ribosomes. In cells, molecular machines first transcribe DNA, copying its information to make RNA "tapes." Then, much as old numerically controlled machines shape metal based on instructions stored on tape, ribosomes build proteins based on instructions stored on RNA strands. And proteins are useful. Proteins, like DNA, resemble strings of lumpy beads. But unlike DNA, protein molecules fold up to form small objects able to do things. Some are enzymes, machines that build up and tear down molecules (and copy DNA, transcribe it, and build other proteins in the cycle of life). Other proteins are hormones, binding to yet other proteins to signal cells to change their behaviour.

Genetic engineers can produce these objects cheaply by directing the cheap and efficient molecular machinery inside living organisms to do the work. Whereas engineers running a chemical plant must work with vats of reacting chemicals (which often misarrange atoms and make noxious by-products), engineers working with bacteria can make them absorb chemicals, carefully rearrange the atoms, and store a product or release it into the fluid around them. Genetic engineers have now programmed bacteria to make proteins ranging from human growth hormone to rennin, an enzyme used in making cheese. The pharmaceutical company Eli Lilly (Indianapolis) is now marketing Humulin, human insulin molecules made by bacteria.

Existing Protein Machines

These protein hormones and enzymes selectively stick to other molecules. An enzyme changes its target's structure, then moves on; a hormone affects its target's behaviour only so long as both remain stuck together. Enzymes and hormones can be described in mechanical terms, but their behaviour is more often described in chemical terms. But other proteins serve basic mechanical functions. Some push and pull, some act as cords or struts, and parts of some molecules make excellent bearings. The machinery of muscle, for instance, has gangs of proteins that reach, grab a "rope" (also made of protein), pull it, then reach out again for a fresh grip; whenever you move, you use these machines. Amoebas and human cells move and change shape by using fibers and rods that act as molecular muscles and bones.

A reversible, variable-speed motor drives bacteria through water by turning a corkscrew-shaped propeller. If a hobbyist could build tiny cars around such motors, several billions of billions would fit in a pocket, and 150-lane freeways could be built through your finest capillaries. Simple molecular devices combine to form systems resembling industrial machines. In the 1950s engineers developed machine tools that cut metal under the control of a punched paper tape. A century and a half earlier Joseph-Marie Jacquard had built a loom that wove complex patterns under the control of a chain of punched cards. Yet over three billion years before Jacquard, cells had developed the machinery of the ribosome. Ribosomes are proof that nanomachines built of protein and RNA can be programmed to build complex molecules.

Then consider viruses. One kind, the T4 phage, acts like a spring-loaded syringe and looks like something out of an industrial parts catalog. It can stick to a bacterium, punch a hole, and inject viral DNA (yes, even bacteria suffer infections). Like a conqueror seizing factories to build more tanks, this DNA then directs the cell's machines to build more viral DNA and syringes. Like all organisms, these viruses exist because they are fairly stable and are good at getting copies of themselves made. Whether in cells or not, nanomachines obey the universal laws of nature. Ordinary chemical bonds hold their atoms together, and ordinary chemical reactions (guided by other nanomachines) assemble them. Protein molecules can even join to form machines without special help, driven only by thermal agitation and chemical forces. By mixing viral proteins (and the DNA they serve) in a test tube, molecular biologists have assembled working T4 viruses. This ability is surprising: imagine putting automotive parts in a large box, shaking it, and finding an assembled car when you look inside! Yet the T4 virus is but one of many self-assembling structures.

Molecular biologists have taken the machinery of the ribosome apart into over fifty separate protein and RNA molecules, and then combined them in test tubes to form working ribosomes again.

To see how this happens, imagine different T4 protein chains floating around in water. Each kind folds up to form a lump with distinctive bumps and hollows, covered by distinctive patterns of oiliness, wetness, and electric charge. Picture them wandering and tumbling, jostled by the thermal vibrations of the surrounding water molecules. From time to time two bounce together, then bounce apart. Sometimes, though, two bounce together and fit, bumps in hollows, with sticky patches matching; they then pull together and stick. In this way protein adds to protein to make sections of the virus, and sections assemble to form the whole. Protein engineers will not need nanoarms and nanohands to assemble complex nanomachines. Still, tiny manipulators will be useful and they will be built. Just as today's engineers build machinery as complex as player pianos and robot arms from ordinary motors, bearings, and moving parts, so tomorrow's biochemists will be able to use protein molecules as motors, bearings, and moving parts to build robot arms which will themselves be able to handle individual molecules.

Designing With Protein

How far off is such an ability? Steps have been taken, but much work remains to be done. Biochemists have already mapped the structures of many proteins. With gene machines to help write DNA tapes, they can direct cells to build any protein they can design. But they still don't know how to design chains that will fold up to make proteins of the right shape and function. The forces that fold proteins are weak, and the number of plausible ways a protein might fold is astronomical, so designing a large protein from scratch isn't easy. The forces that stick proteins together to form complex machines are the same ones that fold the protein chains in the first place. The differing shapes and kinds of stickiness of amino acids—the lumpy molecular "beads" forming protein chains—make each protein chain fold up in a specific way to form an object of a particular shape. Biochemists have learned rules that suggest how an amino acid chain might fold, but the rules aren't very firm. Trying to predict how a chain will fold is like trying to work a jigsaw puzzle, but a puzzle with no pattern printed on its pieces to show when the fit is correct, and with pieces that seem to fit together about as well (or as badly) in many different ways, all but one of them wrong. False starts could consume many lifetimes, and a correct answer might not even be recognized. Biochemists using the best computer programs now available still cannot predict how a long, natural protein chain will actually fold, and some of them have despaired of designing protein molecules soon.

Yet most biochemists work as scientists, not as engineers. They work at predicting how natural proteins will fold, not at designing proteins that will fold predictably. These tasks may sound similar, but they differ greatly: the first is a scientific challenge, the second is an engineering challenge. Why should natural proteins fold in a way that scientists will find easy to predict? All that nature requires is that they in fact fold correctly, not that they fold in a way obvious to people. Proteins could be designed from the start with the goal of making their folding more predictable. Carl Pabo, writing in the journal Nature, has suggested a design strategy based on this insight, and some biochemical engineers have designed and built short chains of a few dozen pieces that fold and nestle onto the surfaces of other molecules as planned. They have designed from scratch a protein with properties like those of melittin, a toxin in bee venom. They have modified existing enzymes, changing their behaviours in predictable ways. Our understanding of proteins is growing daily. In 1959, according to biologist Garrett Hardin, some geneticists called genetic engineering impossible; today, it is an industry.

Biochemistry and computer-aided design are now exploding fields, and as Frederick Blattner wrote in the journal Science, "computer chess programs have already reached the level below the grand master. Perhaps the solution to the protein-folding problem is nearer than we think." William Rastetter of Genentech, writing in *Applied Biochemistry and Biotechnology* asks, "How far off is de novo enzyme design and synthesis? Ten, fifteen years?" He answers, "Perhaps not that long." Forrest Carter of the U.S. Naval Research Laboratory, Ari Aviram and Philip Seiden of IBM, Kevin Ulmer of Genex Corporation, and other researchers in university and industrial laboratories around the globe have already begun theoretical work and experiments aimed at developing molecular switches, memory devices, and other structures that could be incorporated into a protein based computer. The U.S. Naval Research Laboratory has held two international workshops on molecular electronic devices, and a meeting sponsored by the U.S. National Science Foundation has recommended support for basic research aimed at developing molecular computers. Japan has reportedly begun a multimillion-dollar program aimed at developing self-assembling molecular motors and computers, and VLSI Research Inc., of San Jose, reports that "It looks like the race to biochips [another term for molecular electronic systems] has already started.

NEC, Hitachi, Toshiba, Matsushita, Fujitsu, Sanyo-Denki and Sharp have commenced full-scale research efforts on bio-chips for bio-computers."

Biochemists have other reasons to want to learn the art of protein design. New enzymes promise to perform dirty, expensive chemical processes more cheaply and cleanly, and novel proteins will offer a whole new spectrum of tools to biotechnologists.

Second Generation Nanotechnology

Despite its versatility, protein has shortcomings as an engineering material. Protein machines quit when dried, freeze when chilled, and cook when heated. We do not build machines of flesh, hair, and gelatin; over the centuries, we have learned to use our hands of flesh and bone to build machines of wood, ceramic, steel, and plastic. We will do likewise in the future. We will use protein machines to build nanomachines of tougher stuff than protein. As nanotechnology moves beyond reliance on proteins, it will grow more ordinary from an engineer's point of view. Molecules will be assembled like the components of an erector set, and well-bonded parts will stay put. Just as ordinary tools can build ordinary machines from parts, so molecular tools will bond molecules together to make tiny gears, motors, levers, and casings, and assemble them to make complex machines.

Parts containing only a few atoms will be lumpy, but engineers can work with lumpy parts if they have smooth bearings to support them. Conveniently enough, some bonds between atoms make fine bearings; a part can be mounted by means of a single chemical bond that will let it turn freely and smoothly. Since a bearing can be made using only two atoms (and since moving parts need have only a few atoms), nanomachines can indeed have mechanical components of molecular size. How will these better machines be built? Over the years, engineers have used technology to improve technology. They have used metal tools to shape metal into better tools, and computers to design and program better computers. They will likewise use protein nanomachines to build better nanomachines. Enzymes show the way: they assemble large molecules by "grabbing" small molecules from the water around them, then holding them together so that a bond forms. Enzymes assemble DNA, RNA, proteins, fats, hormones, and chlorophyll in this way—indeed, virtually the whole range of molecules found in living things. Biochemical engineers, then, will construct new enzymes to assemble new patterns of atoms. For example, they might make an enzyme-like machine which will add carbon atoms to a small spot, layer on layer. If bonded correctly, the atoms will build up to form a fine, flexible diamond fiber having over fifty times as much strength as the same weight of aluminum.

Aerospace companies will line up to buy such fibers by the ton to make advanced composites. (This shows one small reason why military competition will drive molecular technology forward, as it has driven so many fields in the past.) But the great advance will come when protein machines are able to make structures more complex than mere fibers. These programmable protein machines will resemble ribosomes programmed by RNA, or the older generation of automated machine tools programmed by punched tapes. They will open a new world of possibilities, letting engineers escape the limitations of proteins to build rugged, compact machines with straightforward designs. Engineered proteins will split and join molecules as enzymes do. Existing proteins bind a variety of smaller molecules, using them as chemical tools; newly engineered proteins will use all these tools and more.

Further, organic chemists have shown that chemical reactions can produce remarkable results even without nanomachines to guide the molecules. Chemists have no direct control over the tumbling motions of molecules in a liquid, and so the molecules are free to react in any way they can, depending on how they bump together. Yet chemists nonetheless coax reacting molecules to form regular structures such as cubic and dodecahedral molecules, and to form unlikely-seeming structures such as molecular rings with highly strained bonds. Molecular machines will have still greater versatility in bond-making, because they can use similar molecular motions to make bonds, but can guide these motions in ways that chemists cannot. Indeed, because chemists cannot yet direct molecular motions, they can seldom assemble complex molecules according to specific plans.

The largest molecules they can make with specific, complex patterns are all linear chains. Chemists form these patterns (as in gene machines) by adding molecules in sequence, one at a time, to a growing chain. With only one possible bonding site per chain, they can be sure to add the next piece in the right place. But if a rounded, lumpy molecule has (say) a hundred hydrogen atoms on its surface, how can chemists split off just one particular atom (the one five up and three across from the bump on the front) to add something in its place? Stirring simple chemicals together will seldom do the job, because small molecules can seldom select specific places to react with a large molecule. But protein machines will be more choosy. A flexible, programmable protein machine will grasp a large molecule (the workpiece) while bringing a small molecule up against it in just the right place. Like an enzyme, it will then bond the molecules together. By bonding molecule after molecule to the workpiece, the machine will assemble a larger and larger structure while keeping complete control of how its atoms are arranged. This is the key ability that chemists have lacked.

Like ribosomes, such nanomachines can work under the direction of molecular tapes. Unlike ribosomes, they will handle a wide variety of small molecules (not just amino acids) and will join them to the workpiece anywhere desired, not just to the end of a chain. Protein machines will thus combine the splitting and joining abilities of enzymes with the programmability of ribosomes. But whereas ribosomes can build only the loose folds of a protein, these protein machines will build small, solid objects of metal, ceramic, or diamond—invisibly small, but rugged. Where our fingers of flesh are likely to bruise or burn, we turn to steel tongs. Where protein machines are likely to crush or disintegrate, we will turn to nanomachines made of tougher stuff.

Universal Assemblers

These second-generation nanomachines built of more than just proteins will do all that proteins can do, and more. In particular, some will serve as improved devices for assembling molecular structures. Able to tolerate acid or vacuum, freezing or baking, depending on design, enzyme-like second-generation machines will be able to use as "tools" almost any of the reactive molecules used by chemists—but they will wield them with the precision of programmed machines. They will be able to bond atoms together in virtually any stable pattern, adding a few at a time to the surface of a workpiece until a complex structure is complete. Think of such nanomachines as assemblers. Because assemblers will let us place atoms in almost any reasonable arrangement, they will let us build almost anything that the laws of nature allow to exist.

In particular, they will let us build almost anything we can design—including more assemblers. The consequences of this will be profound/

because our crude tools have let us explore only a small part of the range of possibilities that natural law permits. Assemblers will open a world of new technologies. Advances in the technologies of medicine, space, computation, and production—and warfare—all depend on our ability to arrange atoms. With assemblers, we will be able to remake our world or destroy it. So at this point it seems wise to step back and look at the prospect as clearly as we can, so we can be sure that assemblers and nanotechnology are not a mere futurological mirage.

Conclusions

In everything we have been describing, we have stuck closely to the demonstrated facts of chemistry and molecular biology. Still, people regularly raise certain questions rooted in physics and biology. These deserve more direct answers.

Will the uncertainly principle of quantum physics make molecular machines unworkable?

This principle states (among other things) that particles can't be pinned down in an exact location for any length of time. It limits what molecular machines can do, just as it limits what anything else can do. Nonetheless, calculations show that the uncertainty principle places few important limits on how well atoms can be held in place, at least for the purposes outlined here. The uncertainty principle makes electron positions quite fuzzy, and in fact this fuzziness determines the very size and structure of atoms. An atom as a whole, however, has a comparatively definite position set by its comparatively massive nucleus. If atoms didn't stay put fairly well, molecules would not exist. One needn't study quantum mechanics to trust these conclusions, because molecular machines in the cell demonstrate that molecular machines work.

Will the molecular vibrations of heat make molecular machines unworkable or too unreliable for use?

Thermal vibrations will cause greater problems than will the uncertainty principle, yet here again existing molecular machines directly demonstrate that molecular machines can work at ordinary temperatures. Despite thermal vibrations, the DNA-copying machinery in some cells makes less than one error in 100,000,000,000 operations. To achieve this accuracy, however, cells use machines (such as the enzyme DNA polymerase 1) that proofread the copy and correct errors. Assemblers may well need similar error-checking and error-correcting abilities, if they are to produce reliable results.

Will radiation disrupt molecular machines and render them unusable?

High-energy radiation can break chemical bonds and disrupt molecular machines. Living cells once again show that solutions exist: they operate for years by repairing and replacing radiation-damaged parts. Because individual machines are so tiny, however, they present small targets for radiation and are seldom hit. Still, if a system of nanomachines must be reliable, then it will have to tolerate a certain amount of damage, and damaged parts must regularly be repaired or replaced. This approach to reliability is well known to designers of aircraft and spacecraft.

Since evolution has failed to produce assemblers, does this show that they are either impossible or useless?

The earlier questions were answered in part by pointing to the working molecular machinery of cells. This makes a simple and powerful case that natural law permits small clusters of atoms to behave as controlled machines, able to build other nanomachines. Yet despite their basic resemblance to ribosomes, assemblers will differ from anything found in cells; the things they do—while consisting of ordinary molecular motions and reactions—will have novel results. No cell, for example, makes diamond fiber.

The idea that new kinds of nanomachinery will bring new, useful abilities may seem startling: in all its billions of years of evolution, life has never abandoned its basic reliance on protein machines. Does this suggest that improvements are impossible, though? Evolution progresses through small changes, and evolution of DNA cannot easily replace DNA. Since the DNA/RNA/ribosome system is specialized to make proteins, life has had no real opportunity to evolve an alternative. Any production manager can well appreciate the reasons; even more than a factory, life cannot afford to shut down to replace its old systems.

Improved molecular machinery should no more surprise us than alloy steel being ten times stronger than bone, or copper wires transmitting signals a million times faster than nerves. Cars outspeed cheetahs, jets outfly falcons, and computers already outcalculate head-scratching humans. The future will bring further examples of improvements on biological evolution, of which second-generation nanomachines will be but one. In physical terms, it is clear enough why advanced assemblers will be able to do more than existing protein machines. They will be programmable like ribosomes, but they will be able to use a wider range of tools than all the enzymes in a cell put together. Because they will be made of materials far more strong, stiff, and stable than proteins, they will be able to exert greater forces, move with greater precision, and endure harsher conditions.

Like an industrial robot arm—but unlike anything in a living cell—they will be able to rotate and move molecules in three dimensions under programmed control, making possible the precise assembly of complex objects. These advantages will enable them to assemble a far wider range of molecular structures than living cells have done.

Is there some special magic about life, essential to making molecular machinery work?

One might doubt that artificial nanomachines could even equal the abilities of nanomachines in the cell, if there were reason to think that cells contained some special magic that makes them work. This idea is called "vitalism." Biologists have abandoned it because they have found chemical and physical explanations for every aspect of living cells yet studied, including their motion, growth, and reproduction. Indeed, this knowledge is the very foundation of biotechnology. Nanomachines floating in sterile test tubes, free of cells, have been made to perform all the basic sorts of activities that they perform inside living cells. Starting with chemicals that can be made from smoggy air, biochemists have built working protein machines without help from cells. R. B. Merrifield, for example, used chemical techniques to assemble simple amino acids to make bovine pancreatic ribonuclease, an enzymatic device that disassembles RNA molecules. Life is special in structure, in behaviour, and in what it feels like from the inside to be alive, yet the laws of nature that govern the machinery of life also govern the rest of the universe.

The case for the feasibility of assemblers and other nanomachines may sound firm, but why not just wait and see whether they can be developed?

Sheer curiosity seems reason enough to examine the possibilities opened by nanotechnology, but there are stronger reasons. These developments will sweep the world within ten to fifty years—that is, within the expected lifetimes of ourselves or our families. It would cost many millions of lives, and perhaps end life on Earth. Is the case for the feasibility of nanotechnology and assemblers firm enough that they should be taken seriously? It seems so, because the heart of the case rests on two well-established facts of science and engineering. These are (1) that existing molecular machines serve a range of basic functions, and (2) that parts serving these basic functions can be combined to build complex machines. Since chemical reactions can bond atoms together in diverse ways, and since molecular machines can direct chemical reactions according to programmed instructions, assemblers definitely are feasible.

Nanocomputers

Assemblers will bring one breakthrough of obvious and basic importance: engineers will use them to shrink the size and cost of computer circuits and speed their operation by enormous factors. With today's bulk technology, engineers make patterns on silicon chips by throwing atoms and photons at them, but the patterns remain flat and molecular-scale flaws are unavoidable. With assemblers, however, engineers will build circuits in three dimensions, and build to atomic precision. The exact limits of electronic technology today remain uncertain because the quantum behaviour of electrons in complex networks of tiny structures presents complex problems, some of them resulting directly from the uncertainty principle. Whatever the limits are, though, they will be reached with the help of assemblers. The fastest computers will use electronic effects, but the smallest may not. This may seem odd, yet the essence of computation has nothing to do with electronics.

A digital computer is a collection of switches able to turn one another on and off. Its switches start in one pattern (perhaps representing 2 + 2), then switch one another into a new pattern (representing 4), and so on. Such patterns can represent almost anything. Engineers build computers from tiny electrical switches connected by wires simply because mechanical switches connected by rods or strings would be big, slow, unreliable, and expensive, today. The idea of a purely mechanical computer is scarcely new. In England during the mid-1800s, Charles Babbage invented a mechanical computer built of brass gears; his co-worker Augusta Ada, the Countess of Lovelace, invented computer programming. Babbage's endless redesigning of the machine, problems with accurate manufacturing, and opposition from budget-watching critics (some doubting the usefulness of computers!), combined to prevent its completion.

In this tradition, Danny Hillis and Brian Silverman of the MIT Artificial Intelligence Laboratory built a special-purpose mechanical computer able to play tic-tactoe. Yards on a side, full of rotating shafts and movable frames that represent the state of the board and the strategy of the game, it now stands in the Computer Museum in

Boston. It looks much like a large ball-and-stick molecular model, for it is built of Tinkertoys.

Brass gears and Tinkertoys make for big, slow computers. With components a few atoms wide, though, a simple mechanical computer would fit within 1/100 of a cubic micron, many billions of times more compact than today's so-called microelèctronics. Even with a billion bytes of storage, a nanomechanical computer could fit in a box a micron wide, about the size of a bacterium. And it would be fast. Although mechanical signals move about 100,000 times slower than the electrical signals in today's machines, they will need to travel only 1/1,000,000 as far, and thus will face less delay. So a mere mechanical computer will work faster than the electronic whirl-winds of today.

Electronic nanocomputers will likely be thousands of times faster than electronic microcomputers—perhaps hundreds of thousands of times faster, if a scheme proposed by Nobel Prize-winning physicist Richard Feynman works out. Increased speed through decreased size is an old story in electronics. Disassemblers Molecular computers will control molecular assemblers, providing the swift flow of instructions needed to direct the placement of vast numbers of atoms. Nanocomputers with molecular memory devices will also store data generated by a process that is the opposite of assembly.

Assemblers will help engineers synthesize things; their relatives, disassemblers, will help scientists and engineers analyze things. The case for assemblers rests on the ability of enzymes and chemical reactions to form bonds, and of machines to control the process. The case for disassemblers rests on the ability of enzymes and chemical reactions to break bonds, and of machines to control the process. Enzymes, acids, oxidizers, alkali metals, ions, and reactive groups of atoms called free radicals—all can break bonds and remove groups of atoms. Because nothing is absolutely immune to corrosion, it seems that molecular tools will be able to take anything apart, a few atoms at a time. What is more, a nanomachine could (at need or convenience) apply mechanical force as well, in effect prying groups of atoms free.

A nanomachine able to do this, while recording what it removes layer by layer, is a disassembler. Assemblers, disassemblers, and nanocomputers will work together. For example, a nanocomputer system will be able to direct the disassembly of an object, record its structure, and then direct the assembly of perfect copies. And this gives some hint of the power of nanotechnology.

Assemblers will take years to emerge, but their emergence seems almost inevitable: Though the path to assemblers has many steps, each step will bring the next in reach, and each will bring immediate rewards. The first steps have already been taken, under the names of "genetic engineering" and "biotechnology." Other paths to assemblers seem possible. Barring worldwide destruction or worldwide controls, the technology race will continue whether we wish it or not. And as advances in computer aided design speed the development of molecular tools, the advance toward assemblers will quicken. To have any hope of understanding our future, we must understand the consequences of assemblers, disassemblers, and nanocomputers. They promise to bring changes as profound as the industrial revolution, antibiotics, and nuclear weapons all rolled up in one massive breakthrough. To understand a future of such profound change, it makes sense to seek principles of change that have survived the greatest upheavals of the past. They will prove a useful guide.

6

ELECTROPHORESIS

Capillary electrophoresis (CE) is a special case of using an electrical field to separate the components of a mixture. Electrophoresis in a capillary is differentiated from other forms of electrophoresis in that it is carried out within the confines of a narrow tube. To understand the behavior of molecules under the influence of an electrical field inside a capillary it is essential to understand the phenomena that result from the geometry of a capillary.

FUNDAMENTALS OF ELECTROPHORESIS

Basic Principles

It has long been known that molecules can be either positively or negatively electrically charged. When the numbers of positive and negative charges are the same, the charges cancel, creating a neutral (uncharged) molecule. If given the freedom to move, charged particles will seek regions, such as an electrode, having an opposing charge; in other words, opposites attract. In this example a mixture of ionic substances is dissolved in a suitable solvent, such as water. In the absence of an electrical field, the motion of these ions is essentially random. When the electrical field is applied, charged species begin to move. A crude separation occurs, resulting in a less random distribution of charged particles. Cations (positively charged ions) move toward the cathode (negatively charged electrode), and anions (negatively charged ions) move toward the anode (positively charged electrode).

Figure 6.1 also illustrates another aspect of electrophoresis in solution, the significance of the mass charge ratio (m/z). In this figure there are actually four types of charged particles: large and small positively charged and large and small negatively charged. If each

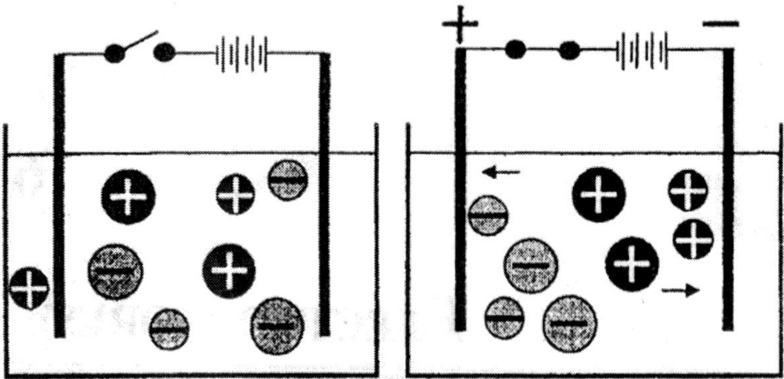

Fig. 6.1. Simple electrophoresis.

particle has only a single charge, then the absolute value of the force on each particle will be the same. The acceleration created by this force can be determined by the relationship Force = mass × acceleration (F = ma). The viscosity of the separation medium opposes the acceleration with the result that a steady velocity is achieved under constant conditions. This means that the system can not only separate particles having opposite charges, but can also separate particles of the same charge if there are other differences between them. The science of electrophoresis is largely concerned with creating systems that exploit the differences between the molecules. Alternatively, the analyst may wish to develop a system that creates differences between molecules. Varying the pH of the separation method is an example. At pH = 10.0, glycine and acetic acid will have the same charge (−1). At pH = 7.0, glycine will have a very small net charge whereas acetic acid will still have a charge of −1. Separation of these two molecules would therefore be different at pH = 7.0 and at pH = 10.0.

Numerous other factors besides pH affect electrophoretic separations. These include the hydrodynamic radius of the molecules, the viscosity of the separation medium, and temperature. In real systems there are other forces, in addition to the electrical field, acting on the charged molecules, e.g., the entire fluid mass may be moving relative to the vessel in which it is contained. Some of these factors can affect the electrophoresis in a very complex manner; for example, the passage of current through a liquid can raise the temperature of that liquid. This change in temperature can influence the electrical resistance of the system (and hence the current), the viscosity, and the velocity of the molecules moving in the field. These factors will be discussed later in the specific context of CE. Janini and Issaq presented a detailed discussion of the theoretical underpinnings of these factors.

Electrophoresis in a Capillary

The electrophoretic process in a capillary has all of the features previously described. In addition, the small diameter of the capillary and its large aspect ratio (length/width) contribute additional factors. What actually defines a capillary? For this chapter the discussion will be limited to capillaries formed from fused silica and having an inner diameter (i.d.) of 100 μm or less. The usual range of inner diameters is from 20–100 μm. Typically the capillaries that are used in CE are circular in cross-section. However, capillaries with square cross-sections have been produced. These may offer some advantages in terms of thermal regulation and detection sensitivity, but to date they have not been widely used.

Advantages of the capillary

Capillaries were introduced into electrophoresis as an anti-convective and heat controlling innovation. In wide tubes thermal gradients cause band mixing and loss of resolution. The use of glass capillaries of 200–500-μm i.d. was reported by Virtanen in 1969. Jorgensen's introduction of 75 μm capillary tubes was the start of modern "*high-performance*" CE.

Figure 6.2 illustrates the separation of a mixture of two components (both negatively charged) in a capillary. A small plug of sample is introduced into one end. When an electrical field is applied, the components begin to move in the field as previously described. The narrow capillary reduces lateral diffusion and insures that temperature

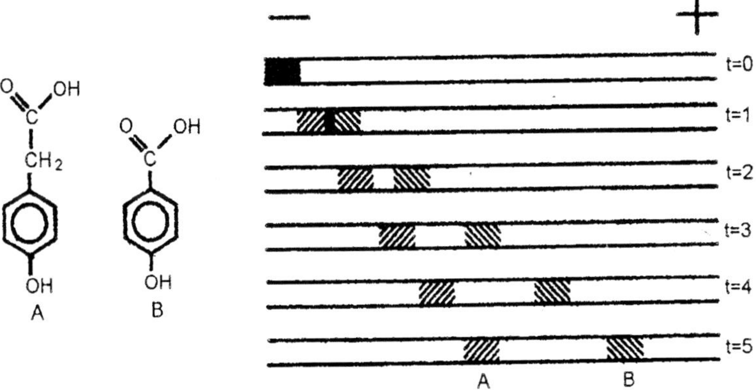

Fig. 6.2. Capillary electrophoresis.

differences between the center of the capillary and the wall are quite small. Because the two components in this example move at different velocities, they can be separated. The geometry and other properties of the capillary electrophoretic separation also lead to a condition known as *plug flow*. Under ideal plug flow conditions, the only factor leading to sample dispersion is diffusion. This contributes to the high efficiency of CE separations.

Electroosmosis

The small diameter of the capillary contributes to another aspect of the separation process. This is the phenomenon known as electroosmosis, electroendoosmotic flow, or simply EOF. EOF exists in any electrophoretic system. It will occur whenever the liquid near a charged surface is placed in an electrical field resulting in the bulk movement of fluid near that surface. Because the surface to volume ratio is very high inside a capillary, EOF becomes a significant factor in CE.

The velocity of the electroosmotic flow through a capillary is given by the Smoluchowski equation (Eq. 1),

$$v_{eof} = -(\varepsilon\zeta / 4\pi\eta)E \qquad \text{...(1)}$$

where ε is the dielectric constant of the electrolyte, ζ is the zeta potential (Volts), η is the viscosity (Poise), and E is the potential applied (Volts/cm). In CE the zeta potential (ζ) is a measure of the charge on the wall of the capillary. This charge arises from both the nature of the material that composes the capillary and the composition of the electrolyte (buffer). The most commonly employed capillary material is fused silica. The surface of a fused silica capillary can be hydrolyzed to yield a negatively charged surface as described previously. The negatively charged wall attracts cations that are hydrated from the electrolyte solution, creating an electrical double layer. In an electrical field they migrate toward the cathode, pulling water along and creating a pumping action. The zeta potential increases with the density of the charge on the surface. For fused silica and many other materials charge density will vary with pH. Bare fused silica behaves much like a weak acid with a pKa of 6.25. EOF also decreases with the square root of the concentration of the electrolyte, i.e., increasing buffer concentration decreases the velocity of EOF. Although the aforementioned discussion has focused on bare fused silica capillaries, other types of surfaces can be chemically created from fused silica. The new surface may be positive, negative, or neutral. The direction of the electroosmotic flow will therefore depend on the sign of the

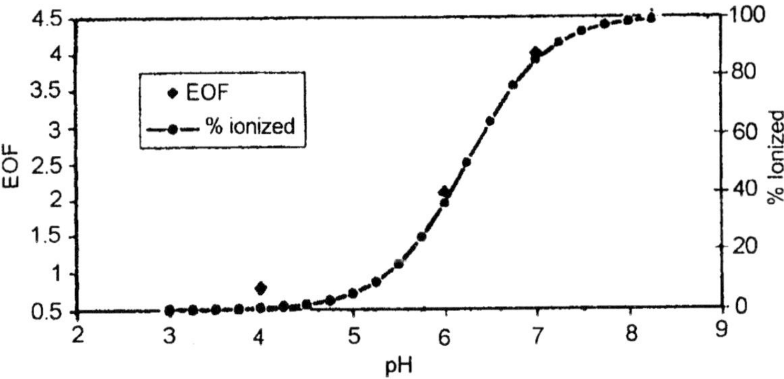

Fig. 6.3. EOF as a function of pH.

charge on the wall of the capillary. Flow is always toward the electrode that has the same charge as the capillary wall. Thus an uncharged wall will have, in theory, no EOF. In reality this is difficult to accomplish.

In the narrow confines of the capillary the velocity of liquid is nearly uniform across the i.d. of the capillary resulting in what has been termed "*plug flow*". This is in contrast to the laminar flow exhibited by pumped systems, which creates a velocity profile across the diameter of the tube. On the other hand, plug flow greatly reduces the band broadening seen in systems (such as *high-performance liquid chromatography* [HPLC]) that are pumped by a pressure differential.

In the example, the EOF is assumed to be zero. In practice it is very difficult to completely eliminate EOF, although it can be reduced

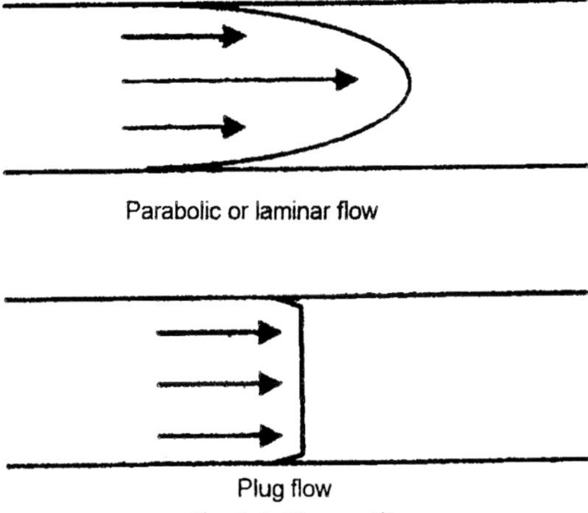

Fig. 6.4. Flow profiles.

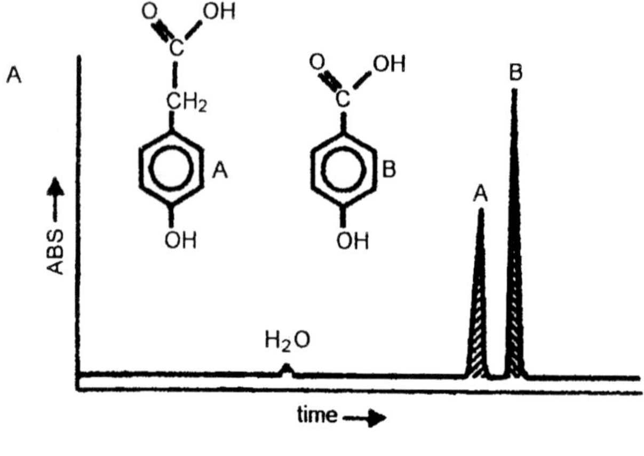

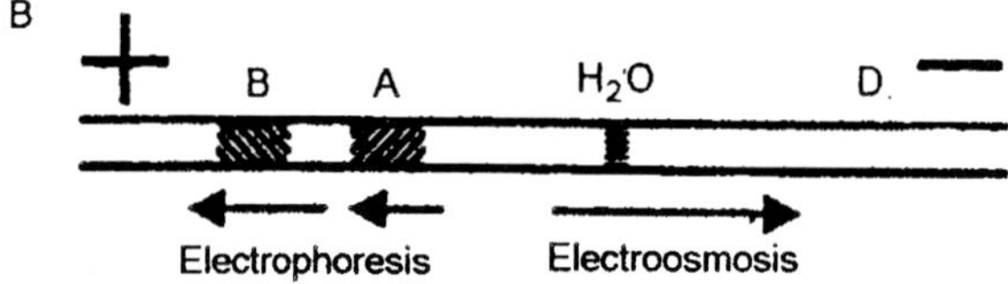

Fig. 6.5. Separation with EOF.

to a value near zero. In many separations it is a significant factor, and it may affect the net movement of the analyte molecules more than does the electrophoretic force. The movement (vectors) due to the electrophoretic and electroosmotic forces may even be in opposite directions. Both analytes have a charge of −1 under these conditions. The EOF in this case is toward the cathode, while the analytes are moved electrophoretically toward the anode. The EOF contribution is larger than the electrophoretic, so that the net movement is toward the cathode. The small peak labeled H_2O represents the water in which the sample was dissolved. This water plug does not have any electrophoretic mobility; the movement of this plug is due solely to electroosmotic flow, and it is the first peak to pass the detector. The next peak to emerge, A, has the higher mass/charge ratio of the two analytes. This is in contrast to the case (no EOF) where the first peak to emerge had the lower mass/charge ratio. Note also that the polarity of the electrodes has been reversed compared. A snapshot of the inside of the capillary during the separation but prior to any peak reaching the detector (D). Because A and B are influenced by both the EOF and the electrophoretic force, their net movement is the sum of the electroosmotic and electrophoretic vectors.

Capillaries

Many materials have been suggested and tested for the construction of capillaries for CE. These include fused silica, borosilicate glass, and polytetrafluoroethylene (Teflon). Fused silica is now the preferred material for the construction of capillaries. The silica used is of very high purity, similar to that used to produce silicon chips for electronic components. Tubes of silica are heated to over 1000°C, then stretched to produce the final dimensions. Fused silica capillaries are extremely brittle. To facilitate handling they are coated with a layer of polyimide 10–25 μm thick, rendering the capillary very flexible. It can easily be wound on a spool 2.5 cm in diameter without breaking. The coating, however, has the disadvantage of being virtually opaque to UV light. It must be removed to create a "*window*" to allow on-capillary detection. This makes an area that is very fragile, warranting careful handling to prevent breakage.

The thickness of the wall of the capillary and the polyimide coating is generally much greater than the i.d. of the capillary. A typical capillary will have an external diameter of 350 μm. If the internal diameter is 50 μm, this leaves a wall thickness of 150 μm. The thickness of the wall is critical, because the heat that is produced when an electrical current passes through the electrolyte must be removed through this wall. Reducing the wall thickness increases the fragility of the capillary at the same time as it increases the heat transfer capacity. External diameter is thus a trade-off between strength and heat transfer.

Variables in Capillary Electrophoresis

Capillary electrophoresis offers many directions from which to approach an analytical problem. This can be advantageous because as the number of options available increases, the likelihood that a solution to the analytical problem can be found also increases. This can also be a disadvantage, particularly to the newcomer, because the wide choice of options can appear daunting. This is compounded by the interactive nature of some of these factors. For example, a change in pH can affect the current; a change in current can affect temperature; a change in temperature can affect pH, the factor we were trying to vary in our experiment. Understanding these key factors will contribute to successful analysis.

Capillary Surface

The inner surface of a capillary is an extremely important factor in CE. The inner wall is in contact with the separation chemistry and the samples. As noted earlier, the capillary wall is the site of the mechanism by which EOF is created. In order to understand how to control the effect of the surface on the separation, it important to understand the behavior of the capillary surface. The inner surface of a CE capillary is more analogous to the surface of the particles packed inside an HPLC column than it is to the HPLC column wall. Unlike the wall of an HPLC column, the capillary wall participates in the separation process. In CE the goal is usually to prevent interaction between the analytes and the inner wall of the capillary while simultaneously controlling EOF. In order to have control over the separation process it is necessary to have control over this surface.

Surface modifications

The nature of the surface of a bare fused silica capillary depends somewhat on the treatment it has experienced. Most of the surface is in a siloxane form; there are few groups present that can create a charged surface. Treatment with base (NaOH or KOH) causes the hydrolysis of the siloxane group. Sometimes referred to as an etched surface, this layer is rich in silanol groups, which are the weak acids referred to in the discussion of EOF. To convert completely the surface of the capillary into this form may require long periods of exposure with 0.1–1.0 M base at elevated temperatures. Incomplete conversion may lead to poor reproducibility in peak migration times.

Not all separations can be optimized using bare silica. However, the silica surface is chemically reactive and can be derivatized in a variety of ways. Silane derivatization reactions are commonly employed

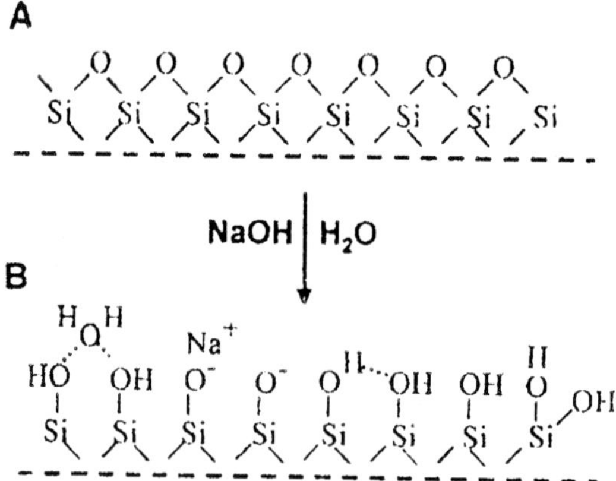

Fig. 6.6. Silica surface chemistry. Surface of fused silica before (A) and after (B) NaOH hydrolysis.

to link other groups to silica. This can result in a linkage through a Si-O-Si-C bond or through a Si-C bond depending on the chemistry employed. The Si-C bond is said to be the more stable of the two at high pH values. Using a molecule that has a silane at one end and a second functional group at the other allows a second layer to bond. This approach allows more complex coatings to be applied in layers. In this way many types of coatings have been created including polyacrylamide, polyvinyl alcohol, polyethylene glycol, and polyvinylpyrolidone. Each of these coatings will have different properties of charge and adsorptivity. Most were designed to reduce protein binding to the capillary wall. Basic proteins, being positively charged, are particularly difficult to analyze in untreated capillaries because they bind tightly to the negatively charged wall. Coatings are also employed to modify the charge of the wall such that EOF is drastically altered or reversed.

Presently, the analyst seldom prepares covalently coated capillaries, as many types are commercially available. With commercially available fixed coatings, it is possible to obtain charges ranging from acidic to neutral to basic and from *hydrophilic* to *hydrophobic*. The analyst may also wish to incorporate noncovalent coatings into a separation method using a bare silica capillary. These transient or dynamic coatings rely on ionic interactions to hold them on the capillary wall. Polyamines will stick to the negatively charged silica, effectively shielding the negatively charged wall from the solution and creating a positively charged surface. In such a capillary, the EOF will be toward the anode. Coatings of this type may be stable for several runs. Cationic detergents such as tetradecyltrimethylammonium bromide (TTAB) can be used to form a positively charged surface as well. Such coatings are actually bilayers. The first layer has the positively charged head of the detergent attached to the negatively charged silanol surface. The hydrophobic tails of the molecules extend radially inward, forming a *hydrophobic* surface. A second layer of detergent forms with the hydrophobic tails in contact with the tails of the first layer and the positively charged heads forming the inner surface of the capillary. Again, EOF will be toward the anode. These detergent bilayers are easy to form but lack stability. They are easily disrupted by organic solvents and usually require that the detergent be added to the run buffer (where it may interact with analytes).

Whenever a capillary is used, the buffer being used modifies the surface. Most workers have come to recognize that capillaries perform best when they are "dedicated" to a specific type of buffer species.

For example, if a capillary that has been equilibrated with a phosphate buffer is used with a borate buffer of similar concentration and pH the migration times will drift from run to run until a new equilibrium condition has been established. Although neither borate nor phosphate would be expected to form a coating on the capillary wall, it is apparent that the capillary does "remember" the buffer it has held. This dedication of a capillary to one type of system is a relatively inexpensive way to improve results.

Capillary regeneration

Regeneration is the process of creating the same surface on the inner wall of the capillary prior to the start of every analytical run with a given method. Although it is only necessary to recreate the same surface condition every time, it is in fact easier in the long run to fully regenerate the surface. There are two goals to regeneration: (1) to clean the capillary of any residual buffer and sample from the previous analysis; and (2) to create the same surface from run to run, thereby enhancing reproducibility. If the method has been properly designed these steps should not be necessary. If sample components are sticking to the capillary wall in sufficient quantity to affect subsequent runs, it is advisable to change the method to prevent the sample from sticking (perhaps by using a coating or by increasing the buffer concentration) rather than incorporating a cleaning procedure. In practice this can be difficult to accomplish, particularly with samples containing protein (such as serum) that is inherently troublesome.

This chapter will focus on the regeneration of bare silica capillaries. The special handling of coated surfaces will be discussed previously. However, when using commercially coated capillaries, it is important to follow the manufacturer's instructions for cleaning and regeneration. This is because coated capillaries are easily converted into bare silica capillaries.

Freshly purchased fused silica capillaries, even those sold as "ready to use," may not be clean when first received. Therefore, some variation of the following procedure should be used whenever a new piece of bare silica capillary is used.

1. Flush the capillary with methanol. This will remove any organic residues (such as oil) and will aid in the subsequent wetting of the surface.
2. Flush with water.
3. Flush with 1 N hydrochloric acid. This removes any residual cations.

4. Flush with water.
5. Flush with 1 N NaOH. This step hydrolyzes the siloxanes to create the charged silanol groups that provide EOF.
6. Flush with water.
7. Flush with electrolyte.

Many variations on this procedure exist. For example, a second hydrochloric acid wash can be added after the NaOH if the removal of residual sodium is desired (as prior to a sodium analysis). The time and temperature of each step is important. A minimum of three capillary volumes should be used over at least 5 min. Static soaking should be avoided in favor of a continuous flow. As expected, the use of elevated temperature (35–40°C) will enhance the cleaning rate. The final flush with electrolyte, although critical to obtain assay reproducibility, is often neglected. The procedure described above creates a virgin surface that may change as analysis proceeds. There is usually an equilibration between the capillary wall and even simple buffers, such as sodium phosphate. Applying voltage to the capillary facilitates the exchange of ions at the surface after filling with the electrolyte.

The regeneration process between sample runs can be milder than the initial cleaning, particularly if the method is not detrimental to the surface and the samples are not sticky. In the best of cases it is only necessary to flush with fresh run buffer. At other times exposure to acid or base, or even an organic solvent, may be required to restore the surface to the original condition. The regeneration procedure, however, should be matched to the separation method. For separations using sodium borate buffer, pH 8.3, regeneration with NaOH is appropriate. For separations using phosphate buffer, pH 2.5, the use of NaOH or other base rinses should be avoided. At low pH, EOF is suppressed because the silanol groups are poorly ionized. Using NaOH to regenerate the capillary will increase the time needed to reequilibrate the surface with the low pH electrolyte (phosphate), thus rinsing with a higher concentration of low pH buffer or even with 1 N acid (i.e., phosphoric) will generally give better results.

Rinsing coated capillaries adds a further complication, in that the coating itself may be labile in the presence of the cleaning solution. Exposure to NaOH will strip many covalently bonded phases from the capillary surface. It is critical to follow the manufacturer's recommendations for cleaning and regenerating coated capillaries. Dynamic coatings such as TTAB should be reconstructed by rinsing with a more concentrated solution of the coating material than is used in the run buffer. For example, if the running buffer contains 0.5 mM of detergent, a pre-rinse with 20 mM detergent followed by the actual running buffer would be appropriate.

Separation Buffers

The significance of the capillary wall in controlling the process of separation in CE cannot be overstated. The separation, however, takes place in the separation buffer. It is here that the conditions are such that the differences in mobility can exist. Buffers can either be made or purchased. Buffer quality is independent of the instrument system used. However, even the best instrument system will not perform properly with a poorly prepared buffer.

Significance of pH

The most common measurement in analytical chemistry is probably the measurement of pH. It is also the measurement that is most often made improperly. In CE it is extremely important to properly control pH since it affects analyte charge, electroosmotic flow, and, by affecting current, heat production. Thus small changes in pH tend to have greater impact in CE than do comparable pH variations in HPLC.

The titration curve of the silica surface behaves as a typical weak acid (e.g., acetic acid). The ionization of acetic acid (pKa = 4.76) is shown in Eq. 2. At the pKa, the analyte will be 50% charged. This does not mean that each molecule carries one-half of a charge, but rather that at any point in time one half of the molecules are charged and the other half are uncharged. A titration curve represents the probability than any given molecule will be charged at a given pH.

$$H_3C-COOH \rightleftharpoons H_3C-COO^{(-)} + H^{(+)} \quad \ldots(2)$$

When the molecule is uncharged it is under the influence of the electroosmotic flow but not the electrical field. When charged the molecule responds to both forces. As the fraction of the time the molecule spends in each state varies with pH, the net migration velocity will change. As discussed earlier, the EOF also changes with pH. Failure to properly control pH is one of the major causes of poor reproducibility in CE.

Buffer species

Buffers are compounds that are used to control the pH of a solution. They are generally weak acid or bases that can accept or donate protons. Buffers reduce the change in pH that is caused by the introduction of additional acid or base; CE buffers are selected on the basis of the pH range that is to be maintained, as well as other factors.

A wide range of buffers has been employed in CE systems. The most commonly employed are phosphate, borate, citrate, acetate, and Tris (trishydroxymethylamino methane). The zwitterionic buffers that have been recommended by several authors have the advantage of carrying less current than do mono-functional buffer molecules. However, the selection of a buffer for CE should be based on several factors, including: (1) pH value desired; (2) operating temperature; (3) charge of the buffer relative to the analytes and the capillary wall; and (4) effects on detection. These factors are further described as follows:

1. For any buffer the effective buffering range is defined as within one pH unit of the buffers pKa. Thus a buffer with a pKa of 5 would be usable from pH 4.0 to pH 6.0. For CE it is much better to have the pH as close to the pKa of the buffer as possible, certainly within ± 0.5 pH units, to reduce the buffer's contribution to variability.
2. Buffer molecules exhibit a temperature coefficient. The pH of a buffered system will tend to change with temperature. In addition, buffers tend to differ in their sensitivity to temperature. For example, phosphate has a low temperature coefficient compared to Tris. The temperature coefficient of the buffer is also important because during electrophoresis the analyst has only limited control of the temperature inside the capillary.
3. Interaction between the buffer and the analytes can change the effective charge on the analyte molecules, changing their migration velocity. Ion pairing can also occur between the buffer molecules and the analytes affecting the separation.
4. Buffers differ widely in their absorbance spectra. This is much less of a problem in CE than in HPLC because of the short path-length in CE detection. However, for high-sensitivity work the background absorbance may become an issue.

Once the buffer composition has been selected, it is essential that it be prepared properly. In order to make a buffer consistent it is critical to make the buffer the same way every time. For example, preparing a 50 mM sodium phosphate buffer by starting with 50 mM sodium phosphate and adjusting the pH with phosphoric acid will result in a buffer with a phosphate concentration higher than 50 mM. Such a buffer is best prepared by titration of phosphoric acid with NaOH. A buffer that has found wide application in CE is sodium borate. However, the preparation of borate buffers can be confusing because boric acid combines with sodium hydroxide according to Eq. 3.

$$4\ H_3BO_3 + 2\ NaOH \rightleftarrows Na_2B_4O_7 + 7\ H_20 \quad ...(3)$$

Four moles of boric acid react with sodium hydroxide to form one mole of sodium tetraborate. A borate buffer will contain both boric acid and tetraborate. In discussing these systems it is customary to refer to the concentration of Boron. Thus, 4 M boric acid and 1 M sodium tetraborate have the same concentration of boron.

Buffer concentration

Increasing the concentration of a buffer will increase the buffering capacity of the system, making it less likely to change pH with the addition of acid or base. In addition, there are other factors in CE that can be affected by the concentration of the buffer. These factors can be critical when trying to achieve optimum resolution and reproducibility.

If the composition of the sample plug is different from the composition of the buffer in the capillary, the phenomenon of stacking may occur. Properly exploited stacking can result in sharper peaks and greater sensitivity. One way to achieve stacking is to keep the conductivity of the sample lower than the conductivity of the running buffer by reducing the concentration of the sample buffer relative to the run buffer. To understand this phenomenon, consider the buffer filled capillary as a wire. If the buffer concentration is uniform from one end of the capillary to the other end, the voltage drop will be linear. Inserting a plug of dilute sample is like inserting a resistor into the wire. The voltage drop now becomes steeper in the region of the plug than it is elsewhere. Since velocity is proportional to the voltage gradient, the velocity of the analytes will be faster in this region. Analytes will move rapidly until they reach the boundary between the dilute sample plug and the more concentrated running buffer. On reaching this point they will stack, increasing the local concentration of analyte and creating a zone that is narrower than the original injection plug.

Because stacking is caused by a difference in conductivity between the buffer and the sample, it might be assumed that dissolving the sample in pure water would create the highest peak efficiencies. In practice, however, this is not usually the case. The resistance across a plug of nearly pure water can be so high that localized heating results. This heating causes convective mixing and de-stacking. There should be always be some buffer ions in the sample plug; keeping samples at about 1/10 the concentration of the running buffer is a good rule of thumb.

One way to maximize stacking would be to increase the concentration of the buffer, particularly if the samples as received are high in salt. As stated previously, EOF decreases with the square root of the buffer concentration. If EOF is to remain constant it will be necessary to increase it in some other way, such as a change in pH, temperature, or voltage. If desired, EOF can be reduced simply by increasing the buffer concentration. However, by increasing the buffer concentration the current that passes through the capillary will also increase. This in turn will increase the heat generated inside the capillary and, if not adequately dissipated, a variety of problems can result. Without adequate cooling the ability to exploit higher buffer concentrations is limited.

When trying to separate basic compounds or proteins that tend to interact with the capillary wall using bare silica capillaries, problems are frequently encountered. In extreme situations they may be totally adsorbed on the capillary wall and never pass the detector. In other cases the wall may provide a drag force that results in peak tailing, loss of resolution, and decreased signal-to-noise ratio. Although coated capillaries are one solution to this problem, merely increasing the concentration of the electrolyte can often reduce these interactions.

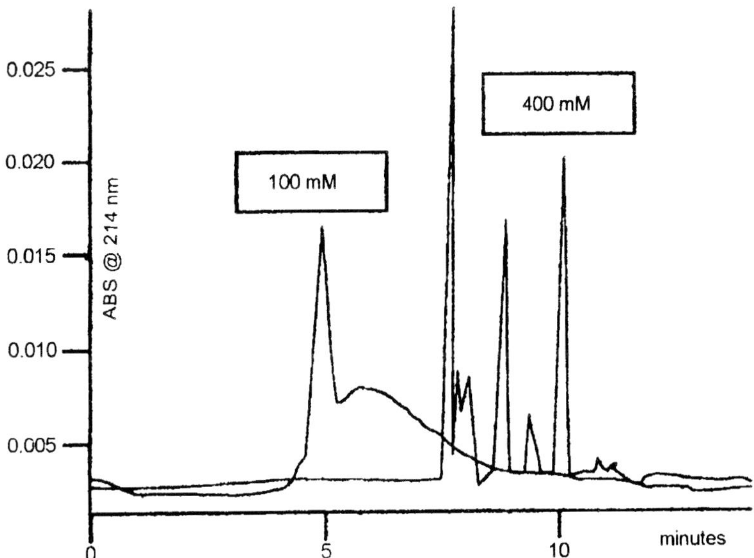

Fig. 6.7. Effect of buffer concentration on peak shape.

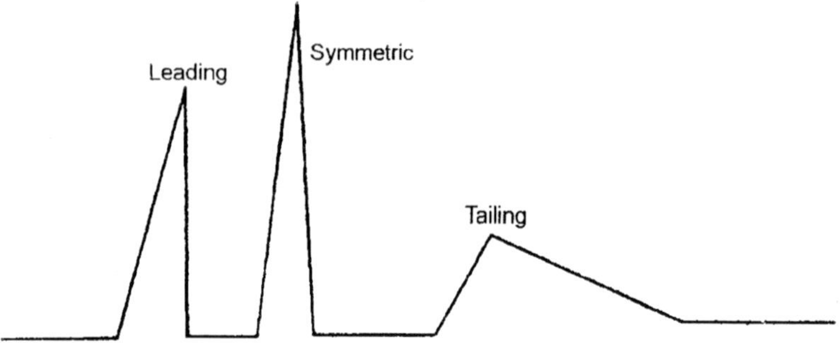

Fig. 6.8. Peak shape in CZE mode.

Peak shape in many modes of CE differs from the bell shape that is commonly seen in HPLC and other chromatographic separations. In open capillary modes (such as *capillary zone electrophoresis* [CZE]), it is not uncommon to see peak shapes. In CE the electrical field influences the buffer ions just as it does the analyte ions. The peak shape in these separations is due to the relative migration rates of the buffer ions and the analyte ions. The degree of asymmetry will increase with increased difference in mobility. Buffer mismatch can so distort the peak shape that analytes may be difficult to quantitate or may even be lost.

Because the buffer ions migrate in the electrical field, the concentration of the buffer ions in the inlet and outlet vials will change over time. Using buffer vials that are too small (i.e., do not contain an adequate volume of buffer) or running too many samples without replacing the buffer can result in ion depletion. This process can result in a change in the sample migration times or resolution.

Additives

Other reagents are frequently added to the buffer systems used in CE. The most common are detergents, such as sodium dodecyl sulfate (SDS), viscosity modifiers, such as linear polyacrylamide, organic solvents, such as acetonitrile, denaturants, such as urea, or combinations of these additives.

The addition of detergent to a buffer used in CE can change dramatically the separation properties of the system. Detergents can aid in solubilizing analytes and in reducing analyte-wall interactions. They may also bind to the capillary wall, affecting the EOF. If the detergents are above their *critical micellar concentrations* (CMC), they can create a pseudostationary phase. Under these conditions the separation mechanism is referred to as *micellar electrokinetic chromatography* (MEKC). Detergents for CE may be anionic, cationic, or nonionic. Protein separations can also be achieved with detergents. By titrating the concentration of SDS in the run buffer it is sometimes possible to achieve protein separations based on differential binding of the detergent to different proteins. This is not a separation by MEKC because the protein molecules are too large to be included in the micelle.

Viscosity modifiers provide a physical retardation to the movement of molecules. Suitable viscosity modifiers include polymers, such as linear polyacrylamide or soluble cellulose derivatives, as well as small molecules like urea. By retarding large molecules more than they do small molecules the dimension of size separation is added to the electrophoretic separation. This is exploited fully in *capillary gel electrophoresis* (CGE) which is discussed earlier. At lower concentrations, viscosity modifiers can aid in the separation of very similar analytes. Viscosity modifiers can also have significant effects on EOF, both by the change in viscosity and by interacting with the capillary wall.

Organic additives routinely used include acetonitrile, methanol, formamide, and dimethylformamide. These solvents improve the solubility of organic molecules that would otherwise be poorly soluble in an aqueous system. At sufficiently high concentrations, they can also reduce the degree of ionization of charged analytes. In MEKC they can alter the relative hydrophobicity of the micellar and non-micellar phases. Organic additives can also affect viscosity and wall charge (and hence EOF). The effect of solvents on viscosity can be complex. Acetonitrile has a lower viscosity than water, and any combination of water and acetonitrile will have a viscosity between that of water and acetonitrile. Methanol also has a lower viscosity than water, but combinations of water and methanol can be substantially more viscous than either solvent alone. In nonaqueous CE organic solvent completely replaces water in the separation solution.

Temperature

Temperature control is crucial to reproducible separations in CE. However, temperature regulation is complicated by several factors. First is that the passage of electrical current through the buffer-filled capillary results in the production of heat. This self-heating effect is inherent in electrophoretic separations and is called Joule heating. Thus temperature control in CE is as much a task of removing heat as it is maintaining a constant temperature environment. Second is that the temperature of the contents inside the capillary is difficult to measure. A layer of silica and a layer of polyimide separate the small fluid volume inside the capillary from the thermostatted medium. Any heat must be transferred to and from the separation buffer through these layers.

Although the temperature inside the capillary cannot easily be measured it can be estimated. Equation 4 is used to estimate the Joule heating (Q). In this equation

$$Q = E^2 \Lambda c \qquad \text{...(4)}$$

Q is the Joule heat generated (Ω/cm^3), E is the voltage gradient (V/cm), Λ is the molar conductivity of the electrolyte ($cm^2\ mol^{-1}\ W^{-1}$) and c is the concentration of the electrolyte (mol/L). Since the conductivity of the electrolyte will vary with temperature, a positive feedback mechanism occurs. In other words, as the temperature increases, the current will also increase. This in turn generates more heat thus generating more current. This feedback loop is moderated by the ability of the cooling system to remove heat and thus limit current.

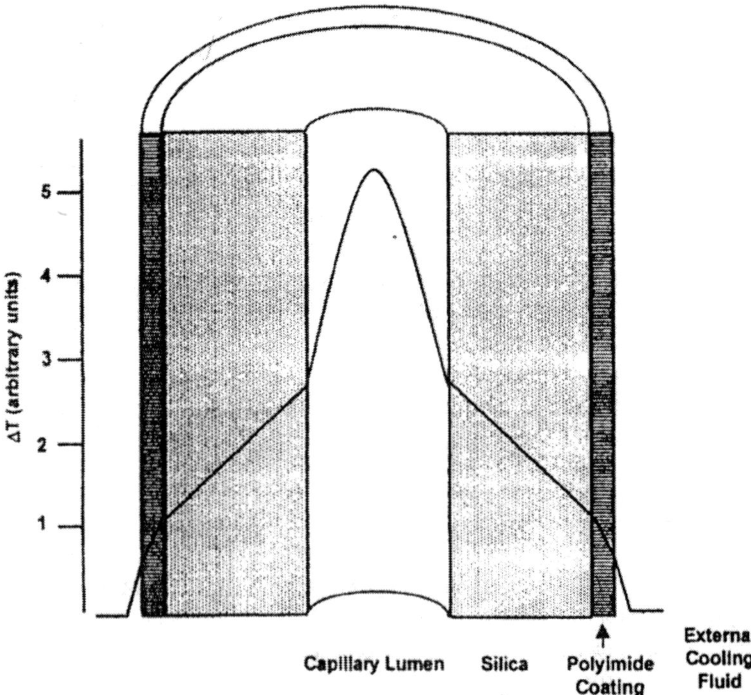

Fig. 6.9. Theoretical radial temperature profile.

Because the cooling system must act at the surface of the capillary and not internally, a temperature gradient exists from the center of the capillary to the outside.

The temperature gradient within the capillary is parabolic. The difference in temperature between the center of the capillary and the wall will increase with increasing capillary internal diameter. Current will also increase as the diameter increases (current α radius^2). The combination of these two factors can lead to the conclusion that heating will cause band broadening and that the band broadening will increase as the internal capillary diameter increases. In systems with adequate temperature control, this thermal band broadening does not become a problem unless the capillary diameter is larger than about 100 μm. A more detailed discussion of heat transfer in CE can be found in an article by Knox.

A second temperature gradient can also exist in poorly designed systems if the entire length of the capillary is not maintained in the same thermal environment. The regions of the capillary that are in buffer vials, passing through cartridge walls, or in detector cells are frequently much less efficiently cooled than are other regions of the capillary. Depending on the system and the capillary length, this uncooled region can account for as much as 25% of the capillary's effective length and can account for much of the variation in performance seen between instruments of different design.

Dynamically cooled CE systems are conveniently divided into two groups: those employing gas and those employing liquid as a heat exchange medium. Liquid cooling is generally considered to be more efficient than gas cooling simply because the heat capacity of liquids exceed that of air. The most effective liquid coolant would be water because of its enormous heat capacity. However, when dealing with electrical equipment, the use of water as a heat exchange medium is not advisable. The requirements of safety has led to the use of perfluorinated organic molecules, having the general formula (F_3C-$[CF_2]_n$-CF_3), because of their very low electrical conductivity. Because of their greater heat capacity, liquids tend to be slower to adjust to changes in temperature than gases. Liquid cooling also necessitates sealed recirculating systems, adding to the expense and complexity of the instrument. Whatever the cooling medium, most modern instrument systems use computer controlled Peltier devices to control the temperature of the cooling fluid. Peltier devices are semiconductor mechanisms that can be made to heat or cool by reversing the direction of current flow through them.

Changes in temperature can also impact other factors inside the capillary. For most liquids, an increase in temperature causes a decrease in viscosity. For example, water has a viscosity of 1.002 centipoise at 20°C. This drops to 0.798 centipoise at 30°C. In the absence of any other changes this change in viscosity will result in a substantial increase in EOF. In addition, the pH of a buffered solution may also change with temperature. A 10°C change in temperature will cause a 0.3 pH unit change in the pKa of Tris. This may be sufficient to change both EOF and the charge on the analyte molecules.

A simple way to determine if a CE system is capable of coping with the heat that will be produced during separation is by creating an Ohm's Law plot. Ohm's Law states that $V = IR$. If we vary voltage (V), the current (I) should change in a linear manner so long as resistance (R) remains constant. An increase in temperature within the capillary generally results in a decrease in resistance (and hence an increase in current at constant voltage). By measuring the current at a variety of voltages a graph of current vs voltage can be created. Regardless of the type of cooling system the current will become nonlinear at some voltage. For the buffer system being considered the point at which the line deviates from linearity indicates the maximum voltage possible for that buffer system/capillary combination on that instrument. In practice it is possible to run at somewhat higher voltage if it is understood that the temperature within the capillary is higher than the thermostatted temperature under these conditions. It is good practice to monitor current during a run because an increasing current during a run may indicate that the heat dissipation capacity of the instrument is being exceeded. In the worst case the heat in the capillary may become sufficient to boil the buffer locally, usually at some poorly cooled location. This usually breaks the electrical circuit (and sometimes the capillary).

Volume Relationships

The dimensions employed in CE are much smaller than those with which most chemical analysts are accustomed to work. Capillary diameters typically are measured in microns and as such the entire volume of a capillary is usually a few microliters. Injected sample volumes are in nanoliters (10^{-9}L). At these scales the tolerances required of instrument systems are extremely small and difference of 1 μm in the diameter of a capillary can create very large differences in the results of an analysis.

Flow Dynamics

Most often the sample being analyzed by CE is injected into the capillary by pressure. To calculate the volume of liquid injected it is necessary to use the Poiseuille equation (Eq. 5), which estimates the flow of liquid through a cylinder.

$$V = (\Delta P d^4 \pi t)/(128 \eta L) \qquad ...(5)$$

In this equation, ΔP is the pressure drop down the length of the cylinder (Pascals), d is the cylinder's inside diameter (m), t is time the pressure is applied (s), η is the fluid viscosity (Pascal-seconds), and L is the total length of the cylinder (m). This equation has wide applicability in fluid dynamics and can be used to calculate fluid flow in CE capillaries, blood vessels, or a water main. It shows that the delivery of fluid into a capillary by pressure is directly proportional to the pressure that is applied and the length of time the pressure is applied. Fluid delivery is inversely proportional to the liquid viscosity and the length of the capillary, and proportional to the fourth power of the diameter. Temperature is indirectly involved in this relationship because the viscosities of most fluids change with temperature.

The volumes typically encountered in CE. This example assumes the viscosity of water at 25°C (1.000 centipoise). Typically injection and rinse pressures are 0.5 and 20 psi, respectively.

As predicted by Eq. 5, the volume of fluid delivered at a given pressure increases dramatically as the diameter of the capillary increases. Doubling the diameter of the capillary increases the delivered volume 16-fold. The implications of this are significant. The volume injected under constant condition changes dramatically when the capillary diameter changes over a narrow range. Because tolerances of 1–2 μm are not uncommon in commercially available capillary, selecting different pieces of capillary nominally 20 μm in diameter could give peak areas that vary substantially. Cutting pieces from a large spool of capillary is no guarantee of uniformity because the diameter may vary along the length of the spool.

The injection plug length is the linear distance within the capillary that is occupied by the injected sample volume. An injection plug that is too long may result in wide bands and loss of resolution, particularly if the analytes cannot focus. Increasing the capillary diameter allows the injection of substantially larger volumes of sample without increasing the plug length. Because the volume of a cylinder increases with the square of the radius a doubling of the capillary diameter will allow the injection of four times as much sample without changing the plug length. The viscosity of the fluid being delivered is the most difficult parameter to know with accuracy. The aforementioned examples have assumed the viscosity of water at 25°C. Within the range of temperatures typically used in CE separations (15–60°), the viscosity of water varies in a nonlinear manner from 1.138 to 0.467 centipoise. Almost anything added to the water will alter both the viscosity and the temperature-viscosity relationship. The additions of macromolecules, such as cellulose derivatives or acrylamide polymers, are extreme cases. These molecules display remarkable viscosity behavior with changes in flow rates. Linear polymers, for example, can extend and align themselves when forced through a capillary. They can show a reduction in viscosity with increased flow. Other systems may increase in viscosity with flow because of polymer entanglement. Even systems as simple as methanol-water can show complex behavior. For example, a 1:1 mix of methanol and water has a higher viscosity than either pure solvent. It is also important to remember that when calculating volumes injected into fluid-filled capillaries, the viscosity of the fluid in the capillary is usually more significant than the viscosity of the sample (unless one is analyzing highly viscous samples).

Entering the Poiseuille equation into a spreadsheet program simplifies fluid delivery calculations such as these. There is also a Windows-compatible computer program called "*CE Expert*" that can perform these calculations.

Sample Handling

Four basic strategies are used to deliver these fluid volumes into capillaries: Positive pressure, vacuum, gravity, and electrophoresis. Positive pressure and vacuum have been the most common methods of filling and rinsing capillaries. Positive pressure up to 100 psi (7 bar) has been used for rinsing. This pressure is delivered either from a source of compressed gas, such as nitrogen, or from an on-board air pump that applies pressure to the headspace of a buffer reservoir. Vacuum delivery is limited to 10 psi or less but can be useful for drawing fluid from containers that cannot be made pressure tight.

Gravity and electrophoresis are not practical for filling and rinsing capillaries, but they are used, along with pressure and vacuum, for injecting samples into capillaries. Gravity injection (sometimes called hydrostatic injection) is accomplished by inserting the inlet end of the capillary into the sample vial and raising the vial and capillary relative to the outlet end. Reproducible gravity injection requires that the vial be raised to the same height for the same duration of time at each injection (gravity itself being a fairly reliable source of motive power). Pressure and vacuum injections are more complicated. Regardless of how the pressure differential is created, a finite time is required for the pressure to reach a steady state. Changes in pressure on the order of 0.05 psi can be significant at the low pressures commonly employed (0.1–1 psi). Systems that rely on pressure or vacuum must have some sort of feedback mechanism to compensate for these variations. These systems either adjust the delivered pressure or the delivery time to maintain the desired product of pressure and time. In well-engineered systems a 3-s injection at 1 psi and a 10-s injection at 0.3 psi should give identical results, because both are 3 psi-second injections. In practice the longer, lower pressure injection usually gives better performance because it allows a longer time for the system to respond to variances.

Electrophoretic or electrokinetic injections do not conform to the Poiseuille equation. In this method of sample introduction, the inlet of the capillary is inserted into the sample and the outlet into a buffer vial. Voltage is briefly applied. Through a combination of electrophoresis and electroosmotic flow sample is drawn into the capillary. This technique is valuable when delivering sample to a gel-filled capillary or when pressure delivery is not possible. There is a possibility of bias when using this technique. Components that migrate more rapidly in the electrical field will be over-represented in the sample compared to slower moving components. To maximize the volume injected, the sample should be at a considerably lower ionic strength than the run buffer. Subsequent injections will show reduced peak areas because each injection delivers salts from the capillary buffer into the sample vial, raising the ionic strength of the sample. This effect can be minimized (and injected quantity increased) by pre-injecting from pure water immediately prior to the sample injection.

Most commercial systems employ some sort of carousel or X-Y-Z robotic system to move the buffer and sample vials to the capillary. These systems may also provide refrigerated storage for labile samples. In this case the sample storage temperature should be regulated independently of the capillary thermostatting system.

Detection

In order to gain useful information from the separation technique, it is necessary to detect and measure the analytes. Detection may be qualitative and/or quantitative. Most CE detection is done on-capillary; that is, a section of the capillary is linked to the detection device and the capillary itself is the detection cell. It is also possible to couple to detectors that are outside of the separation capillary although this does require a specialized interface.

Geometry and Path-Length

The most frequently used method of detection involves absorbance of energy as the analytes move through a focused beam of light. The scale of the detection apparatus and of the signal produced has created some unique challenges for the instrument developer.

On-capillary detection eliminates the problems of coupling the capillary and its power supply to flow cells or other devices. However, detection through the capillary is complicated by the curvature of the capillary itself. The capillary and the fluid it contains make up a complex cylindrical lens. The curvature of this lens must be accounted for in order to gather the maximum amount of light and thereby maximize signal-to-noise ratio. The effective length of the light path through the capillary is actually about 63.5% the stated i.d. of the capillary. Thus a 50-μm capillary has an effective path length of only 32 μm. This can be compared to typical HPLC detectors that have detectors in the 5–10 mm range. Because of this very small light path, the absorbance signal obtained from a CE system is also correspondingly small. Therefore, a peak with an absorbance of 0.002 AU is a significant peak. Noise levels are correspondingly small and are usually measured in microabsorbance units. The maximum absorbance of typical CE detectors is 0.2 AU. The capillary lumen occupies only a small part of the diameter of the capillary; the remainder is transparent silica. This geometry allows for large amounts of stray light to enter the detector so that a fully opaque sample would not reduce the light passing through the capillary to zero. This factor must be considered in the design of effective CE detector hardware and software. Several novel approaches have been taken to increase the path length (and the sensitivity) of CE detectors. One approach has been to use a specially constructed low-volume flow cell. These cells carry the analytes through two right-angle bends. The segment between the bends, which may be over 1000 μm long, is thus at right angles to the direction of the capillary and parallel to the direction of the light beam. Properly designed, these cells can dramatically increase sensitivity but at the risk of some loss of resolution. Another approach has been to create a wide zone or "*bubble*" in the capillary at the window. A 50-μm i.d. capillary may have a window that is 150 μm in diameter, giving a sensitivity increase of approximately threefold.

Absorbance

Absorbance detectors are the most commonly encountered types of detector in CE instrument systems. They rely on the absorbance of light energy by the analytes. This absorbance creates a shadow as the analytes pass between the light source and the light detector. The intensity of the shadow is proportional to amount of material present.

The simplest absorbance detector uses only portion of the available energy. The broad-spectrum light from a source lamp is passed through a filter or diffracted by a grating so that a narrow range of the spectrum is used. In some cases lamps (such as hollow cathode types) are used,

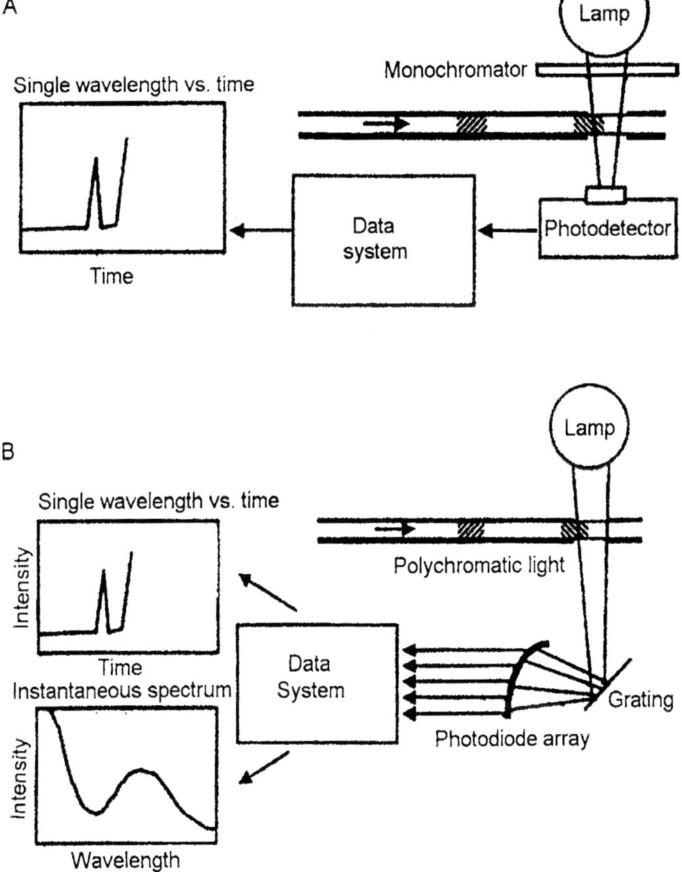

Fig. 6.10. Comparison of single wavelength (A) and photodiode array (B) absorbance detectors.

that produce light at only a few discrete wavelengths. Monochromatic absorbance detectors are relatively inexpensive and rugged. However, because only a part of the spectrum is used, information may be lost from complex samples that have components with differing absorbance maxima.

Another type of absorbance detector looks at changes over a wide range of wavelengths simultaneously. The photodiode array detector (PDA) delivers the entire spectrum of light available from the source lamp to the capillary window. The light passing through the capillary is diffracted into a spectrum that is projected on a linear array of photodiodes. In this manner it is possible to record the entire absorbance spectrum of analytes as they pass by the detector window. PDA-type detectors are more expensive and less rugged than are monochromatic detectors. Because the spectrum can be divided into as many as 512 channels the amount of data acquired in a single run can be very large. The PDA type detector, however, is not suitable for all applications because nearly all the energy of the source lamp is focused onto a very small region of the capillary. Some capillary coatings and buffers will decompose under this onslaught of energy unless some of the energy is filtered out. Despite these limitations the information provided by the PDA detector can be valuable for confirming the identity of analytes. By comparing the change in spectral signature across a peak it is possible to estimate peak purity.

Fluorescence

Fluorescence detectors do not rely on the measurement of shadows. These systems use an external energy source to excite the analyte molecules to a higher energy state. When these excited molecules return to the normal state they emit energy of a lower wavelength, which can be detected and recorded as evidence of the passage of the analytes. Fluorescent detectors in CE systems often use lasers as the source of the excitation energy. Lasers have the advantage of producing intense light at a single wavelength. The intensity of the light contributes to good excitation efficiency. In addition, the monochromatic nature of the laser beam makes it easy to filter out any stray laser light to keep it from interfering with the detection of analytes. Analytes will vary in their excitation and absorbance wavelengths so that a fluorescent detector will not see all the components that may be in a sample. For analytes that are fluorescent or can be made fluorescent by a chemical reaction, the sensitivity of this type of detector can be 10–1000 times better than an absorbance detector.

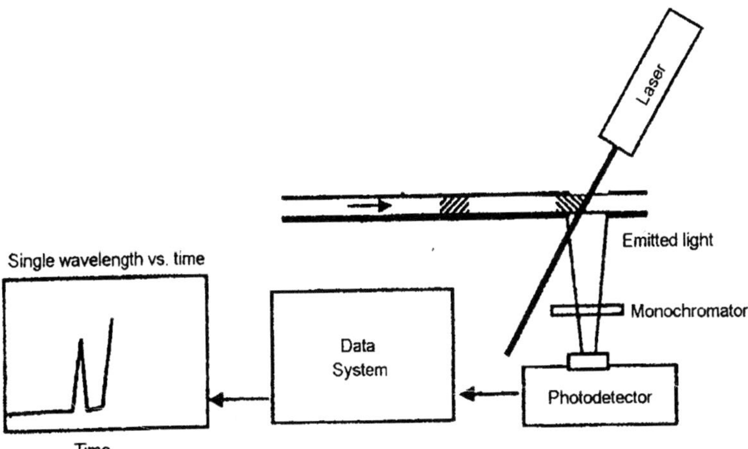

Fig. 6.11. Laser-induced fluorescence detector.

Amperometry and Conductivity

These two detection techniques potentially can offer a high degree of sensitivity and be applied to a wide variety of analytes, including those without appreciable UV absorption. Amperometry and conductivity are difficult to do in practice, and to date these devices are not commercially available. Both of these detection schemes require the use of sensing electrodes that are scaled to the dimensions of the electrophoresis capillary. In addition, these detectors place some limitations on the type of separation buffer employed.

In amperometric detection, an electroactive analyte undergoes an electrochemical reaction inside a detector cell. The CE separation is generally carried out at microampere currents and kilovolt potentials, whereas the detection cell must operate at picoampere currents and millivolt potentials. The two circuits must therefore be isolated, usually by connecting the capillary to the high voltage at a point prior to the end of the capillary where the detection cell is located. This system depends on EOF to carry the analyte past the high voltage electrode to the detection cell. Amperometric detectors have been used successfully for the detection of biogenic amines at levels as low as 10^{-8} m. With continued advances in microfabrication this type of detector should become routinely available.

There are two types of conductivity detector. Both require that two electrodes be placed within the separation capillary. In the first type a short distance down the length of the capillary, separates the electrodes. Because there is a voltage gradient down the length of the capillary a portion of that voltage gradient can be measured between the two sensing electrodes. The presence of analyte zones will change the potential drop in the area of the zone. This change in potential will be sensed as the zone passes the electrodes. In the other arrangement the electrodes are placed opposite one another across the diameter of the capillary. In this case there is no voltage drop between the two electrodes. A circuit is constructed that passes a sensing voltage across the capillary diameter. If the conductivity of the system changes as analytes pass between the electrodes, there will be a change in the current in the sensing circuit. Conductivity detectors work best when there is a substantial difference between the conductivity of the analyte zones and the background buffer. These conditions are not optimized for peak shape and lead to unacceptable peak asymmetry. Various schemes have been described for avoiding this problem. Conductivity detectors have been used for the measurement of inorganic ions and in isotachophoretic separations.

Capillary Electrophoresis-Mass Spectrometry

Hybrid systems such as CE-mass spectrometry (CE-MS) offer an additional dimension of analysis in addition to detection. CE data consists of migration time, quantity, and (using a PDA detector) spectral signature. MS adds the additional data of molecular weight and, using collision dissociation and MS-MS systems, structural information as well. This combination of techniques provides an orthogonal approach to analysis in a single analytical run.

The predominant form of MS that has been coupled to CE has been electrospray MS. The outlet end of the CE capillary is inserted into the electrospray interface. Because the volume of liquid emerging from the capillary is very small, a make-up liquid is pumped through an axial needle. This sheath flow also provides a return connection to the high-voltage power supply of the CE system. The liquid is mixed with a flowing gas stream and nebulized into a spray. The spray vaporizes and the ionized analyte particles are carried into the MS detector. The MS system is usually set to scan across an expected range of mass values. Because the width of peaks in CE can be very small, the MS instrument must be able to scan across the desired mass range very rapidly or peaks may be missed.

Most mass spectroscopists prefer to use buffer systems that are volatile, such as ammonium formate, in order to reduce the accumulation of buffer salts inside the MS instrument. These buffers may not be optimized for the separation of the analyte mixture, although an incomplete separation can be acceptable in CE-MS because the MS systems provide an additional dimension of separation.

Many of the existing commercial CE systems have been interfaced to commercial MS systems. The design of these systems allows the use of UV or other detectors prior to the MS interface, but usually require quite long and awkward reaches of capillary to connect the two systems. Unlocking the true potential of this method will require the development of a CE system that is fully integrated with the MS system.

Indirect Detection

In the previous discussions it has been assumed that detection will be direct, that is, the presence of the sample in the detector cell or window will cause an increase in the output signal. This is not always the case. There are certain applications where the decrease of the background signal provides evidence of the passage of analytes through a detector. In some cases a detector that is useful for indirect detection would not detect the analytes in direct mode. An example of this is the detection of inorganic ions such as sodium or sulfate with a UV detector.

To detect such analytes, the capillary is filled with a buffer that has mobility close to that of the analytes and also has a significant UV absorbency. The buffer/chromophore must carry the same charge as the analyte. For the analysis of sulfate, a buffer containing sodium chromate is used. In the area of the capillary occupied by the sulfate band the chromate is displaced, creating a chromate depleted region. Because chromate absorbs light at 254 nm, by monitoring background electrolyte absorbance at this wavelength the presence of sulfate will be indicated by a negative peak, with an area corresponding to the amount of chromate displaced and hence to the amount of sulfate present. Most CE data analysis packages can invert these negative peaks, producing a normal-appearing electropherogram.

Data Analysis

The collection and analysis of CE data has many characteristics in common with other chromatography-like analyses. Many analysts have utilized data systems developed for HPLC and GC systems. These data systems may not be optimal for CE for three reasons:

1. The signals obtained from CE are usually very small. A typical HPLC peak may have a maximum absorbance of 0.2 AU, whereas a CE detector may have a full range of 0.2 AU. A typical peak in CE may have a maximum absorbance of 0.002 AU.

2. CE peaks can be quite narrow with a width of only a few seconds. Some older data systems (particularly HPLC systems) cannot respond sufficiently fast to deal with this data. If data is to be collected digitally, it must be collected at a high data rate to define adequately the peak shape.
3. CE peaks are often non-Gaussian. Because of the low band broad spreading in CE and the effects of electrofocusing CE peaks tend to be more triangular and less bell-shaped than peaks in HPLC and GC. At times CE peaks may have one edge that is nearly vertical. Some data systems have difficulty locating the peak start and stop times for these shapes.

Two types of peak-detection methods dominate analytical data analysis: slope sensitive algorithm and the moving median filter algorithm. Slope sensitive algorithms look at the slope of the baseline over some interval of time. When the slope exceeds a pre-determined value, a peak is said to have begun. The point at which the slope goes to zero identifies the peak apex, and the point at which the slope returns to the starting value defines the peak end. The slope-sensitive method looks for peaks independently of the baseline shape. Most commercial software packages use this method.

Moving median filter algorithms take a different approach. Peaks are relatively high frequency (impulse) events when compared to baseline drift. These algorithms seek to define how the baseline would look in the absence of peaks by filtering out all impulse events. Whatever differs from the baseline is defined as a peak. In practice, the algorithm makes several passes through the data set (iterations) to determine the best fit. This type of algorithm, commercially available as "Caesar," is more successful at dealing with the abrupt slope changes in CE data than are slope-sensitive algorithms.

A data system for CE needs to include some calculations that can be deemed "*CE specific*." These include the calculation of mobility (a measure of the velocity of an analyte through the capillary) and corrected peak area. Corrected area is necessary because the peaks passing through a CE detector do not all pass through at the same velocity. Early eluting peaks move through more rapidly than do later eluting peaks. This is unlike the situation in HPLC where the velocity through the detector cell is dependent only on the flow rate and not on the retention time. Because later eluting peaks are moving more slowly, they appear to be larger relative to earlier eluting peaks. Corrected peak area normalizes peak area to unit migration time and allows accurate comparison of the components in a mixture.

Fraction Collection

Fraction collection is placed under the heading of detection under the assumption that fractions are being collected for analysis outside the CE instrument. At first glance, fraction collection in CE appears to be analogous to fraction collection in HPLC, but there are significant differences.

The velocity of peaks through the HPLC detector is constant if the flow rate is constant. This is not true in CE where earlier eluting peaks are moving more rapidly than later eluting peaks. The delay time between peak detection and elution (the time for transit from the window to the capillary outlet) will therefore vary, becoming longer with each successive peak. This time can be quite long. Consider a capillary 60-cm long with the window located 10-cm from the outlet end. If a peak passes the detector at t = 5 min it will not emerge from the end of the capillary until t = 6 min. A peak detected at t = 10 min will emerge at t = 12 min. Peaks that pass the detector between 5 and 10 min will all be in the outlet segment of the capillary between t = 6 and t = 12 min. A further complication is that collection requires that the outlet of the capillary be moved to a new collection vial for each peak that is to be collected. At each move the circuit will be broken. The time needed for the power supply to ramp down and ramp up at each move, and the diminished peak velocities during the ramping segments cannot be disregarded. Complex algorithms are needed to do this process smoothly.

A more serious limitation to the collection of fractions from CE is the very small amount of sample that the fraction will contain. Consider a capillary 100 μm in diameter and 50 cm long, with the window located 10 cm from the end. A 5 psi-sec injection into this capillary will inject about 189 nL of sample. If the analyte concentration is 1 mg/mL in the starting sample and the peak is collected with 100% efficiency, the recovered sample will be 189 ng. Repeating this run 10 times would yield less than a two micrograms of product. Although fraction collection in CE is possible, one must question whether it is worth the effort.

Modes of Separation in Capillary Electrophoresis

A CE system can be operated in several different modes. These modes offer the analyst a variety of ways to approach an analytical problem. The choice of mode will be based on the analytical problem under consideration. This section will describe the major modes of capillary electrophoretic separation that are currently in use and provide some applications for each.

Capillary Zone Electrophoresis (CZE)

Mechanism

Section 1 of this chapter dealt almost entirely with this form of CE. CZE is characterized by the use of open capillaries and relatively low viscosity buffer systems. Analyte molecules move from one end of the capillary to the other according to the vector sum of electrophoresis and electroosmotic mobility.

Applications

CZE is the most widespread mode of CE. It has been used for analytes as diverse as sodium ions, drugs, and protein molecules. Analyte species can be separated by CZE if they migrate at different velocities in the electrical field.

Capillary Gel Electrophoresis (CGE)

Mechanism

Capillary gel electrophoresis (CGE) is separation based on viscous drag. In this mode of CE the capillary is filled with a gel or viscous solution. EOF is often suppressed so that the migration of the analytes is solely by electrophoresis. Larger molecules tend to be retarded more by the viscous separation medium than are smaller molecules, so that the separation is effectively based on the molecular size.

Applications

This is the method of choice for molecules that differ in size but not in mass/charge ratio. DNA molecules, for example, can vary greatly in length, but the charge per unit length is quite constant. In a pure CZE separation, all the molecules move at very nearly the same velocity and no separation results. In a viscous medium, the longer molecules are retarded more than shorter molecules. Thus shorter pieces of DNA pass the detector sooner than do larger pieces of DNA.

Protein molecules are composed of more complex subunits than DNA molecules. Even though proteins vary widely in their size, the wide range of charge states possible in a protein sequence complicates size separation. In order to separate proteins by size, it is necessary to mask the native charge and create a more nearly uniform charge-mass ratio. This is done by treating the proteins with the detergent SDS. Although there are exceptions, typically proteins bind to a constant number of SDS molecules per unit length. Because the SDS molecules are highly negatively charged the amino acid subunit charge contributes little to the mobility of the molecules and a separation based on chain length is possible. This technique of SDS-CGE is exactly analogous to SDS-polyacrylamide gel electrophoresis (SDS-PAGE).

Capillary Isoelectric Focusing (cIEF)

Mechanism

Molecules that carry both positively and negatively charged groups exhibit, at a specific pH, an equal number of positive and negative charges. At this pH, known as the isoelectric pH or pI, the molecule, although charged, behaves as if it is neutral because its positive and negative charges cancel each other. The molecule, therefore, has no tendency to migrate in an electrical field. In isoelectric focusing, special reagents called *ampholytes* are used to create a pH gradient within the capillary. These ampholytes are mixtures of buffers with a range of pKa values. In an electrical field, ampholytes will arrange themselves in order of pKa; this gradient is trapped between a strong acid and a strong base. Analytes introduced into this gradient will migrate to the point where the pH of the gradient equals their pI. At this point the analyte, having no net charge, ceases to migrate. It will remain at that position so long as the pH gradient is stable, typically as long as the voltage is applied.

Applications

Capillary isoelectric focusing (cIEF) is used almost exclusively for the separation of closely related protein species. Hemoglobin can be separated into several bands by this technique, whereas separation by SDS-CGE usually results in a single form being identified. In this application, the protein sample is mixed with the ampholyte solution and the mixture is pumped into the capillary. When voltage is applied, the proteins and the ampholytes migrate to their appropriate positions in the gradient. When focusing is complete, the proteins in the mixture are distributed throughout the length of the capillary. In order to detect the proteins it is necessary to mobilize them so that they pass by the detector in turn. There are two ways to accomplish this mobilization. Pressure mobilization utilizes positive pressure applied to one end of the capillary to drive the entire fluid column through the window. In order to prevent band distortion, this pressure must be applied carefully, preferably while the voltage is still being applied. Chemical mobilization requires that one end of the capillary be transferred to a salt solution after focusing has taken place. Upon application of voltage, salt will migrate into the capillary, disrupting the pH gradient and allowing the proteins to migrate past the detector by electrophoresis.

cIEF is also widely used for examining the distribution of carbohydrate isoforms of glycoproteins. In other techniques, such as SDS-CGE or CZE, these proteins tend to move as diffuse bands and to generate broad peaks. cIEF can often resolve these bands into peaks that differ by as little as one charged sugar group.

Capillary Isotachophoresis (cITP)

Mechanism

In this technique the sample plug is introduced between two different buffers. One of these, the leading electrolyte, has the highest mobility in the separation. The second, trailing electrolyte, has a mobility lower than anything else does. The sign of the charge on the analytes and the buffers must be the same. When the voltage is applied the ions in the sample form discrete zones that are not separated into peaks; one zone is adjacent to the next. The concentration of the analyte within a zone is constant within that zone and the length of the zone is proportional to the concentration within the zone. Because the voltage drop is uniform within a zone, isotachophoretic methods are quite compatible with conductance detectors.

Applications

Capillary isotachophoresis (cITP) of peptides and proteins has been used with MS detection. Since this mode of separation delivers zones of uniform concentration to the MS, it may be an ideal separation mechanism for this detection technique.

Micellar Electrophoresis

Mechanism

Electrophoresis is not possible for analytes that are not charged. In order to analyze such analytes, it is necessary to employ some agent in the separation buffer that will transport them through the capillary. The most commonly used mode of CE for these analytes is MEKC. In this technique a suitable charged detergent, such as SDS, is added to the separation buffer in a concentration sufficiently high to allow the formation of micelles. These micelles are arrangements of detergent molecules that have a hydrophobic inner core and a hydrophilic outer surface. Micelles are dynamic and constantly form and break apart. For any given analyte, there is a probability that the molecules of that analyte will associate within the micelle at any given time. This probability is the same as the partition coefficient in classical chromatography. When associated with the micelle, the analyte molecule will migrate at the velocity of the micelle. When not in the micelle, the analyte molecule will migrate with the EOF (if any). Differences in the time that analytes spend in the micellar phase will determine the separation.

Applications

MEKC is useful for a wide range of small molecules such as drugs, pesticides, and food additives that are not charged and are sufficiently hydrophobic to associate with the micelle. While SDS is probably the most widely used detergent for this purpose, cationic detergents such as TTAB can also be employed. Nonionic detergents by themselves do not provide mobility to uncharged analytes, but in combination with charged detergents they will modify the separation. Some detergents are useful in specific applications. For example, sodium cholate is useful in the separation and analysis of a variety of steroids. The micelles formed in this case are not the classical spherical shape but are probably sodium cholate molecules arranged on each other like a stack of coins. Different uncharged steroids differ in their tendencies to participate in these stacks.

Chiral Electrophoresis

Mechanism

Chiral molecules are molecules that can exist in two stereo-specific forms. These chiral forms or enantiomers are identical in molecular weight and chemical formula but differ in the arrangement of the atoms in space. Separation of these enatiomeric forms depends on the tendency to associate differentially with other chiral molecules known as chiral selectors. By incorporating a chiral selector into the CE buffer, it is often possible to separate enantiomers of a chiral molecule. This is analogous to MEKC described previously. The complex of the analyte and the selector will migrate at a different rate than will the analyte alone. Because one of the two enantiomers associates more strongly with the selector than does the other form a separation can be achieved.

Applications

The most commonly employed chiral selector in CE is cyclodextrin, ring shaped carbohydrates made up of 6, 7, or 8 D-glucose subunits. Cyclodextrins may be chemically modified to alter their hydrophobicity or charge. Uncharged cyclodextrins are not suitable for the analysis of uncharged analytes since the complex will move with the EOF.

However, cyclodextrins modified to carry a charge by addition of sulfate groups, can serve both as chiral selectors and as carrier molecules (similar to the detergent in MEKC). Other molecules, such as the antibiotic vancomycin, have also been employed as chiral selectors.

Chiral CE can be used to separate the enantiomeric forms of pharmaceuticals as well as natural substances, such as amino acids. Impurities as small as 0.1% are easily detected by this method.

Nonaqueous Electrophoresis

Mechanism

Electrophoresis usually is considered to occur only in aqueous solutions. However, CE can be performed using nonaqueous systems based on such solvents as acetonitrile, methanol, formamide, and dimethylformamide, to which are added small amounts of anhydrous acid or buffer salts. The separation is by simple electrophoresis as EOF is very low under these conditions.

Applications

There are times when two analytes have the same charge to mass ratio and are not easily separated. In some cases this same analyte pair can be separated in a nonaqueous environment where they may have different pKa values than those expressed in water. The degree of solvation, and hence the radius of the solvated species, may also differ in aqueous and nonaqueous environments. Hence, nonaqueous CE offers an alternative for analytes that are difficult to separate under aqueous conditions. In addition, some analytes are difficult to solubilize in aqueous systems but dissolve readily in organic solvents, thus nonaqueous CE offers an alternative to MEKC for these analytes. Separations by nonaqueous CE have been reported for drugs, dyes, preservatives, surfactants, and inorganic ions.

Capillary Electrochromatography

Mechanism

Capillary electrochromatography (CEC) is a hybrid technique between liquid chromatography and electrophoresis. It is a partitioning technique in which molecules distribute between a stationary and a moving phase. As described for MEKC, different analytes will tend to associate to a greater or lesser extent with the stationary phase, effecting a separation.

CEC capillaries are packed with particles like those used in HPLC columns. Unlike conventional LC techniques, CEC uses electroosmotic flow to drive the mobile phase down the column. The resulting plug flow improves the separation efficiency over that of the laminar flow of pressure driven systems.

Applications

As of this writing, most of the work done on CEC has used model systems, such as polyaromatic hydrocarbons. A great deal of effort is underway to identify applications where this technique can be the method of choice. Because the buffers used are typically high in organic content and hence volatile, CEC may be useful when coupled to mass spectroscopy (CEC-MS).

7

Tools and Techniques

In our culture, the negotiation of novelty is commonplace. Patenting, for example, is a process to decide what counts as novel. Innovations are compared against predecessors and consequential decisions are made on the basis of similarity and difference. The same is true of the Nobel Prizes. One might even argue that all arguments can be recast as a negotiation of similarity and difference.

Nanoscience and nanotechnology are often claimed to be novel and also often claimed to not be so. The *Oxford English Dictionary* is a useful first port of call for this kind of endeavor: The first use of the word was already in 1974 but in an obscure publication, the *Proceedings of the International Conference of Production Engineers*. The second recorded use is Eric Drexler's 1986 *Engines of Creation*, and that is of course the most important locus because this book was widely read. After 1986, one can see the word spread to publications with large readerships: *The New Scientist*, the *Times Higher Education Supplement*, the *Washington Post*, the *Sunday Times*, and *Nature*.

Drexler's *Engines of Creation* is a tremendously successful book, written in an upbeat tone of voice painting a rosy future of tremendous technological ability. Drexler argued that we can now build structures on the nanoscale, meaning that we can move and combine atoms and molecules as we do with Lego™-blocks, as long as the resultant molecules are energetically stable. We can build molecules that have similar functions as the DNA-RNA-protein system found in nature, that is to say our new molecules may be engineered so as to be parts of a self-reproducing system. From this will flow new materials, new drugs, new information technologies, new human tissues, new just about everything. In the introduction to the book, Marvin Minsky, Professor at MIT (and so a credible individual in matters technical), emphasized that Drexler's vision was not fanciful but based on a thorough knowledge of the current science and technology. The vision was compelling for two reasons. (1) The tool for moving individual atoms, the scanning tunneling microscope (STM), became well-known at just this time - it received the Nobel Prize in the same year that *Engines of Creation* was published (1986). (2) The combination of molecular biology, the incipient human genome project, and the understanding of biochemical pathways made it feasible that a slightly different ensemble than the DNA-RNA-protein one could be produced and have the same kind of tremendous power as life. Much of Drexler's book thus addresses the issue of figuring out what kinds of molecules we would want to assemble given our knowledge of molecular biology and biochemical pathways, and how to ensure that the research would be beneficial. At the very same time, in the mid-1980s, the field of artificial life came into being. Nano and A-life are natural bedfellows: one predicts new forms of life created in the laboratory, the other simulates new forms of life on the computer. Both make the creation of new forms of life in the laboratory seem less fanciful.

So, much of the feasibility of the vision depended upon the feasibility of the STM's purported control over individual atoms and upon the feasibility of alternative forms of life. And the novelty of Drexler's vision traded upon the novelty of the STM and A-life. In this paper I will focus upon the novelty of the former. As ever, this was negotiated and renegotiated.

Scanning Tunneling Microscope

What is an STM and how does it work? It was described in the following way in *Scientific American*. A very fine needle is brought very close to a sample surface, for example a crystal surface whose structure is to be examined. When very close, electrons might jump across the gap from sample to tip; especially if an electrical potential is applied (*e.g.* by connecting the tip to a battery and the sample to earth). The jump across the gap is explained within quantum mechanical theory by the phenomenon of tunneling. The electrons tunnel through the vacuum despite the classical, non-quantum mechanical theory predicting that they do not have the energy to surmount the obstacle provided by the vacuum. The tunneling electrons amount to an electrical current that can be measured with great precision. Quantum theory predicts that the tunneling current is very sensitive to the distance

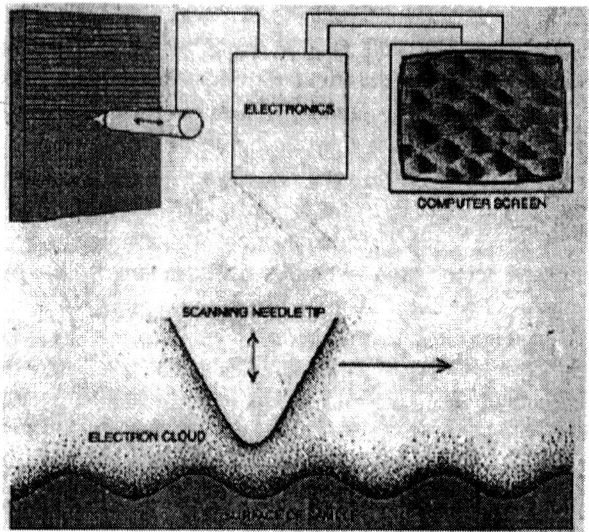

Fig. 7.1. Scanning tunnelling microscopy.

between tip and sample: proportional to the inverse of the distance squared. If one scans the tip across the surface, the distance between tip and sample will oscillate and so will the current. The correlation of tip position and current can thus be used to produce an image on the computer screen giving a rendition of the topology of the sample surface.

The STM was invented by Gerd Binnig and Heinrich Rohrer at IBM Zurich in 1981. Their very first paper was concerned with a tunneling microscope. They argued that they were able to reduce the distance between probe and surface to the dimensions of a single atom. The proof of this lay in the tunneling current measured (inversely proportional to the distance squared, a proportionality that is theoretically explainable only with the quantum mechanical notion of electrons tunneling across the vacuum between probe and surface). The main point here is that they relied on quantum mechanics. They themselves highlighted the fact of atomic resolution: "Surface microscopy using vacuum tunneling is demonstrated for the first time. Topographic pictures of surfaces on an *atomic scale* have been obtained."

Holy Grail of Atomic Resolution

In order to understand this, let us examine the significance of the term "atomic resolution" for the audiences that Binnig and Rohrer addressed. With some hyperbole one might say that atomic resolution had been the holy grail in the natural sciences for at least a hundred years. 19th-century scientists developed a language based on atoms as elementary building blocks with which all sorts of analytical and industrial chemistry was carried out. The concepts of the atom and of Mendeleev's elementary table were tremendously useful. But it was agreed that there was no direct evidence of atoms and many scientists developed a pragmatic attitude, dismissing all discussions of atoms as metaphysical - beyond measurement, beyond our ken.

In the early 20th century, much experimental evidence emerged with radioactivity and x-rays. The visible tracks made by alpha particles in cloud chambers were very powerful, and atom-talk became kosher once more. William Henry Bragg, for instance spent much of his career popularizing such talk, lecturing on BBC radio and at the Royal Institution on individual particles flying through a gas. He also spent much time developing x-rays as an analytical tool in crystallography. A broadside of x-rays will be deflected at a crystal surface, and the many deflected waves combine to produce a pattern on a photographic plate. The power of X-rays lay precisely in their atomic resolution: they yielded information on the average distances between atoms in the crystal lattice.

Many similar techniques were developed to explore surfaces, especially with the growth of the semiconductor industry in the 1950s and '60s. Scientists used light, electrons, or ions of all kinds of wavelengths or energies shooting at all kinds of angles at the surface, sometimes measuring the particles transmitted through the target, sometimes those reflected back. Knowledge of the structure of semiconductor surfaces was obviously of tremendous financial importance and so this armory of techniques became large and very sophisticated. The same techniques were used to examine metallic surfaces for which there was also tremendous industrial interest.

Novelty of the Scanning Tunneling Microscope

So, by the early 1980s, there was a large, if diffuse, social grouping of surface scientists, united by an understanding of, and a commitment to, an array of techniques yielding information about surfaces, often with atomic resolution, but always averaged over many atoms. In the following years, Binnig and Rohrer worked also to explain just what constituted the novelty of their new instrument. In the abstract of one paper (Binnig & Rohrer 1982) they referred to "unprecedented resolution in real space on an atomic scale" (real space in contrast to the conceptual "reciprocal space" used with diffraction techniques). They now explain: "The usual experimental methods to investigate surface structures (*e.g.* LEED, atom diffraction, ion channeling) are indirect in the sense that 'test models' are used to calculate the scattered intensity profile which is then compared with the one measured. In addition, these methods usually require periodic surface structures. The STM, on the other hand, gives 3d pictures of surface structures direct in real space".

Binnig and Rohrer not only advertised their new instrument to a busy but potentially interested audience, they also had to convince them that they were credible. Some scientists directly accused them of fraud and some reviewers rejected their papers. A knee-jerk reaction of many scientists was that the resolution of an individual atom was impossible, due to the uncertainty principle, a fundamental tenet of quantum mechanics. The fact that the quantum mechanical effect of tunneling was centrally involved will have given scientists the immediate association of quantum mechanics and its somewhat different laws for the atomic length scales. The uncertainty principle may be explained in the following way. If one were to determine the position of an individual atom, then one could send out light (a photon) which, if impinging upon the atom, would change direction.

The deflection of the photon would yield information about the atom's position, but unfortunately the deflection of the photon entails the slight movement also of the atom. Thus, some uncertainty will always remain about such issues as the position of individual atoms. Most scientists learning quantum physics will learn about the uncertainty principle with examples such as the one just given. Nowadays STM users will learn that the uncertainty principle does not apply for the case of atoms embedded in a solid and that the examples used to explain the uncertainty principle apply only to free atoms. In other words, while the photon might nudge the atom, the neighbouring atoms will push it back into place. But in the early 1980s, the audience will have consisted of many busy scientists whose knee-jerk reaction when hearing of atomic resolution of individual atoms was to dismiss it.

Some scientists will also have had much investment in the existing techniques and have been reluctant to accept a new one that might render their expertise obsolete. Surface scientists and crystallographers were, generally speaking, proud of their facility to think in terms of both real and reciprocal space. And so Binnig and Rohrer needed to build up their own credibility. For instance, they needed a convincing theory based on quantum mechanics explaining the tunneling process.

According to this theory it is not just a question of "feeling" the topography of the surface but rather a result of the overlap of electron orbitals of the tip and sample atoms with the greatest proximity.

The bottom line is that STM measurements require interpretation according to a theoretical model, and that it is not immediately obvious which model is the most appropriate. On top of all this, it is difficult to get the STM to work properly: proficient users will tell you that it might measure junk for hours and then suddenly yield sensible information. (This phenomenon, so the explanation goes, is due to the chance placement of an atom on the tip that gives it the required sharpness. That is to say, when scanning across the surface very closely, a surface atom might jump from the surface to the tip and sit in such a way as to jut out and give the tip the desired sharpness.)

All of this means that other scientists had plenty of reason to dismiss Binnig and Rohrer's results, and one might expect that those with a career invested in existing techniques would feel threatened by an instrument promising markedly better performance. Many surface scientists thus had both the motivation and the arguments to reject the STM. The politic reaction of Binnig and Rohrer was of the kind: 'okay guys, it's not that novel, really - relax and give us a break'.

They wrote: "we understand the STM as a complement to present microscopy rather than a competitor. For many applications, the STM is best used in combination with another microscope". And indeed everyone used the STM in conjunction with another microscope. The proficient new STM user was able to discern obvious noise from a proper measurement by comparing the result with that obtained from another tool.

The evidence yielded by the STM is mediated through quantum theoretical understanding and a profound pre-existing understanding of surfaces. (For the importance of the pre-existing understanding of surfaces.) And importantly, the novelty of the STM was negotiated: at times it was emphasized, at times downplayed. The novelty sometimes focused on the atomic resolution but it didn't have to. For example, the AFM, the sibling of the STM and much more widely used, does not yield atomic resolution. The utility of the instrument doesn't require atomic resolution. But symbolically, atomic resolution mattered greatly - comparing it to the holy grail is not too much of a hyperbole, after all. As always, a new technique becomes credible only when replicable, and it took years for an STM to be built successfully outside IBM Zurich. Other IBM labs came first and by 1985 there was a small

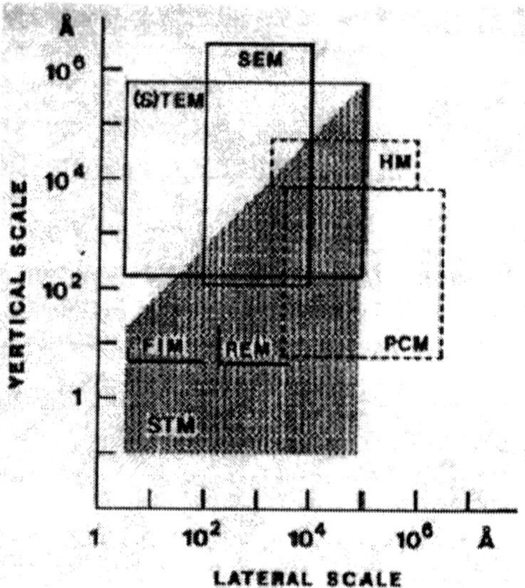

Fig. 7.2. The resolution of various microscopic technique.

community of STM users. At this point, *Scientific American* picked up the story. Binnig and Rohrer wrote the article jointly with the staff of *Scientific American*. The staff of course knew how to address a broader audience than just the surface science community, and so the language shifted importantly. The new kind of microscope enables one to "see" surfaces "atom by atom". The article also advertised the instrument's versatility: it "may extend to investigators in the fields of physics, chemistry, and biology".

The next year, 1986, was the STM's breakthrough year. Binnig and Rohrer received the Nobel Prize, and Eric Drexler published his influential *Engines of Creation* that also popularized the notion of nanotechnology. Drexler does refer to the STM, but not centrally. The manipulation of individual atoms is pretty much taken for granted, and he focuses much more on the implications of that purported ability, thus shifting the discourse towards artificial life and the creation of alternative life forms.

Hyping of the Scanning Tunneling Microscope

The story of the scanning tunneling microscope and its new siblings (collectively called *scanning probe microscopes*, or SPM) after 1986 primarily went off in the direction of immediate utility that is discussed by Cyrus Mody. One might posit a continued disconnect between the actual work done with SPMs and the LEGO™-style construction of life-like molecular systems at the foundation of the Drexlerian vision. Even the historian of science, Jed Buchwald has contributed to this disconnect by rendering an illustration of "Zippenfeld's amazing atomic etcher", purportedly for touching up the family's greeting cards. The illustration is unreferenced and Buchwald in fact made it up himself.

One event has enhanced this disconnect more than others: IBM employees' media stunt, writing IBM with individual atoms. They used "the STM at low temperatures (4K) to position individual xenon atoms on a single-crystal nickel surface with atomic precision. This capacity has allowed us to fabricate rudimentary structures of our own design, atom by atom ... the possibilities for perhaps the ultimate in device miniaturization is evident." The paper made it straight to the front page of the issue of *Nature* in which it was published. The reason for its media success was of course its relevance for the Drexlerian promise/hype. It is of methodological advantage to talk about promise/hype, to retain a Janus-faced ambiguity and not decide in advance whether nanotechnology will succeed or fail. The nature of the promise requires no further explication at this point, whereas the nature of the hype does.

First of all, the IBM experiment worked only at 4K, an extremely low temperature, and at high vacuum. One of Drexler's points was that we would only be able to assemble energetically stable molecules, and IBM's surface with patterns made by xenon atoms is not energetically stable except at these low temperatures. Furthermore, Eigler *et al.* were able to move atoms laterally on a surface, which is rather different from assembling a three-dimensional molecule - DNA, RNA, and proteins are of course not flat. In a word, there is a tremendous disconnect between moving xenon atoms on a surface at 4K, if that is what Eigler actually does, and building large complex bio-molecules LEGO™-style. Xenon, after all, is an inert gas, meaning that it prefers not to bond chemically. Nudging along an atom that skates on the surface without any propensity to engage with the substrate is comparatively easy; picking up a chemically active atom and placing it somewhere in a huge chemically active three-dimensional molecule is completely different.

Don Eigler has continued to popularize this experiment. Visitors may experience the set-up at IBM's Almaden Research Center in San Jose, California, and a virtual art gallery of STM-renditions of xenon atoms on a nickel surface has come into existence. In 1996, Charles Siebert "flew across the country to move an atom" and to write about it in the *New York Times*. That he had to fly from New York City to San Francisco indicates that we are not talking about an experiment that has proliferated greatly. Others have written or drawn other words and images with a similar set-up, but this technique is not being worked on for industrial application. Siebert ignored, and perhaps didn't even understand, the mediated nature of his movement of single atoms. All he did was to "nudge around a single atom of the element xenon, to pick it up and put it back down, to will that atom where I wanted". There's no talk of a hand using a mouse in coordination with an image on a computer screen, and still less talk of what goes into the making of that image. Eigler's program for moving the atom with the mouse even has a chirpy sound, when the atom falls into place, just as LEGO™-bricks click when slotted together. Even more than the *Scientific American* article of 1986, articles like Siebert's elide the disconnect. Obviously, promise/hype sells better than pedantic arguments.

But it is precisely the elision of the pedantic argument that is of interest here, the elision of the differences between atomic resolution, atomic manipulability, and the ability to assemble self-replicating molecular systems LEGO™-style out of individual atoms. The word nanotechnology focuses our attention on the nanoscale, the scale of atoms, and this term covers a multitude of sins. Nano is simultaneously scanning probe microscopy, Eiglerian atom nudging and Drexlerian hype.

Science Fiction

In a very interesting article, Colin Milburn has disclosed the close relationship between Drexler's arguments and the genre of science fiction. Science fiction is identified by the narratological deployment of a novum – a scientific or technological innovation extrapolated from present-day realities – that entails a change in the whole universe of the tale. "Science fiction assumes an element of transgression from contemporary scientific thought that in itself brings about the transformation of the world. It follows that nanowriting, in positing the world turned upside down by the future advent of fully functional nanomachines, thereby falls into the domain of science fiction".

Milburn shows that *Engines of Creation* is composed of a series of science-fictional vignettes, providing a veritable checklist of science-fictional cliches. He finds the same elements in the technical writings of Ralph Merkle, Markus Krummenacker, Richard Smalley, Daniel Colbert, Robert Freitas, Jr., J. Storrs Hall, "and other prophets of the nanofuture [...] Matter compilers, molecular surgeons, spaceships, space colonies, cryonics, smart utility fogs, extraterrestrial technological civilizations, and utopias abound in these papers, borrowing unabashedly from the repertoire of the twentieth-century science-fictional repertoire". Milburn even shows that Feynman's famous 1959 lecture "There is plenty of room at the bottom", which is routinely deployed as an origin myth, belongs in the same category. It too is structured in a series of science fictional vignettes and it too draws on science fiction themes of its time.

The genre is visible in official literature too, for example in that of the munificently endowed National Nanotechnology Initiative – witness its brochure *Nanotechnology: Shaping the World Atom by Atom* the main author of which seems to be Ivan Amato, an author who has also written a book extolling the virtues and promise of materials research. Drexler himself has institutionalized his bolstering role with the foundation of the Foresight Institute. There can be no doubt that the promise/hype of a Drexlerian vision has helped direct funding in a certain direction.

Technological Futures

In a book entitled, *Imagining the Future*, Joseph Corn has assembled half a dozen histories of technological promise/hype. There is for example a story about the early discourse on x-rays for therapeutical purposes (where the promise was to eliminate disease *tout court*), the electrical home (to eliminate domestic labour), or nuclear power (to eliminate war and even social strife). In the epilogue, Corn sums up the imagined technological futures as each fitting at least one of three fallacies. The first is the fallacy of total revolution; that a new technology was expected to herald tremendous change whereas the change in fact turned out to be less significant. The second fallacy is that of social continuity; whereas in fact all new technologies altered the society into which they were introduced. The third is the fallacy of the technological fix or the expectation that the new technologies would strengthen the values of old existing social patterns, whereas they turned out to introduce novel and unintended ones. The Drexlerian vision certainly heralds revolutionary change, and it talks only about the technological changes in store, ignoring and thus not expecting attendant social changes. And the latter part of *Engines of Creation* discusses how to set up an institution of oversight to ensure that the nanotechnological revolution brings only what we desire and none of the technological nightmare conjured up by, say, *Prey* (Crichton 2002).

In this sense, the Drexlerian vision seems to conform to other technological visions.

But why should technological visions have to come true? Their purpose is not to predict but to enroll. They invite other researchers to jump on the bandwagon by depicting an exciting and fruitful field. Whether or not the visions are related to science fiction does not really matter, except to the extent that they help and hinder the political project of bringing allies together. The genre of science fiction explores just what will cause broad excitement, and as such it provides a natural resource for promise/hype. But it is clearly a double-edged sword, especially because of the term "*fiction*". Much of the discourse around Drexler negotiates the proper boundary of reality and fiction. This is Milburn's main concern, along with his argument that the difficulty of maintaining that boundary contributes to a post-modern breakdown of hitherto established identities. At the same time, this negotiation is simultaneously a political dance that makes and breaks alliances.

Role of Visions

Most importantly, visions aim to increase the chances of funding. The National Nanotechnology Initiative's programmatic statement, *Shaping the World Atom by Atom*, is illustrative. The vision is science fictional in the above sense. The argument is then made that R&D funding has been geared to short-term projects with specific goals defined in a cost benefit analysis, but that the promise of nanotechnology couldn't be realized with such funding because the tremendous practical difficulties render the likelihood of short-term marketability unlikely. The role of the government, so the NNI-report, is to step in precisely in such cases as nanotechnology, where the absence of short-term returns prevent investment from private enterprise, but where the promise of long-term benefit makes it worthwhile. No vision, no funding.

In the 1990s physicists in particular have become accustomed to cuts in funding, and this may well be related to the lack of a compelling vision. The National Ignition Facility (NIF) is an interesting contemporary example. This is a facility with the aim of achieving fusion by focusing many high-energy lasers very precisely on a very small area in space, thus providing enough energy to overcome the threshold for fusion. If successful, the system would unlock even more energy than fission, and thus tremendous amounts of energy could be obtained from hydrogen atoms, far more energy than the input to start the fusion process.

The investment for the NIF is several billions of dollars and even if fusion were to be achieved, the engineering task of putting that energy to good use would only just have begun. Thus, in order to attract long-term funding, the vision has to contain much promise. Now, the promise/hype of the NIF is very similar to that for nuclear power in the 1950s. It promises the most powerful weapon ever, and thus a US monopoly, in turn ensuring global peace through deterrence. It will also provide an abundance of energy for civilian use, providing affluence to all, eventually resulting in the end of social strife. Presumably the similarity with the chiliastic nuclear vision is both a source of strength and weakness. The many political alliances that are already in place sustaining nuclear power are likely candidates for enrolment, but for the same reason the well-organized enemies of nuclear power will be enrolled just as easily. Furthermore, the similarity to an older and failed vision makes the NIF project look less than exhilarating. By contrast, the Drexlerian vision's piggy-backing on the promise/hype of fashionable molecular biology gives it sheen and luster.

Technological visions of the future are alive and well, but of course not any vision will do. And as predictions, they are bound to fail: sophisticated notions of the interaction between technological and social change would be counterproductive. The visions are intended to tie together the elements of a heterogeneous network that requires constant maintenance in order to hang together (Latour 2002, 2004). The claim of novelty is essential for technological visions: the elision of the connectedness with practices and theories of the past is as productive as is the claim that success has been shown to be possible in principle, requiring from now on merely developmental labour. The role of promise/hype in motivating researchers and funding bodies discussed here has not been the subject of much research so far. Studies that examine the role of the public sphere instead tend to focus upon the issue of consensus. The topic of nano provides plentiful material for future analysis.

Technological and Scientific Aspects

Since two decades *nanotechnology* has evolved from different scientific fields, such as physics, chemistry, molecular biology, and material science. The nanotechnology aims to study and to manipulate real-world structures with sizes ranging between a nanometer, i.e. one millionth part of a millimeter, and up to one hundred nanometers. The set of typical "*nano*"-objects includes colloidal crystals, molecules, DNA based structures, and integrated semiconductor circuits.

The first scientist who pointed out that "there is plenty of room at the bottom" was Richard Feynman in the year 1959. He envisioned scientific discoveries and new applications of miniature objects as soon as material systems can be assembled at the atomic scale. To this end, machines and imaging techniques would be necessary which can be controlled at the nanometer or subnanometer scale. Fifty years later, the scanning tunneling microscope and the atomic force microscope are ubiquitous in scientific laboratories allowing to image structures with atomic resolution. As a result, various disciplines of the nanotechnology aim towards manufacturing materials for diverse products with new functionalities.

There are two strategies for assembling nanosystems: the *top down* and the *bottom up* approach. The "top down" approach follows the development of the microelectronic industry to miniaturize integrated semiconductor circuits. Modern lithographic techniques enable to pattern nanoscale structures, such as transistor circuits, with highest precision down to several nanometers. In 1965 Gordon Moore, co-founder of Intel corporation, predicted that the number of transistors on a computer chip would double about every eighteen months. The exponential law, also known as *Moore's first law*, has described the development of integrated circuits surprisingly well for decades. As the market for information technology continues to grow, the demand for computer hardware instigates more and more sophisticated "top down" techniques to build more densely packed transistor circuits. *Moore's second law* states that the implementation of a next generation of integrated circuits at minimum cost will be exponentially more expensive as well.

Until all the constraints will finally limit the growth of the semiconductor "*top down*" industry, scientists and engineers assume that the nanotechnology will give answers to most of the technological challenges. For instance, as soon as the feature size of the semiconductor transistors reaches the level that quantum phenomena are important, different concepts for the assembly need to be considered. One possibility is the "bottom up" approach, which is based upon molecular recognition and chemical self-assembly of molecules. In combination with chemical synthesis techniques the "bottom up" approach allows assembling of macromolecular complexes with new functionalities.

Nanoscience and *nanotechnology* will definitely have a strong impact on our lives in many separate areas; e.g. information technology, material sciences, and medicines, to name only a few. The manuscript starts with a paragraph about the most recent developments of lithographic techniques in industry and nanoscience. Moreover, most of the discoveries of the nanotechnologies were initiated by the development of accurate microscopes with atomic resolution. Therefore, the manuscript also gives a short introduction to scanning microscopes. A paragraph about nanoelectronics features approaches and techniques which are supposed to produce successor technologies of the microelectronic industry. Examples include quantum computing and spintronics, two emerging fields of nanoscale electronic circuits. Alternative materials and approaches are currently being investigated for novel products consisting of nanostructures. Aspects of nanoscale materials as well as the impact of the nanotechnology on health sciences are finally described in the manuscript.

Top-down Technique

The optical "*top down*" lithography has been the backbone of the semiconductor industry since 45 years. In this technique a so-called *photoresist*—a liquid, photosensitive chemical that resists etching processes—is spin-coated onto the polished surface of a semiconductor wafer. After hardening of the photoresist, e.g. by heating, the semiconductor sample with the photoresist layer on top is exposed to light. Hereby, patterns can be defined in the photoresist. After a developing process, the exposed (or the unexposed) parts of the semiconductor wafer are bared. By follow-up processes such as etching, metallization, and oxidation the patterns can be translated to conductor paths, logical circuits, or memory cells. Since the eighties of the last century, the demise of the optical lithography has been predicted as being only a few years away. However, each time the optical lithography has reached a limitation, new techniques extended the economically useful life-time of the "*top down*" technique.

The minimum feature size of optically defined patterns depends on the wavelength of the utilized light as well as factors, which are due to e.g. the shape of the lenses and the quality of the photoresist. In 2003, the line-width of semiconductor circuits fell below one hundred nanometers; i.e. the semiconductor industry can be literally seen as being part of the nanotechnologies. For this achievement, Argon Fluoride Excimer lasers are applied with an optical wavelength of 193 nm in the deep ultraviolet. For lithography in this optical range tricks such as *optical proximity correction* and *phase shifting* were invented and successfully implemented. On one hand, the miniaturizing of semiconductor circuits is limited by economic costs for the semiconductor industry, since the implementation of new techniques to realize an ever smaller feature size results in ever increasing costs. On the other hand, physical material properties, such as the high absorption level of refractive mirrors at short optical wavelengths, give a natural limit of only a few tens of nanometers for the miniaturization process. Assuming Moore's laws, the final limit for the optical lithography is supposed to be reached in less than a decade.

At present, there are several possible "top down" successor nanotechnologies: e.g. the *extreme ultraviolet light technique* (EUV), the electron beam lithography with multicolumn processing facilities, the *focused ion beam technique* (FIP), and last but not least the ultraviolet nano-imprinting technique. The implementation of each of the above techniques implies vast technical challenges to overcome. The most promising technique is the one utilizing extreme ultraviolet light of a wavelength of only thirteen nanometers. For this technique, fabrication errors of the "optical" components need to be in the nanometer or subnanometer range. For comparison, the state-of-the-art x-ray telescopes, such as the *XXM Newton* telescope of the Zeiss AG, exhibit a granularity of the mirror surfaces of 0.4 nm. In addition to the "state of the art" optical requirements, all metrological components of the extreme ultraviolet technique need to exhibit subnanometer resolution. As a result, the costs of a stepper machine, by which photoresists in the extreme ultraviolet can be exposed, are fifty millions dollars per system; enabling a line-width of 35 nanometers.

For medium-sized businesses, the *ultraviolet nano-imprinting technique* seems to be the most promising method to fabricate nanoscale circuits. Here, nanostructures are mechanically imprinted into a photoresist. The stamp with the nanoscale patterns is made out of fused quartz; a material which is transparent for ultraviolet light. As soon as the stamp is plunged into the photoresist, a short ultraviolet light pulse causes the photoresist to polymerize according to the patterns on the stamp. In line with the optical lithography, follow-up processes allow defining nanoscale circuits in various geometries. The minimum feature size of the nano-imprinting technique is about ten nanometers. At the same time, the technique is applicable to metals and plastic materials. The costs for an industrial nano-imprinting machine are supposed to be less than one million Dollars. However, the throughput of nano-imprinting machines is much lower than for stepper machines.

Bottom-up Technique

The antipole of the "top down" approach is the so-called *bottom up technique*. Generally, "bottom up" assembly techniques seek to fabricate composite materials comprising of nanoscale objects which are spatially ordered via molecular recognition. The primary examples of the technique are *self-assembled monolayers* (SAMs) of molecules. A substrate, usually made out of metals, is immersed into a dilute solution of a surface-active organic material that adsorbs onto the surface and organizes via a self-assembly process. The result is a highly ordered and well-packed molecular monolayer. The method can be extended towards *layer-by-layer* (LBL) assembly; by which polymer *light-emitting devices* (LED) have already been fabricated. The self-assembly technique also allows positioning of single molecules in between two metal electrodes, and subsequently into an experimental circuit. By this setup, quantum mechanical transport characteristics of single molecules, such as photochromic switching behaviour, can be studied in order to build electronic devices with new functionalities.

Monolayer Techniques

There are several combinations and variations of the self-assembled monolayer techniques. It can be combined with nano-imprinting methods, the atomic force microscope, or the focused ion beam technique; allowing the fabrication of geometrical patterns of molecules with varying friction, chemical functionality, and/or topological characteristics. Since most of the processes are performed in solution, electrical fields give further possibilities to assemble charged compounds in a directed way; e.g. for creation of functionalized sensing electrodes. Electrodes modified by negatively charged gold nanoparticles, and positively charged host molecules have proven highly sensitive for the detection of e.g. adrenaline (as the guest molecule). A further very promising field is the DNA directed assembly of network materials. Here, molecular recognition reactions between single DNA strands are translated into aggregate formation of other materials such as nanoparticles. For instance, the molecular recognition of DNA molecules has been exploited to build so-called nano-tweezers. Last but not least, material networks can be built by so-called block copolymer templates. A typical application of this technique is the formation of networks of metallic nanowires; i.e. metals are vapour-deposited onto a preformed template matrix made out of copolymers. The polymer networks can be used as two- or even three-dimensional templates, while the glass transition of the polymers gives further flexibility to form such porous films.

Generally, there is a wide range of materials which can be lithographically engineered by "bottom up" techniques. The corresponding research has led to numerous sensing, electronic, optoelectronic, and photoelectronic interfaces as well as devices. Prominent examples of nanostructures, which can be assembled with "bottom up" techniques, are presented below in the paragraph entitled "*material sciences*".

Scanning Microscopes

Scanning microscopes are key inventories for the nanotechnology. On one hand they allow probing the characteristics of nanoscale objects with highest resolution. Examples include topology, material configuration, electrical, chemical, magnetic, and optical properties of the studied objects. On the other hand, scanning microscopes allow local manipulation of the nanostructures. In a seminal work in 1982, Binnig and Rohrer invented the *scanning tunneling microscope* (STM). Piezo crystals move a scanning tip across the surface of a sample, while the electric current is recorded between the tip and the sample. If the tip is located very close to the surface of the sample, the electric current comprises of tunneling electrons. To simplify matters, electrons behave as waves in the quantum world. As soon as the distance between the tip and the surface is in the order of the electron wavelength, electrons can tunnel from the tip to the surface. The quantum nature of the tunneling process ensures that the point of the tip, which is closest to the surface, contributes mainly to the current. In principle, this point can be made up of only one atom, which allows atomic resolution.

The scanning tunneling microscope is sensitive to electronic densities on the surface of the sample, which allows imaging of electronic orbits of atoms. At the same time, the scanning tunneling microscope can operate at an atmospheric pressure down to high vacuum conditions at various temperatures, which makes it a unique imaging and patterning tool for the nanosciences. E.g. scanning tunneling microscopes are utilized to pattern nanostructures by moving single atoms across surfaces, while the corresponding change of the quantum mechanical configuration can be recorded *in situ*.

The atomic force microscope operates similarly to the scanning tunneling microscope. Here, the force between the scanning tip and the sample surface is extracted by measuring the deflection of the tip towards the sample. Again the atomic force microscope can be utilized as an instrument to image and to manipulate structures on the nanometer scale. Most importantly, the atomic force microscope is a unique tool to measure forces between two nanoobjects with a resolution of only a few PikoNewtons (a millionth of a millionth Newton). By functionalizing the scanning tip chemically, a single molecule can be utilized for sensing applications; e.g. the binding forces between ligands and receptors can be determined for biological and health applications. Furthermore, magnetic tips can be exploited to image the magnetic topology of magnetic nanostructures which are used for information science purposes. The atomic force microscope can further be used as a miniature groove, e.g. which cuts through nanoscale electronic circuits.

Generally, the resolution of an optical microscope is limited to approximately the wavelength of photons; typically several hundreds of nanometers. In order to resolve smaller structures there are two strategies, either using particles with a shorter wavelength, e.g. electrons, or utilizing a technique called *scanning near-field microscopy* (SNOM). The former possibility gives rise to the *scanning electron microscope* (SEM). Modern scanning electron microscopes can resolve topological variations and chemical configurations of a sample at the subnanometer length scale.

To this end, a sample is located in a high-vacuum chamber, and a beam of electrons is focused onto the surface of a sample. The reflected electrons are then detected as a function of the position of the initial beam. In order to study the optical properties of nanoscale structures, which are substantially smaller than the photon wavelength, the *scanning near-field microscopy* (SNOM) is an excellent method. Here, glass fibers with apertures of only hundred nanometers or less are utilized. The size of the aperture defines the spatial resolution of the technique, while the intensity of the gathered light is recorded as a function of the position of the glass fiber.

NANOELECTRONICS

The field of *nanoelectronics* aims towards developing devices comprising of nanostructures with new overall functionalities for the information technology. By exploiting electronic, optical, and magnetic properties of nanostructures, integrated circuits and memory devices are developed for faster and more reliable information processing schemes at lower heat dissipation and with a greater portability of future computers. Since the seventies of the last century, so-called quantum wells in semiconductor heterostructures have been subject of intensive scientific research. Heterostructures are semiconductor crystals with composite monolayers of different atoms. Most importantly, heterostructures can be grown in which electrons are confined solely to the plane of the *quantum well*. These investigations on nanoscale monolayers of composite semiconductors yielded to applications such as the CD drives, the laser printers, and the input-amplifiers of cell phones; investigations which were honored with the Nobel Prize in 2000.

Since approximately fifteen years, so-called nanocrystals or colloidal quantum dots are promising compounds of a successor nanoelectronic technology. Generally, the colloidal quantum dots are semiconductor crystals in which an electronic quantum state is localized within a tiny volume. Most importantly, the optical and the electronic properties of quantum dots are defined by the quantum mechanical characteristics of the electron state. A nanoelectronic device, which is very similar to a quantum dot, is the so-called *single electron transistor* (SET). The transistor relies on the switching of a single electron. The advantage of such devices would be the combination of minimum size with a minimum power dissipation, both of which become very important issues for densely packed logical circuits.

At the same time, optical processing schemes are expected to become important parts of the future information technology. Only recently Raman lasers, optical waveguides, and fast optical switches have been realized on silicon chips. The realization of an all-silicon based optical processing scheme is an important step to combine fiber optics with the present information technology at low cost. So far optical signals in fibers can be switched by micro-mechanical elements, and/or optical information is translated into electronic signals e.g. by photo diodes. Subsequently, the information is processed via electronic transistor elements and in turn, it is transferred again to optical information by a laser. This very inefficient scheme is expected to be replaced by the silicon-based technology mentioned above. In future devices, further nanostructures, such as *photonic crystals*, are expected to play an important role to control light within a semiconductor crystal directly. In "photonic crystals" the optical constants of a material are modulated in three dimensions by nanostructuring the (semiconductor) host crystal.

As a result, only photons with a defined wave vector and energy are allowed within the periodic structure. The incorporation of nanoscale defects into a photonic crystal further allows the deflection of certain optical wave modes within the photonic crystal. Moreover, photonic crystals can be designed, in which well-defined optical modes can interact with nanostructures. Recently it was demonstrated that by positioning a quantum dot at the center of an artificially designed optical mode volume, the electron states of the quantum dot strongly interact with the electromagnetic fields of photons. Latter experiments give rise to the feasibility of an information processing schemes purely based on quantum mechanics.

In principle, the atomic *granularity* defines a natural limit to the miniaturization of electronic circuits. The electron orbits of atoms are the smallest building blocks of a possible "nano-computer". In other words, a conductor path can not be narrower than a single atom. Therefore, it is reasonable to ask whether a computer, which exploits the quantum nature of the electron states, is technically feasible. The quantum mechanical tunneling effect is one of the more obvious effects which will arise as soon as the dimensions of the electronic circuits fall below a certain value. Since electrons are waves, they may tunnel from one conductor path to another if the paths are located in a distance which compares to the wavelength of the electrons.

Typical electron wavelengths in semiconductor crystals are in the range of several nanometers. Albeit the resulting leakage currents are small, the total effect can be significant for millions of densely packed transistors on a chip. On one hand, tunneling electrons give rise to additional heat dissipation. On the other hand, dislocated charges have to be considered as logical errors and/or they influence the general functionality of the integrated circuit via the Coulomb interaction; e.g. via *cross-capacities* between adjacent conductor paths. Furthermore, an effect called *electro migration* can derogate conductor paths made out of e.g. Aluminum; i.e. the electron current *flushes away* single metal atoms.

The quantum mechanical nature of electrons comprises properties and corresponding technical possibilities which go beyond the functionality of a classical computer. In principle, the classical desk top computer is based upon semiconductor physics with foundations in quantum mechanics as well. However, the functionality of the classical computer can be simply described by "classical" physics. The quantum computer aims towards exploiting the physical laws of quantum mechanics to make certain calculations more efficient. As far as the quantum computer is concerned, the equivalent to the binary digit (BIT) of the classical computer is the so-called *quantum bit* or *qu-bit*. In quantum mechanics, particles such as electrons are described by a set of independent parameters, which are called *quantum numbers*. For instance, the magnetic moment of an electron, the so-called *spin*, is a possible quantum number, which can be mathematically represented by a plus or a minus one half.

A quantum bit consists out of the sum of independent quantum numbers, while continuous prefactors define the weight of a specific quantum number. The superposition of quantum numbers also includes so-called entangled states. Latter are physical states of a quantum mechanical particle, where one quantum number depends logically on another in an intrinsic way. The entanglement is unknown to classical physics (and the most intuitive way to accept its reality is via mathematics). Most importantly, the superposition and the entanglement of quantum bits give rise to fast and parallel information processing schemes at a minimum heat dissipation.

A further discipline of nanoelectronics is the field of spintronics, where both charge and spin degrees of freedom of an electron are exploited to realize new electronic devices. As mentioned above, the "spin" describes the magnetic moment of an electron. It is classified by two values, which are usually referred to as "up" and "down" (the two expressions depict the fact that the magnetic moment of an electron orientates in the directions up or down with respect to an external magnetic field). The magnetic interaction of an electron spin with the crystal environment is much lower than the Coulomb interaction felt by electron charges. Therefore, heat dissipation effects are predicted to be significantly smaller for "spintronic" circuits than for "electronic" circuits. However, there are ubiquitous spin relaxation mechanisms in semiconductor crystals, by which the orientation of the spin (and thus the corresponding classical information) is lost within several hundreds of nanoseconds.

The precise fabrication of nanoscale crystalline layers has also given rise to the development of magnetic layer systems, the resistance of which alters significantly when small magnetic fields are applied. Latter effect is known as the *giant magnetoresistance* (GMR). The technology has produced the prime example of a spintronic device: the so-called *magnetic random access memory* (M-RAM). The memory concept was originally proposed by IBM, and it is now introduced into the marked of information storage technology. In this technology, the information is stored by magnetic domains, while the orientation of the domains is read out via the *giant magnetoresistance*.

In principle, the M-RAM technology combines all advantages of the present memory technologies: it is a fast, nonvolatile information storage technology, and it allows information storage at high density and minimum heat dissipation. The magneto-mechanical hard drive, present in most of the current computers, provides also a medium for high density, nonvolatile information storage. However, it is comparably slow. Hard drives, based upon a concept called *dynamic random access memory* (D-RAM), are fast, however, they need to "refresh" the information iteratively by current pulses. The "flash" memories, which are utilized in MP3-players, cell phones, and digital cameras, are quite slow and they can only be used approximately one million times. Therefore, M-RAM devices seem to be promising candidates for a "*nanoelectronic*" successor technology.

Last but not least, nanomaterial sciences give further possibilities for new architectures of integrated circuits and memory devices. For instance, a memory concept, usually referred to as *phase change random access memory* (phase change RAM), utilizes the phase transition between "*crystalline*" and "*amorphous*" states of materials to encode information. The phase transition is triggered by electrical impulses, while the resistance difference between the crystalline and the amorphous state allows identifying the binary information.

The crystalline state is re-initialized, if the material is homogeneously melted by the application of long electrical current. In principle, the method allows nonvolatile information storage at highest densities; i.e. a Terabit on the area of a stamp. A corresponding method is already implemented in *digital versatile disks* (DVDs) made out of polymers, where a short laser pulse melts small areas of the disk; i.e. the spots undergo a phase transition from crystalline to amorphous. Generally, for memory device with highest information densities, material sciences will play an important part in nanoelectronic engineering.

Material Sciences

In recent years, the material science and technology have explored a variety of new materials in which nanoscale components give rise to better properties of well known products. To this end, optical, electric, thermal, mechanical, and chemical characteristics of nanostructures are exploited to enhance certain properties of the composite systems. Examples of "nano"-products include paints, inks, cosmetics, lubricants, grinding pastes, ceramics, luminescent materials, glues, and protective lacquers; to name only a few. The new materials are envisioned to fulfill tasks, such as the ability to decompose pollutants at higher rates or to converse light into current more efficiently. For such and more complex tasks novel materials are based on several nanoscale components whose spatial organization is engineered at the molecular level. The macroscopic behaviour arises from the combination of the novel properties of the individual building blocks and their mutual interaction.

Materials with nanoscale pores, such as zeolites, are utilized for catalyzers in chemical processes or as ten-sides in cleaning detergents. Metallic alloys show a certain degree of mechanical memory. New terms such as nanotubes, nanowires and quantum dots are now common jargon of scientific publications. These objects are among the smallest, man-made units that display physical and chemical properties which make them promising candidates as fundamental building blocks of novel transistors.

Quantum dots are the ultimate example of a solid, in which all dimensions shrink down to a few nanometers. The electronic and optical properties of quantum dots are a consequence of their dimensions. At the same time, colloidal quantum dots can be synthesized from a wide range of materials. The dimension of these particles makes them ideal candidates for the nano-engineering of surfaces and the fabrication of functional nanostructures. Moreover, semiconductor quantum dots are probably the most studied nanoscale systems at the moment.

The surface of the lotus flower is *hydrophobic* due to nanoscale structures. In a biomimetic ansatz, these properties have been transferred to paints which are strongly "*soil resisting*". Investigations of the nanoscale architecture of shells and bones gave rise to new developments in the field of layered materials which are extremely robust. Carbon nanotubes have the topology of a graphite sheet which is rolled up to form a tube. These wires with diameters of only a few nanometers and a controlled chemical make-up show intriguing electrical properties and extremely high strength and stiffness. Due to their properties, carbon nanotubes are proposed to be utilized e.g. as field emitters in flat screens. At the same time, carbon nanotubes exhibit a unique ratio of inner volume to surface, which makes them ideal candidates for extremely light hydrogen storage devices. Moreover, compounds made out of carbon fibers are light and robust materials which are used in a broad range of application.

An important technique of the material sciences is the sol/gel technique, by which nanostructures are formed and controlled by the application of so-called *colloids*. Generally, colloids are systems in which droplets of one substance are formed and in turn, resolved within a second substance. Daily life examples of colloids are the "sauce bearnaise", in which droplets of vinegar are resolved in butter. Other examples are cosmetic cremes and paints. In material sciences colloids are exploited to form nanoscale structures with chemically controlled composition and/or nanoscale structures which tend to build networks as soon as the solvent is removed.

By utilizing the sol/gel technique, protective lacquers of a variety of substances have been produced. E.g. silicon nanostructures can be assembled in the liquid phase, and in turn, they are sprayed onto the surface of arbitrary surfaces. After the solvent has vapourized, appropriate heating gives rise to the formation of a ceramic cover layer which is extremely thin and hard at the same time. Due to the large surface to volume ratio of nanostructures, such ceramics can be formed at rather low temperatures. The sol/gel technique can be even utilized to fabricate optical components such as glass fibers and frequency doublers.

A whole class of nanomaterials is defined by so-called *aerogels*, which are highly porous materials consisting mainly out of "air". Again, a daily life example of such a material can be found in the French cuisine. The "Baizer" consists of white egg in which thousands of microscale air bubbles are incorporated. The air bubbles in "*Baizer*" have a size of several microns. As a result, visible light is scattered at the air inclusions and "Baizer" appears as a white substance, while white egg is transparent to visible light. If the enclosed air bubbles have a submicron diameter, aerogels appear to be transparent.

At the same time aerogels are very good heat insulators, since heat can not circulate in the air bubbles. Hereby, window glasses with remarkable insulator properties can be fabricated. Aerogels with nanometer inclusions can be produced by the sol/gel technique; i.e. spheres are formed within a colloid by utilizing networking nanostructures. A gel is formed as soon as the solvent is removed while the volume of the sol is kept constant.

The large inner surface of an aerogel gives rise to several applications in the energy sector. Lithium batteries with enhanced storage characteristics have been built. Electrical capacitors of up to 2500 Farads have been fabricated by aerogels. Last but not least, better fuel cells can be envisioned by this technique.

Health Sciences

There are several applications of *nanotechnologies* which are already implemented in commercially available cosmetic products. For instance, nanoscale spheres, made out of apatites and proteins, are implemented into toothpastes in order to enable *biomineralization* of the teeth, because the corresponding nanospheres consist out of the same material as teeth. Other well-known examples are sunscreens which contain colloids of zinc oxide. Such a sunscreen gives an excellent protection against ultraviolet radiation, which is simply reflected by the nanoscale structures, while the sunscreen appears transparent to human eyes. In addition, aerogels can be utilized for a very efficient drug delivery. Nanoscale capsules with functionalized molecular linkers can be utilized to deliver drugs to metastases via molecular recognition.

Recently, a method called *magnetic fluid hyperthermia* was introduced to oncology, which is based on heating tissues for therapeutic purposes. Generally, tumour cells exhibit a hypersensibility to heating (*hyperthermia*). At the same time, surfaces of magnetic nanoparticles can be functionalized in a way that they accumulate only/mainly in tumour tissue. Therefore, if an alternating magnetic field is applied, the dissipation losses, induced due the movement of the magnetic colloids, heat up and, in turn, destroy the tumour cells. By solving the biochemical and physiological specificity problem, cancer-specific hyperthermia protocols have been developed.

A well-known vision of scientists working in the field of nanotechnology is the "lab on a chip". The chips would have a size of roughly a square centimeter, while millions of nanoscale instruments would analyze droplets of liquids for biological, medical or forensic purposes. One possibility to move the liquids across the surface of the chip is given by so-called *surface acoustic waves* (SAW). The latter technology has recently been realized for commercially available products.

To conclude, nanoscience and nanotechnology have a great potential to solve several challenges of the modern society. Moreover, there exists a great economic expectation for growth related to nanoscale products. At present, there are about 450 companies alone in Germany, which manufacture products related to nanomaterials. To this end, the manuscript intends to give an introduction to some of the nano-technologies which will become important to the daily life in the near future; in particular these are the information technology, material sciences, and health sciences.

8

INDUSTRIAL BIOTECHNOLOGY

Yeasts are eucaryotic organisms that exist in nature predominantly as single cells. Traditionally, they have played an important role for man, being used for thousands of years in bread-making, brewing, and making certain foods palatable and nutritious. The commercial importance of yeast has grown considerably over the past few decades, and they are now being used in a variety of fermentative processes for the synthesis of simple sugars and ammonium nitrogen as well as certain fats, vitamins, and proteins. Although there are about 350 recognized species of yeast, only a relatively small number are important commercially, notably members of the genera *Saccharomyces* and *Candida*. Species of *Saccharomyces*, such as *S. cerevisiae* and *S. uvarum* (which is sometimes referred to as *S. carlsbergensis*) are used in the manufacture of breads, wines and beers, while some species of *Candida* are sources of animal food and fodder. One species in particular, *S. cerevisiae*, has received an enormous amount of attention from generations of biologists. This is not so much due to its commercial importance, but rather because it is amenable to most of the cultural techniques and genetic manipulations used with laboratory bacteria. It has been extremely well-characterized genetically and defined mutations in most biochemical pathways have been established.

The genetic engineering of yeast using recombinant DNA techniques has been made possible by the development, recently, of a transformation system and the construction of DNA cloning vectors which mediate the introduction of DNA into yeast cells. These technological advances, combined with classical genetic studies, have provided a wealth of knowledge of yeast molecular biology. The ability to introduce individual genes into yeast and apply selection by complementation to identify a clone carrying a particular gene now permits the isolation of virtually any yeast gene provided that a suitable recipient yeast strain is available. Furthermore, gene cloning in yeast is likely to be of great industrial importance, not only for the improvement of existing strains used in fermentation processes, but also for the bulk production of commercially important proteins not selectable in or compatible with bacterial systems.

Transformation

Gene cloning is a method whereby DNA molecules are joined together *in vitro* and introduced into living cells where they are able to be maintained and replicated. An integral part of this method is transformation which is a process whereby cells take up and maintain exogenously-supplied DNA and express genes on the DNA to alter the phenotype of the recipient cells. In common with many organisms, yeast does not have a natural system for DNA uptake and all procedures for yeast transformation rely on artificial methods to introduce the products of *in vitro* DNA manipulation into cells. They are based on the effects of polyethylene glycol on the surface of spheroplasts, and on the reasonably efficient regeneration of normal cells from spheroplasts.

Procedures

Yeast cells are surrounded by a thick cell wall which is composed mostly of the polysaccharides mannan and glucan, with some protein and lipid components, and a small amount of chitin. The preparation of spheroplasts involves enzymatically removing the cell wall with mixtures of β-glucanases. Some commercial preparations of this mixture that are widely used for this purpose are Glusulase and Helicase, both crude extracts of snail gut. Partially pure preparations, such as Zymolase and Lyticase are now widely used. However, Glusulase and Helicase have the advantage that overdigestion with these enzyme mixtures is virtually impossible, whereas Zymolase and Lyticase treatments must be carefully monitored because long exposure to the enzymes seriously disrupts cell metabolism leading to poor regeneration of spheroplasts. Prior treatment of yeast cells with a reducing agent, such as 2-mercaptoethanol or dithiothreitol, is sometimes used to sensitize the cell wall to subsequent enzymatic degradation by reducing disulfide bridges in the protein components of the wall thus making the glycosidic bonds available for cleavage.

Table 8.1. A Procedure for yeast transformation.

Materials	
SED	SOS
1 *M* sorbitol	10 ml 2 *M* sorbitol
25 *mM* EDT A, pH 8.0	6.7 ml yeast nutrient medium
50 *mM* dithiothreitol	0.13 ml 1 *M* calcium chloride
	27 μl 1% solution of amino acid to be selected
SCE	3.17 ml water
1 *M* sorbitol	Regeneration agar
0.1 *M* sodium citrate, pH 5.8	182 g sorbitol
10 *mM* EDTA	20 g agar
STC	6.7 g Difco yeast nitrogen base
1 *M* sorbitol	20 g glucose
10 *mM* calcium chloride	amino acids as required (to 40 μg/ml)
10 *mM* Tris-HCI, pH 7.5	water to 1 liter
PEG	All solutions are sterilized
20% (w/v) polyethylene glycol, M.W. 4000	
10 *mM* calcium chloride	
10 *mM* Tris-HCI, pH 7.5	

Procedure

1. Inoculate a flask containing 100 ml of yeast nutrient medium with an overnight culture of yeast to give a starting O.D.600 of 0.1 (equivalent to 5×10^7 cells/ml).
2. Grow the cells with shaking at 30°C until the culture has reached an O.D.600 of 0.4 (about 2×10^7 cells/ml).
3. Harvest the cells by centrifugation at 5,000 rpm for 5 min. Wash the cell pellet once with 10 ml of 1 *M* sorbitol.
4. Resuspend the cells in 10 ml of SED. Incubate at 30°C for 10 min with occasional gentle shaking.
5. Collect the cells by low-speed centrifugation. Resuspend the cells in 10 ml of SCE. Add 0.1 ml of Glusulase and incubate at 30°C for 20 to 30 min, occasionally gently shaking the suspension.
6. Monitor for speroplast formation by diluting 10 ml of the cell suspension in a drop of 5% (w/v) sodium lauryl sulfate on a microscope slide and observing "ghosts" under the phase contrast microscope.
7. Collect the spheroplasts by spinning at 3,000 rpm for 4 min. Wash twice with 10 ml each of 1 *M* sorbitol.

8. Resuspend in 1 ml of STC. Pipette 0.2 ml aliquots of the suspension into 10 ml Falcon tubes.
9. Add 1 to 5 micrograms of DNA. Incubate the tubes at room temperature for 5 min.
10. Add 2 ml of PEG to each tube and leave at room temperature for a further 10 min.
11. Centrifuge the tubes at 3,000 rpm for 4 min. Resuspend the spheroplasts in 0.5 ml of SOS and incubate at 30°C for 30 min. At this point the cells may be plated or held at 4°C for a few days.
12. Add 6 ml of regeneration agar (kept at 45°C) to each tube and pour the contents immediately onto pre-warmed plates. Incubate the plates at 30°C for 2 to 7 days to allow transformants to grow.

Spheroplasts are maintained in an hypertonic medium provided by 1 *M* sorbitol or 0.6 *M* KCI. Uptake of naked DNA is promoted by co-precipitating calcium-treated spheroplasts and transforming DNA with polyethylene glycol. Regeneration of the cell wall and selection of transformed colonies is done simultaneously by embedding the spheroplasts in a solid matrix composed of 3 percent agar and selective medium which permits only the growth of transformed cells. Colonies develop in the agar within two to seven days depending upon the particular DNA and recipient strain used in the transformation.

A few variations have been developed on this basic method. In one procedure, *Escherichia coli* protoplasts carrying recombinant plasmids can be fused directly to yeast spheroplasts and yeast transformants selected by plating the entire fusion mixture in yeast regeneration agar. This technique, known as yeast transfusion, is relatively inefficient compared to conventional yeast transformation but does have the advantage of eliminating the time and labor spent in the isolation of recombinant plasmid DNA from *E. coli*.

An even more rapid procedure for yeast transformation has been developed recently that completely eliminates the need to make spheroplasts, an often tedious and time-consuming procedure. Intact yeast cells are made competent by treating them with alkali metal ions such as lithium and rubidium salts. Incubation of competent cells with polyethylene glycol promotes the uptake of naked DNA molecules and transformants can be grown directly on selective agar plates without using regeneration agar. The transformation efficiency obtained with this method is generally lower than that obtained with conventional spheroplast techniques.

Table 8.2. A Procedure for yeast transformation of intact yeast cells.

Materials		
TE	LC	PEG
10 *mM* tris-HCI, pH 7.5	0.1 *M* lithium chloride	50% (w/v) polyethylene glycol, M.W. 4000, in water
1 *mM* EDTA	10 *mM* tris-HCI, pH 7.5 1 *mM* EDTA	

All solutions are filter-sterilized.

Procedure

1. Inoculate 100 ml of yeast nutrient medium with an overnight culture of yeast to give a starting O.D. 600 of 0.1. Grow the cells at 30°C with shaking to an O.D. 600 of 0.4.
2. Harvest the cells by centrifugation and wash once with 10 ml of TE.
3. Resuspend the cell pellet in 1.5 ml of LC. Incubate the suspension at 30°C for 60 min.
4. Pipette 0.2 ml of the suspensicm into 10 ml Falcon tubes. Add 1 to 10 μg of DNA. Incubate at 30°C for 30 min.
5. Add 1.2 ml of PEG to each tube. Incubate at 30°C for a further 50 min.
6. Heat shock the suspension at 42°C for 5 min. Pellet the cells by low-speed centrifugation and wash twice with 10 ml of sterile water.
7. Finally, resuspend the cells in 0.2 ml of sterile water and plate directly onto selective agar plates. Incubate the plates at 30°C for a few days to allow transformants to grow.

Genetic Markers for Yeast Transformation

Selection of successful transform ants is based on the ability of genes carried on the transforming DNA to complement auxotrophic mutations in the recipient strain. The first yeast genes characterized were responsible for the synthesis of enzymes involved in amino acid biosynthesis, and were isolated from yeast genome libraries by their ability to complement mutations in *E. coli*. Subsequently, these genes have been used extensively in the construction of yeast cloning vectors. Several host-vector systems have been developed by combining these vectors with the corresponding auxotrophic yeast strain as recipient. However, since selective pressure for successful transformants depends upon the reversion frequencies of these mutations (which is usually of the order of 10^{-6} to 10^{-7}), the appearance of a large number of revertants may obscure the identification of transformed cells, especially when transformation has been inefficient. This problem has been largely overcome by making stable mutations for most of the commonly used auxotrophic markers. These include introducing double or triple mutations into a gene, or by the *in vitro* construction of deletions inserted into the yeast genome by replacement of the wild-type gene.

Constructing recipient strains with suitable genotypes to be used in transformation is difficult and time-consuming. Antibiotic resistance determinants carried on the transforming DNA should be generally useful as selectable markers and could be used with any sensitive yeast strain as recipient. However, although some antibiotic resistance genes of bacterial origin are expressed in yeast, they cannot be used as selectable markers in yeast transformation because yeast cells are naturally resistant to most of these antibiotics. An exception is G418, an aminoglycoside antibiotic which inhibits the growth of a wide range of procaryotic and eucaryotic organisms, including yeast. Resistance to G418 is encoded by a bacterial transposable element, *Tn*601, and yeast cells transformed with the element are resistant to high levels of the antibiotic. The use of G418 for yeast transformation selection, by eliminating the need for specially constructed recipient strains, should be particularly valuable in the genetic engineering of industrial yeast strains.

Yeast Cloning Vectors

Nearly all systems for isolating and analyzing gene sequences to be cloned in yeast require the initial isolation of recombinant DNA molecules in *E. coli*. This gives the experimenter several practical advantages. *E. coli* offers a high efficiency of transformation, powerful hybridization screening methods for the detection of cloned DNA sequences, and good amplification of recombinant DNA molecules to be used in yeast transformation. The ability to maintain cloned genes in two alternative hosts also allows the experimenter several options carrying out manipulations on the cloned DNA by taking advantage of the particular properties of each host and its mutants. The vectors currently used in gene cloning in yeast are therefore hybrids composed of bacterial plasmid or bacteriophage DNA and yeast DNA sequences.

The particular purposes of cloning experiments generally dictate which type of vector system is used. For instance, if a small segment of DNA is to be purified for the genetic manipulation of yeast, then plasmid systems are generally employed. Plasmid vectors offer the advantages of high purification of cloned DNA sequences and simplicity of manipulation of plasmid DNA, and are by far the most popular choice of vectors for gene cloning in yeast. However, if random DNA fragments are to be screened by physical methods in which completeness of representation and large-sized inserts are desired, then other types of cloning vectors must be used, the most generally useful being hybrid plasmid-bacteriophage vectors called cosmids.

The different modes by which DNA can be maintained in yeast following transformation has led to the development of two main classes of cloning vectors: integrating vectors and autonomously-replicating vectors. Each type of vector has different characteristics and uses, but all have the following properties in common:

1. They contain suitable genetic markers that can be selected in yeast recipient strains defective for that gene. For some genes (such as *LEU2* and *TRP1*) selection can also be carried out in suitable *E. coli* hosts as well as yeast. However, most of the vectors in general use carry one or more antibiotic resistance genes, and selection for the presence of these markers can be carried out in virtually any *E. coli* strain.
2. All vectors carry an origin of replication for maintenance in *E. coli* cells which often results in a high copy number.
3. They contain single sites for one or more restriction endonucleases suitable for the cloning of foreign DNA fragments.

Integrating Vectors

Yeast integrating vectors (often referred to as YIp) consist of a bacterial cloning vector which can be either a plasmid, bacteriophage, or cosmid, and a suitable yeast gene which can be used for the selection of transformanl cells. Transformation is accomplished by integration of the vector into the recipient yeast genome and is mediated by recombination between the yeast DNA sequence on the vector and the homologous DNA sequence in the nuclear DNA of the recipient cell. The transformation efficiencies obtained with these vectors are very low, being of the order of about 1 to 10 transformants per microgram of vector DNA (equivalent to 1 transformant per 10^6 to 10^7 viable, regenerated cells). The low number of transformants obtained does not normally present a major problem to the experimenter, since high yields of vector DNA can be obtained by initial amplification in *E. coli*. Much higher transformation frequencies are obtained by linearizing the vector DNA through cleavage in the yeast moiety. This also has the effect of directing the integration of the transforming plasmid at a

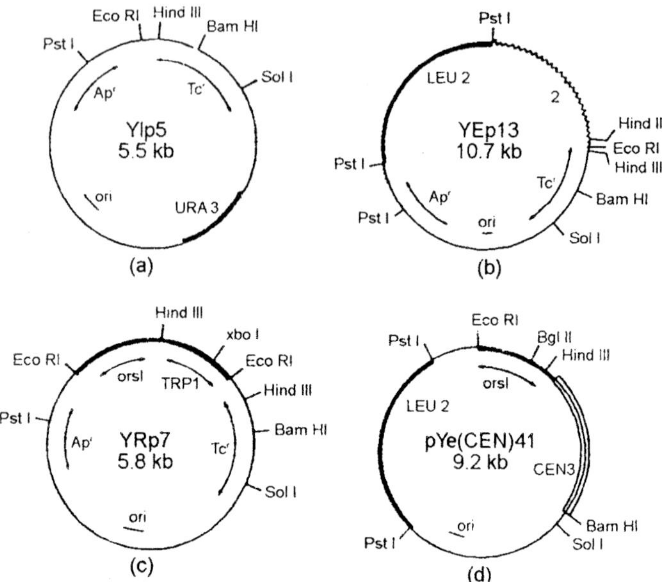

Fig. 8.1. Physical maps of yeast cloning plasmids. Ap^r and Tc^r are resistance determinants for ampicillin and tetracycline, respectively; ori is the bacterial origin of replication.

specific site on the chromosome. The free DNA ends are apparently recombinogenic and act to increase the efficiency of integration and, thereby, transformation. The efficiency of transformation may be increased still further by cotransforming yeast with a YIp vector and an extrachromosomally-replicating vector. If the vectors have DNA sequences in common, recombination can take place between them leading to a single extrachromosomally-replicating molecule. This feature of cotransformation has other practical advantages. Only one of the two vectors needs to have a yeast selectable marker and since recombination can take place between shared bacterial sequences it can provide a convenient way of introducing genes into yeast which have been cloned on a purely bacterial vector.

The transformed phenotype obtained with YIp vectors is very stable, the rate of segregation of integrated vectors being less than 1 percent per generation of growth in non-selective medium. The segregation frequency seems to be correlated with the amount of homology between recipient and vector DNA, and extensive homology results in higher transformation as well as higher segregation frequencies. Normally, there is only a single copy of the transforming vector DNA in each

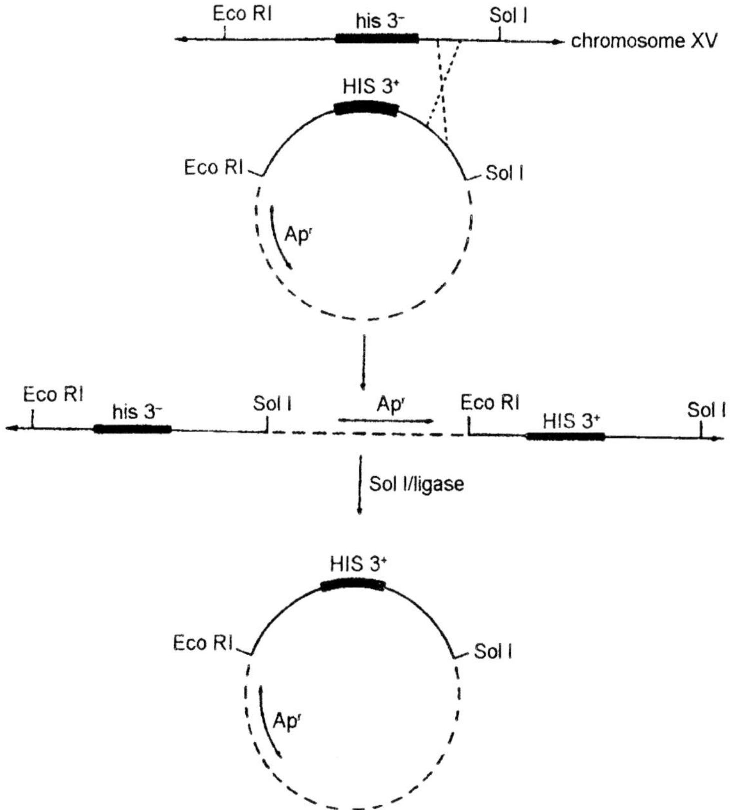

Fig. 8.2. A scheme illustrating the integration of a YIp vector into yeast nuclear DNA and recovery of the YIp DNA.

stably transformed cell, the integrated vector replicating with nuclear DNA and being inherited in the same manner as any regular chromosomal gene.

One major problem associated with integrating vectors is the tedious procedure needed to recover them from transformed cells. It is necessary to prepare total DNA from yeast transformants and digest it with a restriction endonuclease which cuts the vector DNA once within or at the boundary of the yeast DNA insert. The mixture of DNA fragments is then ligated at low DNA concentration and the recircularized vector recovered by transforming *E. coli* cells and selecting for an antibiotic resistance gene on the vector DNA. Depending upon the location of the integrated vector DNA with respect to the chromosomal marker gene and the sites of restriction endonuclease cleavage, either the mutant allele (originally from the recipient genome) or the wild-type allele (originally from the vector) of the yeast marker gene will be recovered.

Yeast integrating vectors are exceptionally useful tools for genetic analysis and gene manipulation in yeast. The specific integration and excision mechanism of YIp vectors has been used to map cloned genes and provides the possibility of mutating yeast structural genes in a defined manner. For instance, they can be used for the construction of strains carrying gene duplications or substitutions by virtue of their integration into nuclear DNA. Sophisticated methods have been developed whereby cloned genes, mutated using *in vitro* techniques, can be recombined into recipient yeast DNA to replace the wild-type allele. These methods should have potentially wide applications for constructing yeast strains carrying fully characterized mutations in virtually any yeast gene.

Extrachromosomally-Replicating Vectors

The ability of cloning vectors to replicate extrachromosomally in yeast is conferred by special yeast sequences derived from either the endogenous yeast 2-micron plasmid or nuclear DNA. Those vectors that include a segment of 2-micron plasmid DNA for maintenance are called yeast episomal plasmids (YEp); vectors that contain chromosomal replicators or autonomously-replicating sequences (*ars*) are called yeast replicating plasmids (YRp); and, finally, vectors that contain a functional centromere are referred to as yeast centromere plasmids (YCp).

Yeast episomal plasmids

Many, if not all strains of *S. cerevisiae*, contain an endogenous plasmid often referred to as the 2-micron plasmid, or *Scp*1, that replicates independently of the chromosomes. 50 to 100 copies of the plasmid are present in each cell and it appears to replicate under cell nuclear control. The entire nucleotide sequence of the plasmid has now been established and reveals two inverted repeat sequences, 598 base-pairs in length. Recombination between the inverted repeat sequences results in the coexistince of two forms of the plasmid molecule which differ in the respective orientation of the two non-repeated sequences.

Cloning vectors that contain certain segments of this plasmid also behave like autonomously-replicating units and transform yeast with a high efficiency (10^3 to 10^5 transformants per microgram of DNA) that is several orders of magnitude greater than integrating vectors. The copy number of the plasmids after transformation is quite high, ranging from 25 to 100 copies per cell, which may be important where a high level of expression of a cloned gene is required. They can be recovered as plasmids from yeast cells, but the stability of the transformed phenotpye varies depending upon the size of the transforming plasmid and which segment of the 2-micron plasmid is present in the vector. The smaller and more versatile cloning vectors, such as YEp13 contain only a small portion of the 2-micron plasmid which provides high transformation efficiency, multiple copies of the vector and relatively stable maintenance in yeast cells.

Some disadvantages, however, are apparent with these cloning vectors. Sequence rearrangements, resulting in complex recombinant molecules, sometimes occur at low frequencies and are due to homologous recombination between 2-micron DNA on the vector and the resident 2-micron plasmid in recipeint cells. Also, instability of the transforming vector seems to increase with additionally inserted DNA and may be due to replicative competition with the resident plasmid. The obvious solution to these problems is to use 2-micron plasmid-free mutants as host cells. Vectors containing the entire 2-micron plasmid transform these strains with high efficiency and transformants are extremely stable. However, it appears that most YEp vectors which are relatively stable in yeast strains containing the 2-micron plasmid are rapidly lost in plasmid-free strains. This is presumably due to the absence on these vectors of certain 2-micron sequences required for efficient replication.

Yeast replicating vectors

Yeast replicating vectors contain segments of yeast chromosomal DNA carrying a biochemical genetic marker closely associated with an autonomously-replicating sequence which functions as an origin or replication in yeast. Similarly to yeast episomal vectors, they transform yeast with high efficiency (10^2 to 10^4 transformants per microgram of DNA), but the yeast transformants obtained contain only one or two copies of the plasmid per cell. This may be advantageous for cloning genes with deleterious effects. A major disadvantage of these vectors is their high mitotic and meiotic instability in transformed cells, the plasmid being lost at a rate greater than 1 percent per generation of growth in nonselective medium. Sometimes, however, rare stable transformants can be obtained in which the transforming vector has integrated at a homologous site on the yeast genome in an identical manner to yeast integrating vectors. This property of yeast replicating plasmids may be useful since it combines a high frequency of transformation with the possibility of stable integration through homologous recombination into yeast nuclear DNA.

Yeast centromere vectors

Artificial minichromosomes have been constructed recently which consist of a yeast vector containing a chromosomal or plasmid origin of replication and DNA sequences apparently derived from yeast centromeres. These plasmids behave in mitosis and meiosis as stably inherited additional linkage groups. More so than yeast replicating vectors, these plasmids represent potentially useful genetic systems, combining the advantages of efficient transformation with stable propagation of a single gene dosage carried on the vector.

DNA Cloning in Yeast Plasmids

The procedures for introducing foreign DNA fragments into yeast plasmid vectors to generate viable recombinant molecules are straightforward in principle, but in practice care is taken in order to minimize the subsequent effort required to identify and characterize recombinant clones. Insertion of foreign DNA is usually by simple ligation of plasmid DNA cut with a restriction endonuclease in the presence of excess foreign DNA cut with the same enzyme. In a ligation mixture, those molecular species that are most effective in *E. coli* transformation and subsequent stable plasmid propagation are circular molecules. Circle formation is optimum when the concentration of DNA fragment ends is low, the only major undesirable product of the reaction being circularized plasmid containing no foreign DNA insert. Recircularization of plasmid DNA may be limited to some extent by adjusting the relative concentrations of vector and foreign DNAs during ligation. However, this problem is dealt with most efficiently by using either biochemical methods, such as treatment of vector DNA with an alkaline phosphatase to remove 5'-phosphoryl groups, to prevent the recircularization reaction, or genetical methods which involve the identification or selection of plasmids that contain foreign DNA inserts.

Cloning vectors able to replicate autonomously in both *E. coli* and yeast are generally referred to as "shuttle vectors." They are easy to recover from yeast by a simple DNA extraction followed by transformation of *E. coli*. This makes it possible to transfer, on a routine basis, genetic information back and forth between *E. coli* and yeast. Nearly all of the widely used plasmid vectors are based on the bacterial plasmid, pBR322, which carries genes for resistance to ampicillin and tetracycline. A typical vector of this type, YEp13, can be selected in *E. coli* for these antibiotic resistances or by complementation of *leu*B mutations, and selected in yeast for complementation of *LEU2* lesions. There are several restriction endonuclease sites available for cloning, the most useful being the *Bam*H1 cleavage site which is situated in the tetracycline resistance gene; insertion of foreign DNA at this site inactivates this gene and provides a method for distinguishing recombinant molecules from recircularized plasmid DNA.

Some cloning vectors have been modified to accept DNA fragments made with almost any restriction endonuclease producing cohesive ends, and others have been constructed for cloning genome fragments at random using partial digests with restriction endonucleases of low specificity that produce cohesive ends complementary to those generated by an enzyme that cleaves the vector DNA just once (for example, *Sau*3A for *Bam*HI vectors). These vectors are generally used to construct and amplify DNA libraries from small genomes in *E. coli*. The library can be transferred to yeast and the high frequency of transformation obtained facilitates the screening of large numbers of transform ants for specific gene sequences. Individual transformants are then returned to *E. coli* for preparative isolation and genetic manipulation. These methods have been widely used not only to isolate and characterize many yeast genes which cannot be isolated by cloning in bacterial systems, but also some genes from other eucaryotic organisms that are able to complement certain yeast mutations.

DNA Cloning in Yeast Cosmids

Yeast cosmids have been designed for the convenient cloning of large DNA fragments in *E. coli* and their transfer to yeast. The recovery of recombinant clones carrying large DNA inserts is sufficiently high to establish gene libraries of complex eucaryotic genomes from microgram amounts of DNA. Yeast cosmids consist of (a) bacterial plasmid DNA sequences for replication and selection in *E. coli*; (b) yeast 2-micron plasmid DNA sequences or chromosomal replicators and yeast markers necessary for maintenance and selection in yeast; and (c) the cohesive end fragments (*cos*) of bacteriophage lambda which allow packaging of large recombinant molecules into lambda phage heads. Restriction and ligation of large DNA fragmenls into restriction endonuclease-cleaved cosmid DNA results in long concatenated DNA structures that resemble the natural DNA substrate of the lambda packaging system. Hybrid concatamers with *cos* sites

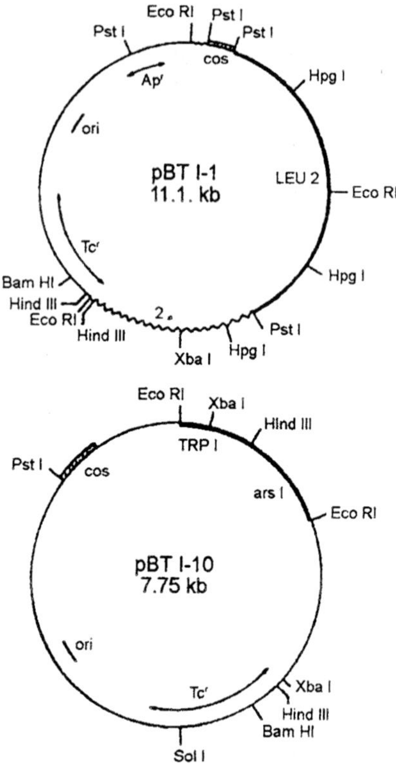

Fig. 8.3. Physical maps of yeast cosmids, pBTI-1 and pBTI-10. Ap^r and Tc^r are resistance determinants for ampicillin and tetracycline, respectively; ori denotes the bacterial origin of replication.

situated approximately 36 to 50 kilobase-pairs apart, and in the same orientation, are packaged using an *in vitro* system into lambda particles which can then be transduced into *E. coli* with high efficiency. The small size of yeast cosmids (about 10 kilo base-pairs) and the size selection imposed by the packaging reaction, allows the cloning of DNA fragments 25 to 40 kilobase-pairs in size.

The use of restriction endonucleases with six-base-pair recognition sites to cleave genomic DNA is usually avoided because they can lead to the exclusion of very large or very small DNA fragments during cloning. Instead, foreign DNA is partially digested with a restriction endonuclease recognizing a four-base-pair sequence that cuts DNA more frequently to produce a more random fragment population. Examples are *Sau*3A and *Mbo*1 which cleave DNA to produce cohesive ends complementary to those generated by *Bam*HI which cleaves most vector DNAs just once. In order to ligate DNA fragments into the vector in an authentic gene arrangement without obtaining a random assortment of insert fragments, the endonuclease-cleaved DNA is fractionated by ultracentrifugation on linear sucrose gradients to purify DNA fragments between 20 and 40 kilo base-pairs in size. Finally, the ligation reaction is carried out at high DNA concentrations to ensure the formation of concatameric DNA.

Transduction procedures, in which an *in vitro* system is used to package recombinant cosmid DNA into infectious lambda phage particles, is an extremely efficient method for the transfer of large DNA molecules to *E. coli*. The *in vitro* packaging procedure uses a combination of extracts of two *E. coli* lysogens each of which is defective in a different step of lambda morphogenesis. In one lysogen which contains a prophage mutation in gene *A*, empty precursor particles accumulate within the bacterial cell upon induction. This mutation can be complemented by the addition of the missing *A* protein supplied by an induced lambda *E* lysogen. This mutation affects the synthesis of the main capsid protein; in its absence, all other head proteins are available in soluble form. Ligated DNA is added to a concentrated, lysed mixture of the two lysogens, and in the presence of ATP and polyamines, DNA is packaged, the phage head matures, and tails, which are made by both lysogens, are added. The process results in DNase-resistant, infectious virus particles which can be used to transduce sensitive *E. coli* cells like any *in vivo* produced bacteriophage. Transduced, recombinant cosmids circularize via the lambda cohesive ends within host cells, and since cosmids do not contain all the essential lambda phage genes, they do not go through a lytic cycle. Instead, recombinant molecules are maintained in their hosts as large, autonomously-replicating plasmids that can be amplified and used directly to transform appropriate recipient yeast cells.

The efficiency of yeast transformation by recombinant yeast cosmids seems to be unaffected by increase in size which is in sharp contrast to the strong dependence on plasmid size in *E. coli* transformation. They can be recovered from yeast either by transformation of *E. coli* with total yeast DNA or by *in vitro* packaging.

Fusion Plasmids

Gene fusions have been very useful for investigating regulatory mechanisms in *E. coli*. In particular, gene fusions in the *lac*Z gene of *E. coli*, which specifies β-galactosidase synthesis, have been widely used because the enzyme retains its activity when up to 30 amino acid residues are removed from its amino terminus and replaced with other polypeptide residues. The ability of the *lac*Z gene to be functionally expressed in yeast has been widely used to investigate regulatory phenomena in yeast by fusing this gene to yeast genes. Yeast plasmids capable of replication in both *E. coli* and yeast and having selectable markers for both organisms have been constructed containing the *lac*Z gene which is missing the first few amino acid codons. In order to be expressed, *lac*Z must be ligated *in vitro* to a segment of DNA that encodes the regulatory signals and amino-terminal region of a gene in the correct reading frame. This is achieved by first manipulating the insert DNA with exonuclease *Bal*31 and synthetic oligonucleotide linkers which allow linkages in different translational reading frames at the restriction endonuclease cleavage site at the boundary of the *lac*Z gene. When introduced into yeast, successful fusions result in the expression of β-galactosidase enzyme activity which can be detected by including a chromogenic substrate in agar plates or by enzyme assay of yeast transformants. This kind of genetic tool has been used to study the mechanisms involved in the regulation of yeast genes and could also be used to indicate and measure the function of exogenously-derived promoters.

Vectors for High Level Gene Expression

Many attempts have been made to isolate particular genomic sequences from other eucaryotic organisms by complementing defined yeast mutants affecting genes common to both organisms. This approach has met with limited success suggesting that there may be many differences in the mechanisms of gene expression between yeast and higher eucaryotes. Specialized yeast vectors are now available, however, which maximize the chances of functional expression of defined heterologous (i.e., non-yeast) eucaryotic genes in yeast by using endogenous yeast promoters. These expression vectors are composed of selectable yeast and bacterial genes, a yeast replication origin (plasmid or chromosomal), and transcription initiation and termination specifying sequences derived from highly expressed yeast genes that flank a restriction endonuclease cleavage site where foreign DNA is inserted. Powerful promoters are derived from the alcohol dehydrogenase I gene (*ADH*1) and the 3-phosphoglycerokinase gene (PGK).

In many gene expression experiments, the coding region of the gene to be expressed is joined to the yeast promoter fragment in such a way that the protein made has no amino terminal extension beyond that of the natural protein. Therefore, it is necessary to remove all

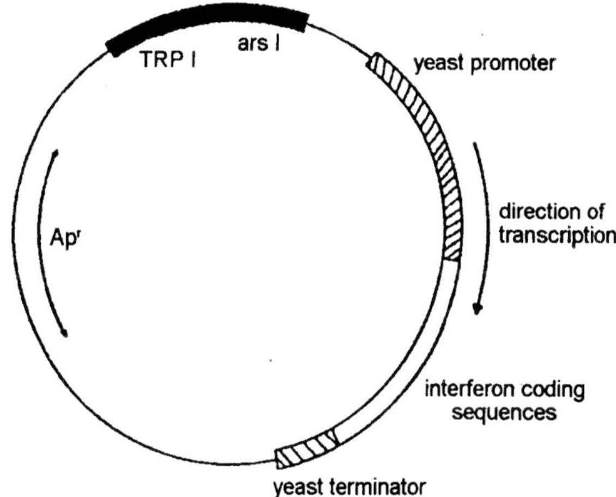

Fig. 8.4. Schematic representation of a plasmid for the expression of an interferon gene in yeast.

the protein coding units, including the initiation codon (ATG), from the yeast promoter fragment, and replace them with a natural or synthetic ATG initiator as part of the coding sequence to be expressed. This is usually done by making a series of deletions with exonuclease *Bal*31 that extend to varying distances into the untranslated leader mRNA of the endogenous promoter, then attaching synthetic linkers and cloning the deleted plasmids. By using this set of deletion plasmids, the coding region for any heterologous protein can be attached to the promoter region with a variety of spacings between the transcription and translation starting points. The different deletion plasmids mostly allow appreciable levels of heterologous gene expression indicating that a large part of the leader region may not be essential for the synthesis of a translatable mRNA.

Expression plasmids have been used for the synthesis in yeast of a number of proteins of biological and commercial importance which will be discussed in greater depth in the next section. There are, however, several requirements necessary for heterologous expression in yeast besides the need for a promoter sequence from yeast to initiate correct transcription. These are:

1. The coding sequences of the protein to be expressed must contain no intervening sequences of the type not processed in yeast. Thus, genomic sequences that are free of intervening sequences or complementary DNA (cDNA) synthesized *in vitro* from the mature mRNA of the gene are used.

2. The untranslated leader sequence just preceding the AUG initiation codon must have a specific sequence to support efficient translation initiation by yeast ribosomes. Highly expressed yeast genes contain only C, A, and an occasional T residue in this region, and the presence of G residues in the mRNA just before the AUG may adversely affect the initiation of translation.
3. The codon usage in all yeast genes which have been sequenced is biased towards a limited number of preferred codons. The general differences in codon usage patterns between different organisms may limit the expression of an heterologous gene in yeast whose codon usage patterns are inappropriate to the yeast intracellular tRNA population.

Industrial Applications

The techniques described here have made possible the genetic engineering of yeast to obtain the expression of eucaryotic gene products having pharmaceutical and biological importance. There are, in fact, several features which make yeast an attractive organism for the large-scale production of biologically-active proteins. Yeast can be produced in large bulk using the technology of the brewing and fermentation industries, enabling a large quantity of the product to be isolated at low cost and in a system giving high purity. Yeast is capable of withstanding hydrostatic pressure and cells do not lyse after death even in cultures grown to high cell density. Furthermore, yeast will be very useful for the synthesis of glycoproteins since the addition of the inner core of carbohydrates occurs in yeast by the same mechanism as in animal cells.

The expression of several types of human interferon has been achieved in yeast using the expression vectors described above. The gene sequence coding for mature leucocyte interferon has been incorporated in extrachromosomally-replicating yeast vectors under the control of endogenous promoters derived from either the *ADH*1 gene or the *PGK* gene. The expression of 6×10^6 units of interferon per liter has been obtained, constituting a potential alternative method for the large-scale production of interferons. Much attention has been focused recently on the production of γ-interferon, which has been reported to have a potential antitumoral activity. Expression of the structural gene for human γ-interferon has now been achieved in several heterologous systems, including yeast. In this system, the mature interferon coding sequence was inserted behind the promoter for *PGK* when up to 10^8 units of protein per liter were obtained, the product having identical properties to authentic human γ-interferon. In addition, the natural secretory pathway of yeast has now been altered to allow the natural secretion into the growth medium of interferons synthesized on expression vectors.

Other foreign genes which have been expressed in yeast using this system are the genes for human serum albumin, human growth hormone, epidermal growth factor and some insulin-like growth factors. Recently, techniques have been described whereby expression vectors have been used to synthesize the surface antigen of Hepatitis B virus in yeast. The protein produced in yeast is assembled into particles very similar to those secreted by human cells. This not only promises the large-scale production of the protein for possible use as a vaccine, but also illustrates the versatility of the yeast cloning system for the production of complex structures.

9

SPECTROPHOTOMETRY

Because of the similarity of function and use, both spectrophotometers and photoelectric colorimeters will be presented here. These instruments that are similar in design with respect to having many identical components can be markedly different in application and versatility. The colorimeter, which utilizes a filter as a monochromator, has proven to be a most rugged and stable instrument requiring very little or specialized usage because of the fixed wavelength of the monochromatic system. This wavelength can be changed only by changing the filter. The spectrophotometers, on the other hand, are more versatile but require much more maintenance. This is because the condition of the light source and the dispersing element is directly related to the quality of the monochromatic light; therefore this system requires closer monitoring and more maintenance. However, the dispersion element enables the selection of a continuously variable source of monochromatic light with either a simple manual or automatic adjustment.

It seems that as the efficiency of the monochromatic system to produce a narrow half-bandpass with a relative high percentage of light transmitted increases, the complexity of the associated optics and light source also increases. As the sensitivity of the detectors increases, the requirements for a more stable power supply and meter system also increase. Because of the interrelation of the aforementioned, it must understood that a good knowledge of the component parts, and their function as a part of that system, is a prerequisite to understanding the system. With a knowledge of these systems, principles of operations, and of the laws that govern the absorption of light, the student is more able to cope with the problems of photometry.

PRINCIPLES

In relation to analytical chemistry, photometry refers to the measurement of the light-transmitting power of a solution in order to determine the concentration of light-absorbing substances present within. Photometry can, of course, be applied to measure the transmission of energy in the ultraviolet, infrared, and visible regions of the radiant energy spectrum. Instruments that are used to measure transmittance at various wavelengths are called spectrophotometers or photoelectric colorimeters, depending upon certain essential differences in construction, particularly the method of producing monochromatic light.

In all such instruments, monochromatic light is passed through an absorbing column of an often coloured solution of a fixed depth and directed upon a photosensitive device which converts the radiant energy into electrical energy. The current produced under these conditions is measured by means of a galvanometer or a sensitive voltmeter.

Absorbance as measured in photometers involves not only the absorbance of the solute in the solution being evaluated, but also all of the molecules of the liquid through which the light passes. It is necessary, therefore, to adjust the instrument by means of a "blank." This blank is prepared by placing in the absorption cell (cuvette) all of the constituents of the unknown solution (solvents, reagents, and so forth), but under conditions that will not permit the colour reaction to take place. The blank solution is used to set the meter of the instrument at a fixed point (100% T or zero absorbance). After adjusting the meter, the cuvette containing the unknown is placed in the instrument and read. By this means, the absorbance of the reagents used can be cancelled. It follows that the reading of the meter with the unknown cuvette in place is a measure of the amount of absorbance of

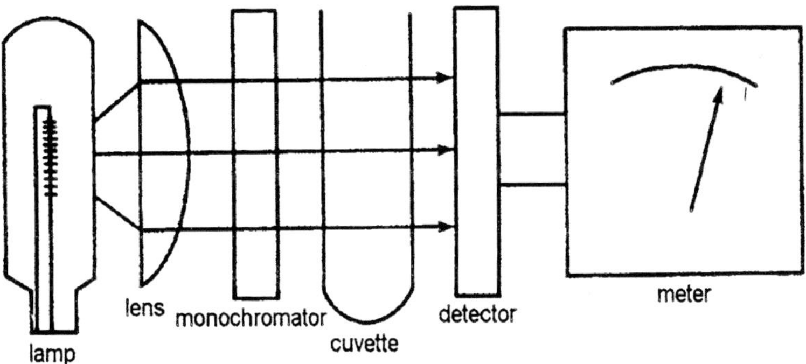

Fig. 9.1. Schematic of single-beam system.

monochromatic light by the unknown. The greater the number of molecules or ions of absorbing substance present, the greater is the absorption of light. In other words, the deeper the colour, the greater is the deflection of the galvanometer from its original setting. Thus, the concentration of absorbing component present in a solution may be accurately measured by a photometer, provided that the necessary monochromatic light is used.

BEER'S LAW

Although the laws of photometry have been referred to as "Beer's Law" (for convenience sake), they are actually laws derived by five different individuals. Bouguer showed (1729) that when light passes through an absorbing body, there is always the same difference between the logarithms of the amount of the light entering and light leaving any section of the same thickness; or, as Lambert expressed it (1760), if one-half the amount of light entering leaves the first section, then $\frac{1}{2} \times \frac{1}{2}$ of the original light will leave the second section. Beer (1852) found that one solution of copper sulfate will absorb the same amount as another if the concentration of the first is twice that of the second and the length of the light path of the first is one half that of the second.

These laws then state that light, in passing through a coloured medium, is absorbed in direct proportion to the amount of the coloured substance in the light path. Thus, the strength of the observed "colour" is directly proportional to the concentration of the absorbing chromagen in solution.

If we let P equal the transmitted energy and Po the incident energy, then the ratio P/Po represents the transmittance (T) of the absorbing material. If the material does not absorb at all, then P and Po are the same, and P/Po=T, and %T=100. By using the negative logarithm of T, measurements are transformed to the energy absorbed 1/T. The equation now becomes A = ¾log T; since 2 is the log of 100, the % T formula becomes A=2¾log %T. In practice, we adjust the photometer to read 100% T with a cuvette containing a blank solution. We then substitute a standard or unknown solution and read the meter. Thus, we may use % T and semilog paper or A with regular (Cartesian) graph paper to get a straight line when plotting against concentration as the abscissa. When a straight line is obtained on one of the above graphs, we may also calculate the unknown from the formula: Concentration of the unknown equals A of unknown divided by A of standard times the concentration of the standard.

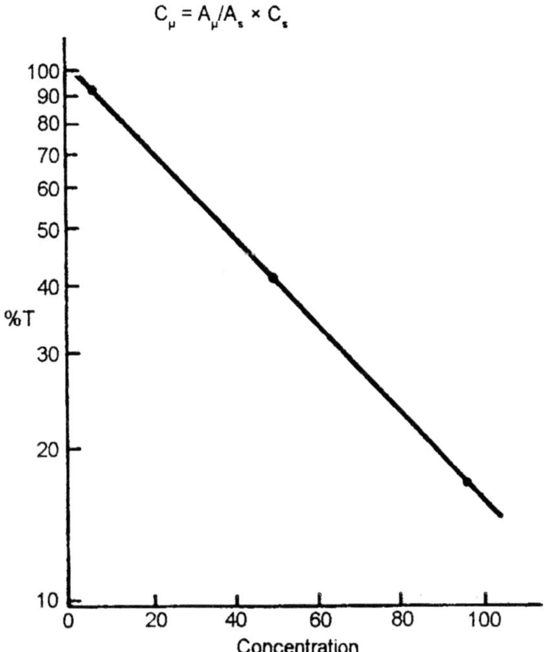

Fig. 9.2. Calibration curve: %T vs concentration plotted on semilog paper.

COMPONENT PARTS

Power Supplies

The commercial power supply is adequate for most general uses and the usual fluctuations of voltage and current do not significantly upset the equilibrium of lamps, heaters, or motors.

However, no stable operation of light-measuring instruments is possible without better control of the power source. Therefore, a power supply capable of furnishing adequately regulated electrical energy for the particular need of the instrument must be utilized. These power supplies fall into three general categories: batteries, voltage-regulating transformers (Sola), and electronic power supplies.

Batteries, either wet cell or dry cell, produce quite stable voltages and are relatively inexpensive. Dry cell batteries are mainly used for the operation of some small photometers. Wheatstone bridges, and radiation detectors, or wherever portability is necessary and a low direct current supply can be used.

The wet cell (lead-acid) battery, as used in automobiles, also produces a fairly stable current supply. However, it is also limited to

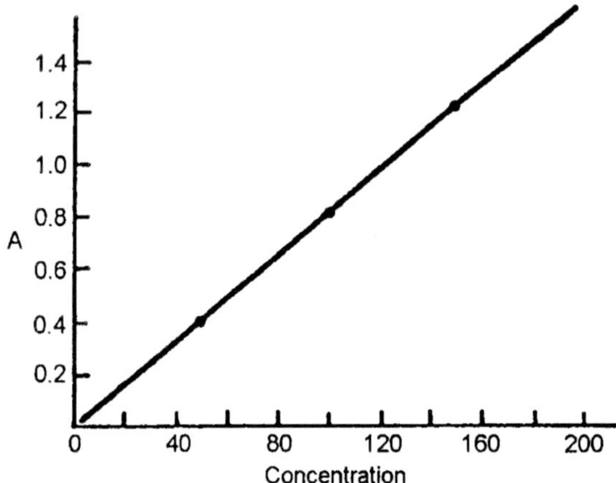

Fig. 9.3. Calibration curve: A vs concentration.

being a source of low voltage and direct current. In addition, it requires regular maintenance consisting of frequent recharging and addition of water. After charging, the battery must go through a short period of discharging before stable current is obtained.

The constant-voltage transformers (such as Sola) will regulate power output very closely as long as the input remains between 95 and 125V and at 60 Hz (cps). These devices are essentially trouble-free and have no moving parts.

The electronic power supplies will give more precise regulation than can be easily obtained from the "Sole-type" transformer. Situations may be encountered where frequency variations must be taken into account. This necessitates the use of an electronic device to regulate and supply the desired voltage. These devices are very complex and quite expensive. They utilize either vacuum tube or solid-state regulators. The latter are becoming much more common and it is claimed that they are more dependable. However, it should be remembered that just about anyone can change a vacuum tube, but no inexperienced person should even try to tinker with solid-state components.

Radiant Energy Sources

The function of the light source is to provide incident light of sufficient intensity for measurement. For work in the visible, near infrared, and near ultraviolet regions, the most common source is the glass-enclosed, tungsten filament, incandescent lamp. These lamps with prefocused bases are most useful with respect to easy replacement in an optical system. However, when used with a grating or prism monochromator, wavelength calibration must be checked when lamps are changed. Although these lamps produce a continuous spectrum over a wide range, they do have some shortcomings. Most of their energy is emitted in the near-infared region of the spectrum, about 15% in the visible, when operated normally. Operating temperatures can be increased to produce a greater percentage of short-wavelength energy, but this drastically shortens lamp life.

For work in the ultraviolet region, the high-pressure hydrogen (or deuterium) discharge lamp is adequate from 200 nm to an upper limit extending to about 375 nm. At longer wavelengths the emission is no longer continuous. When deuterium is used instead of hydrogen, the light intensity is tripled.

For very high levels of ultraviolet illumination, the xenon arc or high-pressure mercury vapour lamp provide a large amount of continuous radiation plus high energies at the spectral lines of these elements. These lamps become very hot in operation and may even require thermal insulation, with or without auxiliary cooling, to protect the surrounding components.

Although these special lamps are available and can be used, they usually require special power supplies and mountings. When planning to use these types of energy sources, it is better to purchase the equipment as a unit so that the light source will be properly set up and the necessary insulation will be in place.

Monochromator

When measuring the absorption or emission of radiant energy by an absorbing solution, it is necessary to be able to isolate the desired wavelength of that energy and exclude the rest. In other words, by restricting the band of wavelengths passing through the sample to those absorbed by the substance of interest, the sensitivity of the instrumental measurements to concentration changes is greatly enhanced. Thus the important characteristics of a monochromator (dispersing device) are its bandpass width, the nominal wavelength, and peak transmittance.

There are numerous ways of isolating the desired spectrum. The simplest device is a *filter* (glass, Wratten, or interference) placed just in front of the sample holder. The usual glass and Wratten filters are of relative wide bandwidths and low peak emission. Since the function of the filter is to absorb unwanted energy, the dissipation of the heat produced in the filter must be considered.

Narrower bandwidths are obtained with *interference filters*. This type consists of an evaporated coating of a transparent dielectric spacer

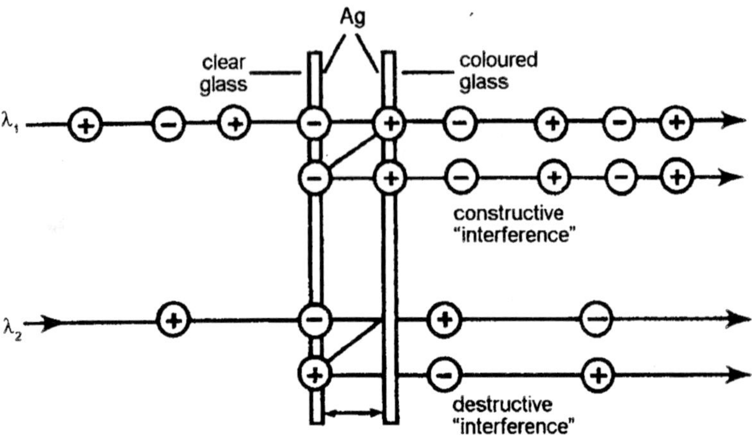

Fig. 9.4. Interference filter.

of low refractive index sandwiched between semitransparent silver films. Sharp cutoff filters are placed on each side of these films to eliminate other than first-order effects. These filters have a bandwidth of 10-17 nm and a peak transmittance of 40-60%.

Multilayer interference filters consist of successive layers of high and low refractive index dielectrics on alternating layers. These filters are characterized by a bandpass width of 8 nm or less and a peak transmittance of 60-90%.

These multilayer interference filters will transmit only light with a wavelength two times the distance between the silver films (in phase), that is, with complete constructive interference. All other wavelengths will show partial or complete destructive interference as the reflected light rays are retransmitted out of phase with those of the same wavelength which are transmitted directly.

These interference filters can be used with high-intensity light sources since they remove unwanted radiation by transmission and reflection and not by absorption.

Many instruments use *prisms* or *diffraction gratings* as monochromators. These devices separate the various wavelengths of radiant energy as produced by a tungsten lamp by refraction or diffraction, and present them as a spectrum from which the desired wavelengths may be selected.

The action of a prism in dispersing a polychromatic beam of radiation into a spectrum depends on the variation of the index of refraction with wavelength. As used in spectrometers, the light from

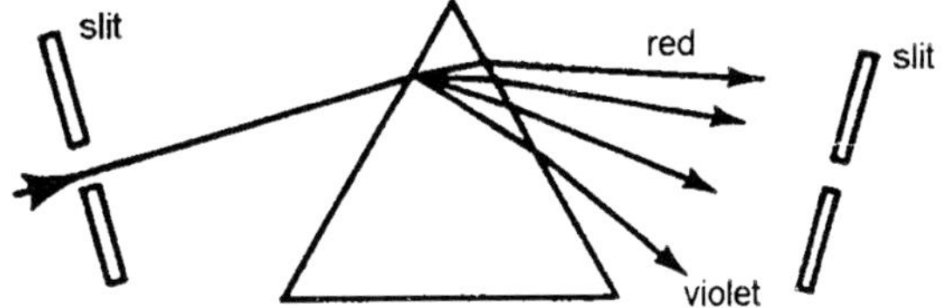

Fig. 9.5. Light dispersion by prism.

the source is directed through a convergent lens into an entrance slit at the focal point of the lens, then through the prism and a second convergent lens. At an air-prism interface an entering light ray at an angle of incidence will be bent toward the vertical to the surface and, at the prism-air interface, it is bent away from the vertical. The image of the entrances slit is projected onto the exit slit as a series of coloured images arranged next to each other. Violet light is refracted (bent) farther than red.

The actual separation between two wavelengths depends on the dispersive power of the prism and the apical angle of the prism. A nonlinear wavelength scale results. The longer wavelengths are not refracted as much and thus are crowded. Dispersion by glass is about three times that of quartz. Quartz or fused silica is mandatory for inclusion of the spectra below 350 nm.

Diffraction gratings, on the other hand, produce a linear noncurved spectrum. A diffraction grating is based on the principle that light rays of radiant energy will bend around a sharp corner, the degree of bending depending on the wavelength. Thus, with a beam of energy containing many wavelengths striking a grating, many tiny spectra are produced, one for each line of the grating. As the wavelengths move past the corners, wavefronts are formed. Where these cross, those that are in phase reinforce one another, and those that are not cancel out and disappear, thus leaving a complete spectrum to display upon the exist slit, much the same as with the prisms. Some grating monochromators have a half-bandwidth as high as 35 nm, while other more expensive units have half-band-widths of less than 0.5 nm.

Despite the good dispersive qualities and the linear spectrum presentation, diffraction gratings have problems with stray light. The stray light may be defined as radiant energy at unwanted wavelengths reaching the detector. One problem is slight imperfections in the ruling. Another is caused by second-order effects of the grating. Part of this trouble can be eliminated by using a double monochromator or by a special mounting such as the Ebert system.

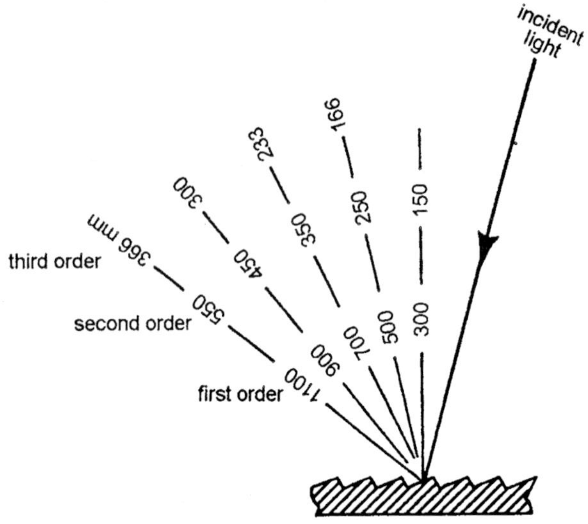

Fig. 9.6. Light dispersion by diffraction grating, showing possible second and third order interference.

Sample Cell (Cuvette)

Absorption spectrophotometry usually evaluates the absorption of a solute in a liquid solvent. The square or rectangular absorption cells have plane parallel faces and a light path of constant length. They are free of optical aberrations. Sets of cuvettes may be matched to very close tolerances. This type of cuvette is rather expensive and is usually utilized in the more expensive instruments. For ultraviolet work (below 340 nm) it is mandatory that this type of cell, constructed from silica or quartz, be used.

For most work in the visual range, the cylindrical test tube type is sufficiently accurate. These cuvettes are prone to variable reflection and refraction errors as well as lens effect. Glass tubing (from which cuvettes are made) is rarely round and is not polished, so three are considerably more surface irregularities. For this reason, round tube-typed absorption cells require close calibration and segregation into matched groups. Because these cells are generally somewhat oval, it is necessary that they be marked in front for proper light-path orientation.

Another solution to this problem would be to use a flow-through cuvette. Since the same cuvette is in the light path at all times, cuvette errors would be compensated for by the blank. Highly reproducible results can be obtained, provided the cuvette does not become dirty and no bubbles are introduced and/or trapped in the cell. A source of vacuum is needed to empty the cell. Filling is often a very real problem because of the small tops. This may lead to spillage into the instrument and resultant corrosion.

Associated Optics

The range of transmittance of materials for construction of windows and lenses is a critical factor. The absorbance of any material should be less than 0.2 at the wavelength of use. Ordinary silica glass transmits satisfactorily from 350-3000 nm. Special Corex glass will extend the ultraviolet range to about 300 nm. Quartz or fused silica must be used below this.

Beam reduction is accomplished by condensers that can reduce the beam size by a factor of 25 without loss of appreciable energy.

In colorimeters and economically priced spectrophotometers, simple lenses are used to focus or collimate the light beams. In the more expensive instruments, front-surfaced mirrors are used to reduce energy loss. These mirrors are aluminized on their front surfaces are other metallic surfaces show selective absorption at certain wavelengths. The surfaces of these mirrors are coated with a thin film of magnesium fluoride to reduce light scattering.

Detectors

Any photosensitive device may be of use as a detector of radiant energy, provided that it has a linear response in the part of the spectrum to be used and is sensitive enough for the task at hand. These devices must first convert electromagnetic energy to a different type of energy, namely electrical energy. The electrical energy produced can then be measured.

The most common of these is the photocell. These require no external voltage source, but rely on internal electron transfers to produce a current in an external circuit. These cells are composed of an iron back plate and a layer of crystalline selenium or cadmium on one surface. Electrical contacts are made with the iron plate and an electrical conductive transparent film on the front active layer. The cells are then sealed to prevent physical or chemical damage.

A photon striking one of the selenium or cadmium atoms transfers its energy to an electron and raises it into a conduction band. This electron now travels from the front to the back of the photocell and through the external circuit for measurement. These cells are simple, quite inexpensive, and in general are very dependable. However, they

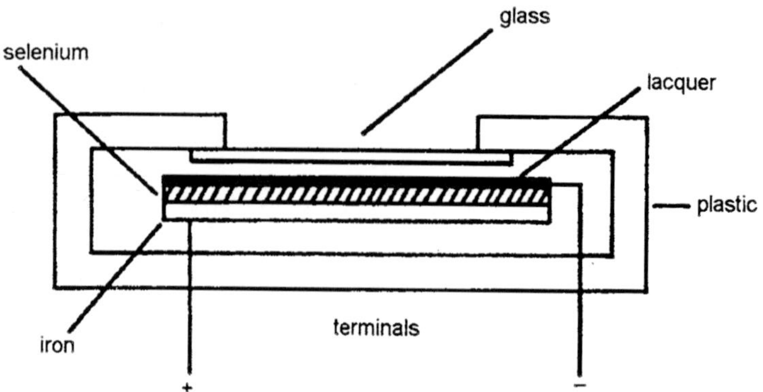

Fig. 9.7. Photocell.

have certain faults that should be understood. These cells can become "light blinded," a condition similar to that of the human eye. They need time to rest. Their response is temperature sensitive, requiring some warm-up time for temperature stability. Their low internal resistance and low output of electrical energy are not easily amplified.

Phototubes are constructed of a negatively charged cathode and a wirelike positively charged anode. The cathode is coated with a photoemissive substance such as cesium oxide. When a photon strikes their layer, electrons are emitted and jump through the vacuum over to the anode, where they are collected and return via the external circuit.

The output from these tubes can be amplified and then fed into external circuits for greater sensitivity. For this reason, this type of detector is used on all precise instruments that have a close restriction on the slit and thereby the wavelength presentation.

The photomultiplier tube combines photocathode emission with multiple cassade stages of electron amplification of primary photocurrent within the tube itself. The tube is constructed so that the primary photoelectrons from cathode are attracted and accelerated to several succeeding dynodes. These dynodes are constructed of a material that will give off several secondary electrons when hit by other high-energy electrons. Thus the photomultiplier tube can measure light intensities about 200 times weaker than the ordinary phototube.

Readout Devices

A galvanometer is an instrument used to detect or measure currents. A flat suspended coil lies in the plane between the poles of a permanent magnet. When the current to be detected or compared passes through the coil, it sets up a magnetic field whose poles are at the front and back, or 90° away from the permanent magnet poles. In trying to set its lines of force in line with those of the magnet, the coil turns as far as the magnetic forces can twist the wire. The stronger the current, the stronger the magnetic field and the farther the coil turns. The iron core makes a uniform filed for the coil to turn in; a beam of light reflected from the mirror on the coil serves as a pointer.

When an arbitrary readout device, such as percent scale or optical density scale, is incorporated into this system, we have what is called a *direct readout meter*. Actually, this readout could be in millivolts, milliamps, mg%, mEq/liter, or any other arbitrary unitage, provided the scales take into consideration the linear relation of milliamps and % T and the log relationship between millivolts and absorbance. Therefore, this type of readout is generally considered to be the fastest, simplest, and the easiest to use.

If, on the other hand, an electrical force governed by the action of a variable resister (potentiometer) is used to bring the meter to

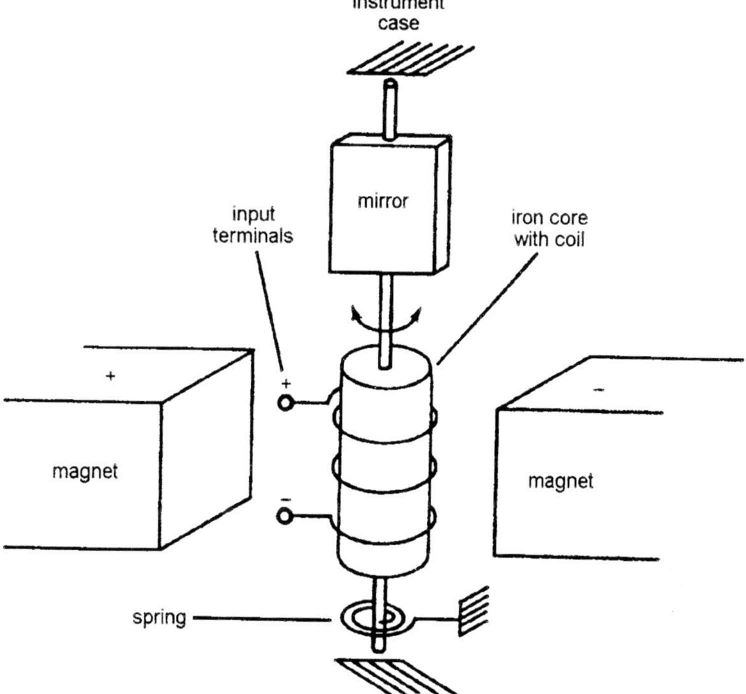

Fig. 9.8. Galvanometer schematic with terminals.

null point, the system is said to be a *null-point system*. In this case, an arbitrary scale is fitted to the potentiometer scale. This scale may carry the same arbitrary type of unitage as the first.

This slide wire potentiometer may also be connected to a servomotor of a digital readout system or of a recorder.

Types of Instruments

Although the filter photometer is not to be considered as a spectrophotometer, because it does not give a continuous source of monochromatic radiant energy, they must be considered together when it comes to function and system design. Colorimeters that use the regular glass filter will also use a simple low-energy light source. In fact, this type of filter dictates the use of this type of source, as well as eliminating the need for even a single lens. Therefore, the simplest system would require a power supply, light source, filter, sample holder, detector, and a meter in that order. However, when an interference filter, prism, or diffraction grating is use, a more elaborate system is necessary. These types of monochromators require that radiant energy be focused into the system in parallel lines. Diffraction gratings and

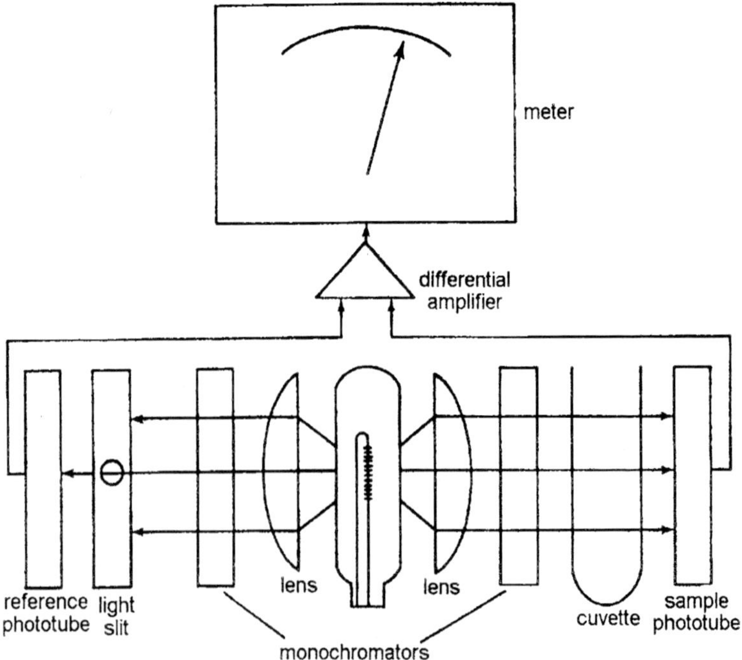

Fig. 9.9. Double-beam system with optics.

prisms also require the use of entrance and exit slits to *limit* the spectrum and exclude stray light. In fact, most grating monochromators require that sharp cutoff filters be placed at the entrance to eliminate second-order spectrum. When you add to this a sample holder, a detector, and a meter, you have then a single-beam spectrophotometer.

In a double-beam system monochromatic light from either a single or double monochromator is focused through both a reference and a sample compartment. The intensity of these two beams of light is then measured by either one or two detectors, and the sample beam is compared to the reference side as a ratio. This ratio may then be fed into a meter or directly into a ratio recording and a beam splitting and reuniting system of choppers to overcome the problem of balancing two phototubes to an exactly equal spectral response.

Instrument Quality Control

The most insidious problems of instrument malfunction have to do with the certainty of wavelength, bandwidth, linearity, stray light, and photometric accuracy. These terms probably mean very little to the laboratory workers who rely completely on the colorimeters to perform, without a question as to how well they did or could function; but quality in instrument performance cannot be assumed, it must be proven. The monitoring of instrument performance is even more important in the spectrophotometer. This is especially true in the narrow bandpass units. These systems with their greater potential accuracy and sensitivity demand closer attention to their functional performance.

Electrical malfunction will not be considered here, even though it is a part of instrument quality control. This type of problem is usually covered in detail by the instrument handbook and requires a specially trained individual to make repairs.

Wavelength Calibration

Wavelength calibration is a means of insuring that the radiant energy being emitted from the exist slit of the monochromator is the same as that specified by the wavelength selector and within the tolerances of the instrument. This simple check should be made whenever a new lamp is installed and routinely thereafter. This is to insure that the slight changes in lamp filament position during usage are corrected and do not lead to gross analytical errors. Minor errors in wavelength settings are of relatively little importance in wide bandpass spectrophotometers, and these units may be checked less often. Wavelength calibration of the more sophisticated narrow bandpass instruments are of the utmost importance, and no scan of any substance is worth doing if the monochromator is not in calibration.

Didymium and holmium oxide glass filters have been used for many years on the general laboratory instruments. More recently, the combination of the didymium glass filter and an acid solution of nickel sulfate has been used. This solution gives a peak transmission at 510 nm. A solution of cobalt chloride has a peak absorbance at this wavelength. This solution may be helpful as a secondary check on the nickel sulfate when it is not in agreement with the didymium filter.

Precision instruments require a more precise means of wavelength calibration. The emission bands of mercury arc or deuterium lamps are good sources of radiant energy with specific emission spectra. The quartz mercury arc lamp, with its multiple emission bands, provides the best source for accurate wavelength calibration from 205—1014 nm. The deuterium lamp in ultraviolet instruments has two major emission lines, 486 and 656.2 nm, that provide a good source of wavelength calibration in the visual range. This lamp provides the most practical means of wavelength calibration of a grating instrument, for only two lines are needed when light dispersion is linear.

Stray Light

Stray light is any radiant energy measured by the detector that is outside the spectral region isolated by the monochromator of the instrument. It can be caused by dirty optics, poor baffling, or faulty grating in the monochromator, as well as fluorescence of the sample itself. Because stray light may be independent of sample concentration, its relative effect is usually greatest at low transmittance, where instrument errors are also greater.

The American Society of Testing Material E-13 Committee Standards have been adopted for the measurement of stray radiant energy in spectrophotometers and provides a good means of instrument evaluation. However, for ease of understanding and routine checking, a common sense approach may be more practical. Scattered light can be of any wavelength, and in grating instruments it may also be of second-order reflection. This may also be true of old and poorly constructed interference filters. Thus, this light may well be of a wavelength far removed from that selected and will not be absorbed by the test solution. From this information, it is easy to reason that scattered light can cause nonlinearity and insensitivity.

Sharp cutoff filters may be used to directly assess the amount of stray light at the extremes of the instrument range. These filters have the unique ability to stop all light above or below a certain range. Thus, by the proper selection of filters, it is possible to check the apparent stray light of any instrument. This may be done by setting the wavelength selector at the desired wavelength (400 or 700 nm) and then adjusting the instrument meter to read 100% transmission. Place a filter in the sample compartment and not the meter reading. This reading is an indication of stray light and must be considered as an instrumental error.

If one is performing time rate analyses, one of the simplest checks to make is to measure the absorbance of NADH at 340 nm with the tungsten and deuterium light source. If the absorbance of the solution is greater with the deuterium lamp, stray light was present when the tungsten lamp was used and this has been corrected by employing the deuterium radiation source. This is because the deuterium lamp is a discontinuous radiant energy source in the visual range, and therefore does not emit radiation at 680 nm, the wavelength of light that is of primary importance in second-order effect. Also very little energy is emitted by this lamp above 400 nm, so that scattered-light problems are greatly diminished when working in the ultraviolet range.

Nickel sulfate solution has also been used for this purpose over the visual range. This solution shows maximum absorption at 400 and 700 nm, and under certain circumstances no light should be transmitted at these wavelengths. Any deviation from this on repeated checks is an indication of stray light.

Photometric Accuracy

Provided that the stray-light checks is negligible, the linearity check tests the ability of the photocell to produce a signal proportional to the light intensity and of the instrument's meter system to measure this signal accurately. A linearity check can be made by reading the absorbance of standard solutions of various salt solutions that have major absorption peaks at certain wavelengths or by using neutral density filters that absorb energy over a broad range.

Solutions of potassium chromate and potassium dichromate have been used for evaluating photometric accuracy in the ultraviolet range. A solution of 0.04 g/liter of potassium chromate dissolved in 0.05N potassium hydroxide will show maximum absorbance at 273 and 373 nm. The absorbance range reported for this solution is 0.189—0.199A and 0.247—0.249A, respectively, for these two wavelengths. When measuring absorbance below 260 nm, this solution should not be more than 6 months old. This may be the reason for the wide range in absorbance values at 273 nm. A solution of potassium dichromate,0.05 g/liter dissolved in 0.01N sulfuric acid, also exhibits two absorbance peaks that may be used. These are at 257 and 350 nm and should read 27.6% transmission at the latter wavelength. Absorbance patterns of this solution are varied by changes in acidity: therefore, the pH should be carefully controlled. Cobalt ammonium sulfate and Thompson's solution have been used for this evaluation in the visual range. These solutions require great care in preparation and handling. The absorbance cells used must be cleaned and matched before use. Although these materials are relatively stable, they may react with contaminants and deteriorate. The solvent solution must also be checked for absorbance in the ultraviolet range when these solutions are to be analyzed. Frings, et. al., has proposed using French's green food colouring for monitoring detector response at three wavelengths. This dye solution has absorbance maxima at 257 nm, 410 nm, and 630 nm. The advantage of this solution is that it provides a means of response verification in both visual and ultraviolet ranges with one solution at a very modest cost.

The National Bureau of Standards developed a set of neutral density glass filters for the specific purpose of checking the photometric scale of spectrophotometers. These filters are premounted in holders ready for use in the standard 10-mm cell compartment. They are not readily affected by temperature change nor do they require a critical wavelength adjustment. These standard reference materials (SRM) come as a set of three individual filters calibrated and certified to ± 0.5% transmittance over four different wavelengths of the visual range. These filters were the forerunners of other sets manufactured by Bausch and Lomb as well as Chemetrics Corporation. All of the later filter sets have a filter that will fit a standard 10-mm cell compartment. These sets provide a means of checking for stray light, bandpass, and wavelength as well as photometric accuracy.

10

BIONANOMACHINES

The most basic theory relevant to chemistry is called *quantum electrodynamics*, or *QED*. In simple terms, QED tells us how photons and electrically charged particles, like electrons or atomic nuclei, behave. Physicists, of course, are interested in deeper stuff like how the nucleus behaves internally, but we can get along just fine without that. In fact, we can get along fine without a lot of QED itself. This is good, because calculations using it are quite difficult. It would be a bit like using the theory of relativity to calculate where a baseball is going. There is no practical difference between the results of the much simpler calculation using *Isaac Newton's laws* of motion and the more difficult relativistic one. In the same way, for purposes of chemical interactions, we can almost always get away with using the *Schrodinger equation*. This still captures most of the quantum effects.

The quantum world is strange and hard to understand. Our instincts about the way things work are formed in the larger world, where objects exist, and waves have to be waves in something, like sound in air, or at the surface of water. In the quantum world, both of these descriptions apply to some degree to everything. Something can seem to be an object, like an electron, but can act as if it were in many places, or moving in many different directions, or had many different energies at once. This odd property is called superposition of states and is the thing that will let quantum computers be able to calculate many different things at the same time, thus being vastly more powerful than ordinary computers.

By the same token, when you take an ordinary computer and try to simulate something in the quantum world, the calculations can easily get out of hand. You have to keep track of all the superposed states and the interactions between all the particles. The amount of work blows up exponentially with the number of particles. Therefore, even when doing the quantum mechanics of a chemical interaction, the analysis is limited to a relatively small number of atoms, and something called the *Born-Oppenheimer approximation* is used. This approximation treats the nuclei as if they were ordinary objects and calculates the electrons quantum mechanically around them. Nuclei, it turns out, are heavy enough to act like ordinary objects most of the time.

Even so, a study such as that done by Robert Freitas and Ralph Merkle of the reactions involved in depositing carbon atoms to build up diamond machine parts required 50,000 CPU-hours of calculation.

We know that the Born-Oppenheimer approximation doesn't always hold. Physicists can do experiments that cause superposition with entire atoms being in two places at once, or meld together in overlapping states called *Bose-Einstein condensates*. But this is where the distinction between science and engineering is important. A can design machines that don't do those experiments, and only do the things we can analyze (relatively) straightforwardly instead. We know that atoms act like objects under most normal conditions—we can image them with scanning probes, push them around, put them somewhere, and find them still there when we look back.

Molecular Mechanics

If you are an automotive engineer, you need to analyze the chemistry in the fuel-air explosions in the cylinders of your motor, but you don't need to simulate the chemistry inside your structural steel to know that it remains steel from moment to moment. You use a higher-level approximation that involves weight, tensile strength, coefficients of elasticity, and so forth. The higher-level approximation that is appropriate for the parts of nanomachines where chemistry is not being done is called *molecular mechanics*.

Molecular mechanics treats atoms as objects and condenses the descriptions of what the electrons do around and between their into a number of simpler forces. First is the *van der Waals force*, which is a small attraction between atoms up to a point, and a larger force resisting their being pushed together after that. The van der Waals force makes atoms act like objects, somewhat soft and springy, but with a definite size.

Next are the *ionic forces*, which occur when atoms gain or lose electrons (or part of one, in a molecule). The resulting electric charge

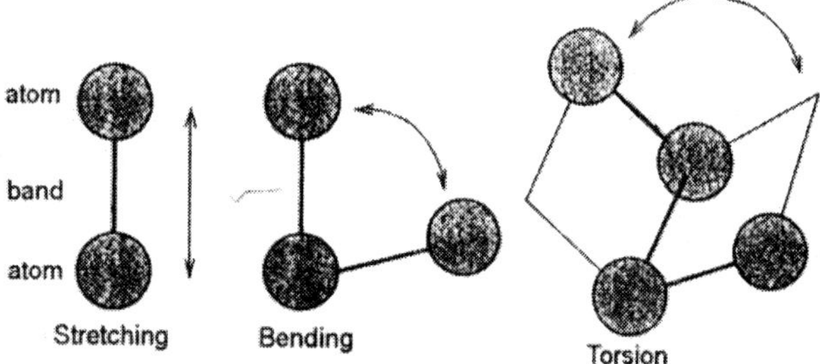

Fig. 10.1. Molecular mechanics. The forces along and between covalent bonds that are accounted for in a typical molecular mechanics force field.

causes a longer-range attraction or repulsion between atoms. For example, ordinary table salt is a crystal of sodium and chlorine ions in which the sodium have lost electrons and the chlorines have gained them. (*Ion* simply means an atom that has lost or gained one or more electrons.) The electrostatic forces that result make the atoms form a very strictly alternating pattern in the crystal. In the *water molecule*, H_2O the oxygen atom pulls the electron from each hydrogen in dose, so that the hydrogen's nucleus (a single proton) is left sticking out a little. The result is that the hydrogen act positively charged and the oxygen acts negatively charged, though not as strongly as if the electron had been pulled all the way off. Thus water molecules stick to each other, hydrogen to oxygen, and stick to the ions of salt, which is why salt dissolves so well in water.

Finally comes the force associated with a *covalent bond*. This is where, instead of one atom grabbing the electron and the other letting go, they both hold on more or less equally. (Actually, the electrons are shared in pairs.) A covalent bond acts as if there were a strong spring between the atoms, holding them much closer together than the van der Waals force would let them, because they actually overlap where the shared electrons are.

Covalent bonds are much stronger than ionic forces, and ionic forces in turn are much stronger than van der Waals ones. Diamond, for instance, is a crystal held together with covalent bonds, and salt is one held together ionically. For a van der Waals crystal, you'd have to look at solid xenon, which melts at −169° Fahrenheit. Stronger solids are held together by van der Waals forces between molecules, but they get help by having the molecules themselves held together by covalent bonds. Ordinary ice is a crystal held together by the weaker ionic bonds between the hydrogen of one molecule and the oxygen of others (which are often called *hydrogen bonds*).

Molecular mechanics describes what goes on in and between molecules at this level of detail. A particular set of parameters describing the strengths and behaviours of the forces and bonds in molecular mechanics is commonly called a force field or sometimes a potential, short for "*potential energy*." (The potential energy of two atoms with a force between them varies with their distance, just as the potential energy of a heavy object increases with its height above the ground.) Potential and force are two sides of the same coin; force is just the derivative (steepness of slope) of potential. In studying the same molecule, a chemist is likely to be more interested in the potential and a mechanical engineer more concerned with the force.

Van der Waals and ionic forces are pair wise and radial; that is, they cause forces between two atoms to push them directly toward or away from each other. Covalent bonds are more complex. They have a radial, pairwise component that is typically called "*stretching*."

Then a component called "*bending*" depends on the angle between two bonds on a given atom. A component called "*torsion*" depends on the angle between the planes formed by three bonds (connecting four atom). The actual magnitude of the force depends on which elements ire involved, and sometimes on what other elements they are bonded to. For example, the length of a carbon-carbon bond depends on whether the carbon is bonded to three other carbons, as in graphite, or four carbons, as in diamond. Various formulations of force fields have other bells and whistles, but these are the main features.

Most force fields were developed by chemists for different reasons, using different methods to arrive at their parameters. Many are line-tuned to work best in the aqueous environment of a cell's interior and are used for analyzing DNA, proteins, drugs, and so on. Others are optimized for the study of crystals, metals, plastics, and so forth. The reason tuning matters is that the force field is, after all, an approximation to the true quantum mechanical situation, and in a different situation a different approximation may be a little closer.

To a chemist, "a little closer" makes a big difference. For example, a very small difference in the potential of a protein folded into shape A as compared with shape B can mean that in the cell or test tube, virtually all the protein molecules will be in shape A and almost none in shape B. Protein molecules are floppy, with many places where just one bond joins one part of the molecule and the next. The amount of energy necessary to cause that bond to turn one way rather than another is very small. The nanomechanical engineer, however, can use the difference between science and engineering to good effect. We can design stiff molecules where there is at least a tripod of bonds holding each atom in place. The amount of energy necessary to change the shape of such a molecule is much greater. This means that if there is a small inaccuracy in the force field, it hardly changes the shape of the molecule at all, where it would make the protein completely different.

For example, the molecular mechanical engineer can use, for preliminary work, a force field that only describes bond stretching and bending. *Bond torsion* is a correction of about 1%, and thus can be ignored for a first approximation. Any molecule where bond torsion would make a significant difference is nowhere near stiff enough for use as a molecular machine part. In contrast, solution chemists often use an approximation where the bond stretch is not modeled, but an average length used as if it were completely rigid. They can do this because in solution, there are no forces strong enough to change the average length significantly.

Imagine simulating a steel chain. If you are a chemist, you are trying to predict what shape of a heap it will make when you drop it on the floor. You need to be very exact about how much friction there is between links, how they bounce off one another as they clatter down, and so forth. You can assume each steel link will remain the same size and shape, though, and worry only about the smaller forces that move them around. If you are a molecular engineer, however, you are simulating the chain being used to lift a ten-ton weight. You can bet that the links will hang in a straight line from the winch to the weight; but you had better be concerned about how they stretch and deform as the weight nears their holding capacity.

The molecular mechanics force held, together with the description of the particular molecule you are interested in, lets you calculate a lot of things about it. It will tell you what shape it will take, and how hard it will be to bend, and what its vibrational frequencies will be. The other thing you can do is simulate it in what is called molecular dynamics.

In molecular dynamics, you take a molecule or a set of them, and give each atom a velocity, chosen at random, to simulate the kind of velocities they'd have due to heat at, say, room temperature. Then you use the force held to calculate the force on each one, and figure where it would have moved to a femtosecond or some traction thereof later. Now make a movie, doing this entire calculation once per frame.

Watching this movie lets you see if the machine you designed does what you thought it would. Molecular dynamics is peculiarly suited to simulating nanomachines as contrasted with many of the other things, like protein activity, that it is used for. The reason is that things happen much faster in a nanomachine, and the glimpse you get with a simulation is brief.

A femtosecond is a unimaginably short amount of time. If you made a movie in which each successive frame represented a femtosecond of simulated time, it would take a million years to watch one second of simulated time. The speed of sound in diamond is faster than ten miles per second, though, so it would only take two seconds of viewing time for a vibration to move across a one nanometer part.

Most existing molecular dynamics software, and the force fields it implements, is aimed at chemists and materials scientists. That is, it typically treats the group of atoms it is simulating as a sample of a much larger amount of material, and it is used to determine things like heat capacity, pressure-volume-temperature relationships, and so forth. For simulating nanomachines, however, we want to treat molecules as machine parts and measure such things as stiffness, torque, and power. New molecular mechanics and dynamics software is being developed specifically for the purpose of simulating and analyzing mechanical nanomachines.

Nanodesign

Nanomachines are going to be among the most complex things that humans have ever designed. Grains of beach sand range in size from about 100 microns to about a millimeter. A nanomachine the size of a small grain of sand might easily contain five trillion parts, many of which would be moving parts. If you spread the parts out on a table to get a look at them before trying to put the machine together, and you magnified them only to the size of grains of sand, your table would have to be bigger than 500 square feet. If you magnified them to the size of typical machine parts in a car, the table would have to be bigger than 100 acres.

VLSI (very large scale integration) chips such as microprocessors are among today's most complex designed machines, with millions of working parts. Even these are much too complex to design by hand, even for a large, well-organized engineering team. Instead the team relies heavily on design and simulation software. There are a number of strong parallels between microchips and nanomachines. First, you can simulate them at many different levels of detail and accuracy, with more accuracy being more computationally expensive. On a chip, you can simulate the transistors and wires as three-dimensional pieces of material and simulate the electron density flowing through them. Once you have transistors that work the way you want, you tend to have simpler models for them and the wires. You can simulate the generation of electromagnetic waves by the flowing current or ignore it. You can treat the transistors as simple on/off switches, and either simulate or not the delay times for signals to move through them. You can simulate just the logic, where small groups of transistors are represented as Boolean functions such as AND and OR, the primitive one-bit operations of which the computer logic is built at this ultra simplified level, a microprocessor is fiendishly complicated.

Each of these levels of simulation will properly represent the machine if the science/engineering distinction is observed. In other words, each level of simulation makes some assumptions about the design at the levels below it and only gives valid results if those assumptions are true. It is up to the designer to make sure that they are.

In electronics design, the assumptions are ensured in a couple of ways. First, designers often make use of predesigned parts that have been carefully analyzed at the lower levels. For example, someone doing logic design will use logic gates that have been carefully checked out as electronic circuits to have guaranteed limits on signal rise and fall times, propagation delay, and so forth. Then the logic designer need only be concerned with the Boolean logic and not have to worry about electronic engineering as well.

Second, the CA software that designers use contains subsystems called *design-rules checkers*. These are very similar to the spelling checkers that are of text—editing programs. If you are drawing a circuit as a direct on your screen, a design rule may say you cannot put two wires closer than some given distance apart. You do not need to know the low—level analysis that showed inductive coupling if they were too close, or that the diffraction limit of the fabrication process just won't let you actually build them closer. All you need to know is that when you draw them too close, they get outlined in red boxes and you must move them apart.

With libraries of prepackaged modular parts and automatic design rules checking, you can be confident that circuits designed at quite an abstract level will work as intended. Exactly the same techniques will be used to design nanomachines.

Today, when people design physical mechanical machines, they pay a lot of attention to each part, optimizing function, manufacturability, and cost. This is very much like how integrated circuits were designed in the 1960s and 1970s. There was enormous attention to detail, and lots of analysis of the effects of each part on every other part. Then circuits began getting more and more complex, as integrated circuits (ICs) became large scale integration (LSI), which became very large scale integration (VLSI). The complexity of a chip rapidly hit and surpassed the limits of what a human brain could understand. A semiserious joke in the industry in those clays was that any electronics engineer had two chips in him—after he'd designed two, he was burned out and fit only for management.

There are three broad ranges in the complexity of machines. The first is like a person living alone, with a bedroom, wardrobe, kitchen, garage, and car. Everything simple and straightforward: when you need the can you drive it; when you need to cook, you go to the kitchen. The second stage is more like having a large family living in the same circumstances. Now everything is being used more efficiently, at the cost of keeping track of trade-offs and interactions. Someone is always in the kitchen, and junior is doing chemistry experiments in the sink between meals. Dad gets driven to work because Mom needs the car during the day. You almost need a reservation to use the bathroom.

The third stage is like a large company. Each person gets facilities that are appropriate to his task. Sharing and interaction are planned and are nowhere as helter-skelter as in the family. There are what seem to be large inefficiencies: the office building sits unused all night, the cafeteria is unused much of the day, where with the family, homework is done in bedrooms and on the kitchen table. Still, every thing works well enough.

In electronics, the first stage is the *discrete transistor* in its own package. In mechanical engineering, it is the craftsman building a single tool. The second stage is the *IC* of the 1970s or the mechanical system of today. Lots of ingenuity goes into getting everything into the constrained resources and making it work together. The giant VLSI chips of today, and the astronomically complex nanomechanical systems of tomorrow, represent the third stage. In the third stage, the bottleneck is not the physical resources of the system, but the ability to design and direct. In the company, it's often cheaper to get extra rooms, machines, whatever, than to buy fewer ones but hire another manager to decide how they to be shared efficiently. In *VLSI*, modules are pre designed and used as is, even though each one might have been further optimized for its specific function. Chip layouts are regularized and laid down by automatic programs, even though carefully (and time-consuming) human attention might have clone it a little better. Indeed, chip designs today look more like programs, written in languages like VHDL and Verilog, than they do any graphical picture of the chip.

Stage three for nanomachines will follow the same route. A mechanical engineer of today; looking at (any small part of an advanced nanomachine, would consider its design quite sketchy and pedestrian. The reason is that it will be generated automatically as part of a vast system, probably described in something like a programming language instead of anything three-dimensional. The designer will have written, essentially, what she wanted the machine to do, rather than trying to describe the low-level mechanism to do it.

Indeed, the designer might not even know (or care) whether she is describing hardware or writing software. Suppose she is designing a machine that will perform a certain motion. She describes the motion she wants, and the design software decides whether it will be a clever linkage, a more straightforward linkage driven by cams, a general arm driven by stepper motors controlled by a microprocessor, or some combination.

Several kinds of current-day software have capabilities that promise the kind of design capabilities that we will need for big nanomechanical systems. There are the VLSI design systems.

Another, more common example is software development systems. Back in the 1970s, Fortran and Cobol were the tools many developers used in scientific and business programming, respectively, but a preponderance of work was still done in assembly language, specifying low-level machine instructions one at a time. Today, a vast array of software synthesis techniques is available. Specialized programming languages for almost any imaginable purpose are available, which cut the effort to develop a program by orders of magnitude compared to assembly language. For highly stylized applications, systems that are essentially automated questionnaires let you describe the application you need, and generate it automatically. And there are methods at virtually every point between these extremes.

Simulation and Testing

Today, the best we can do to test a nanomechanical design is simulate it. In the future, however, it will he possible to build and test systems of considerable complexity, like the trillion-parts grain of sand, in seconds. Design practice has tended to move from physical testing to simulation over the past several decades, as computers have become cheaper and faster. In the next few decades, however, the ability to make things physically will follow the same track. For certain applications anyway, there might be a swing back in the opposite direction.

Imagine a complete machine shop, factory, and engineering laboratory in a building with ten acres of floor space. It has every conceivable kind of construction and test equipment, overhead cranes, mobile robots—the works. Reduced to the scale of nanomachines, this fits in a 40-micron square. It would sit comfortably on the cut end of the finest human hair. Your nanoengineering design workstation could have thousands of these facilities built right into the processor. Having specified your design, the workstation could build and test thousands of minor variants in the time it took to press the return key.

Nanotubes

Nearly 10 years ago Sumio Iijima, first noticed odd *nanoscopic threads* lying in a smear of soot. Made of pure carbon, as regular and symmetric as crystals, these exquisitely thin, impressively long macromolecules soon became known as *nanotubes*, and they have been the object of intense scientific study ever since.

Just recently, they have become a subject for engineering as well. Many of the extraordinary properties attributed to *nanotubes*—among them, *superlative resilience*, *tensile strength* and *thermal stability*—have fed fantastic predictions of microscopic robots, dent-resistant car bodies and earthquake-resistant buildings. The first products to use nanotubes, however, exploit none of these. Instead the earliest applications are electrical. Some General Motors cars already include plastic parts to which nanotubes were added; such plastic can be electrified during painting so that the paint will stick more readily. And two nanotube-based lighting and display products are well on their way to market.

In the long term, perhaps the most valuable applications will take further advantage of *nanotubes*' unique electronic properties. Carbon nanotubes can in principle play the same role as silicon does in electronic circuits, but at a molecular scale where silicon and other standard semiconductors cease to work. Although the electronics industry is already pushing the critical dimensions of transistors in commercial chips below 200 nanometers (billionths of a meter)—about 400 atoms wide—engineers face large obstacles in continuing this miniaturization. Within this decade, the materials and processes on which the computer revolution has been built will begin to hit fundamental physical limits. Still, there are huge economic incentives to shrink devices further, because the speed, density and efficiency of microelectronic devices all rise rapidly as the minimum feature size decreases. Experiments over the past several years have given researchers hope that wires and functional devices tens of nanometers or smaller in size could be made from nanotubes and incorporated into electronic circuits that work far faster and on much less power than those existing today.

The first *carbon nanotubes* that lijima observed back in 1991 were so-called *multiwalled tubes*: each contained a number of hollow cylinders of carbon atoms nested inside one another like Russian dolls. Two years later lijima and Donald Bethune of IBM independently created single-walled nanotubes that were made of just one layer of carbon atoms. Both kinds of tubes are made in similar ways, and they have many similar properties—the most obvious being that they are exceedingly narrow and long. The single-walled variety, for example, is about one nanometer in diameter but can run thousands of nanometers in length.

What makes these tubes so stable is the strength with which carbon atoms bond to one another, which is also what makes diamond so hard. In diamond the carbon atoms link into four-sided tetrahedra, but in nanotubes the atoms arrange themselves in hexagonal rings like chicken wire. One sees the same pattern in graphite, and in fact a nanotube looks like a sheet (or several stacked sheets) of graphite rolled into a seamless cylinder. It is not known for certain how the atoms actually condense into tubes, but it appears that they may grow by adding atoms to their ends, much as a knitter adds stitches to a sweater sleeve.

However they form, the composition and geometry of *carbon nanotubes* engender a unique electronic complexity. That is in part simply the result of size, because quantum physics governs at the nanometer scale. But graphite itself is a very unusual material. Whereas most electrical conductors can be classified as either metals or semiconductors, graphite is one of the rare materials known as a *semimetal*, delicately balanced in the transitional zone between the two. By combining graphite's semimetallic properties with the quantum rules of energy levels and electron waves, carbon nanotubes emerge as truly exotic conductors.

For example, one rule of the quantum world is that electrons behave like waves as well as particles, and electron waves can reinforce or cancel one another. As a consequence, an electron spreading around a nanotube's circumference can completely cancel itself out; thus, only electrons with just the right wavelength remain. Out of all the possible *electron wavelengths*, or *quantum states*, available in a flat graphite sheet, only a tiny subset is allowed when we roll that sheet into a nanotube. That subset depends on the circumference of the nonatube, as well as whether the nanotube twists like a barbershop pole.

Slicing a few electron states from a simple metal or semiconductor won't produce many surprises, but semimetals are much more sensitive materials, and that is where carbon nanotubes become interesting. In a graphite sheet, one particular electron state (which physicists call the *Fermi point*) gives graphite almost all of its conductivity; none of the electrons in other states are free to move about. Only one third of all carbon nanotubes combine the right diameter and degree of twist to include this special Fermi point in their subset of allowed states. These nanotubes are truly metallic nanowires. The remaining two thirds of nanotubes are *semiconductors*. That means that, like silicon, they do not pass current easily without an additional boost of energy. A burst of light or a voltage can knock electrons from valence states into conducting states where they can move about freely. The amount of energy needed depends on the separation between the two levels and is the so-called *band gap* of a semiconductor. It is semiconductors' band gaps that make them so useful in circuits, and by having a library of materials with different hand gaps, engineers have been able to produce the vast array of electronic devices available today.

Carbon nanotubes don't all have the same hand gap, because for every circumference there is a unique set of allowed valences and conduction states. The smallest-diameter nanotubes have very few states that are spaced far apart in energy. As nanotube diameters increase, more and more states are allowed and the spacing between them shrinks. In this way, different-size nanotubes can have band gaps as low as zero (like a metal), as high as the band gap of silicon, and almost anywhere in between. No other known material can be so easily tuned. Unfortunately, the growth of nanotubes currently gives a jumble of different geometries, and researchers are seeking improvements so that specific types of nanotubes can be guaranteed.

Fat multiwalled nanotubes may have even more complex behaviour, because each layer in the tube has a slightly different geometry. If we could tailor their composition individually, we might one day make multiwalled tubes that are self-insulating or that carry multiple signals at once, like *nanoscopic coaxial cables*. Our understanding and control of nanotube growth still falls far short of these goals, but by incorporating nanotubes into working circuits, we have at least begun to unravel their basic properties.

Several research groups, including our own, have successfully built working electronic devices out of carbon nanotubes. Our *field-effect transistors* (FETs) use single semiconducting nanotubes between two metal electrodes as a channel through which electrons flow. The current flowing in this channel can be switched on or off by applying voltages to a nearby third electrode. The nanotube-based devices operate at room temperature with electrical characteristics remarkably similar to off-the-shelf silicon devices. We and others have found, for example, that the gate electrode can change the conductivity of the nanotube channel in an FET by a factor of one million or more, comparable to silicon FETs. Because of its tiny size, however, the nanotube FET should switch reliably using much less power than a silicon-based device. Theorists predict that a truly nanoscale switch could run at clock speeds of one terahertz or more—1,000 times as fast as processors available today.

The fact that nanotubes come with a variety of hand gaps and conductivities raises many intriguing possibilities for additional nanodevices. For example, our team and others have recently measured joined metallic and semiconducting nanotubes and shown that such junctions behave as diodes, permitting electricity to flow in only one direction. Theoretically, combinations of nanotubes with different band gaps could behave like light-emitting diodes and perhaps even *nanoscopic lasers*. It is now feasible to build a *nanocircuit* that has wires, switches and memory elements made entirely from nanotubes and other molecules. This kind of engineering on a molecular scale may eventually yield not only tiny versions of conventional devices but also new ones that exploit quantum effects.

We should emphasize, however, that so far our circuits have all been made one at a time and with great effort. The exact recipe for attaching a nanotube to metal electrodes varies among different research groups, but it requires combining traditional lithography for the electrodes and higher-resolution tools such as atomic force microscopes to locate and even position the nanotubes. This is obviously a long way from the massively parallel, complex and automated production of microchips from silicon on which the computer industry is built. Before we can think about making more complex, *nanotube-based circuitry*, we must find ways to grow the nanotubes in specific locations, orientations, shapes and sizes. Scientists at Stanford University and elsewhere have demonstrated that by placing spots of nickel, iron or some other catalyst on a substrate, they can get nanotubes to grow where they want. A group at Harvard University has found a way to merge nanotubes with silicon nanowires, thus making connections to circuits fabricated by conventional means.

These are small steps, hut already they raise the possibility of using *carbon nanotubes* as both the transistors and the interconnecting wires in *microchip circuits*. Such wires are currently about 250 nanometers in width and are made of metal. Engineers would like to make them much smaller, because then they could pack more devices into the same area. Two major problems have so far thwarted attempts to shrink metal wires further. First, there is as yet no good way to remove the heat produced by the devices, so packing them in more tightly will only lead to rapid overheating. Second, as metal wires get smaller, the gust of electrons moving through them becomes strong enough to bump the metal atoms around, and before long the wires fail like blown fuses. In theory, nanotubes could solve both these problems. Scientists have predicted that carbon nanotubes would conduct heat nearly as well as diamond or sapphire, and preliminary experiments seem to confirm their prediction. So nanotubes could efficiently cool very dense arrays of devices. And because the bonds among carbon atoms are so much stronger than those in any metal, nanotubes can transport terrific amounts of electric current—the latest measurements show that a bundle of nanotubes one square centimeter in cross section could conduct about one billion amps. Such high currents would vapourize copper or gold.

Carbon nanotubes have a second interesting electronic behaviour that engineers are now putting to use. In 1995 a research group at Rice University showed that when stood on end and electric field, carbon nanotubes will act just as lightning rods do, concentrating the electrical field at their tips. But whereas a lightning rod conducts an arc to the ground, a nanotube emits electrons from its tip at a prodigious rate. Because they are so sharp, the nanotubes emit electrons at lower voltages than electrodes made from most other materials, and their strong carbon bonds allow nanotubes to operate for longer periods without damage.

Field emission, as this behaviour is called, has long been seen as a potential multibillion-dollar technology for replacing bulky, inefficient *televisions* and *computer monitors* with equally bright hut thinner and more power-efficient *flat-panel displays*. But the idea has always stumbled over the delicacy of existing field emitters. The hope is that nanotubes may at last remove this impediment and clear the way for an alternative to *cathode-ray tubes* and *liquid-crystal panels*.

It is surprisingly easy to make a high-current field emitter from nanotubes: just mix them into a composite paste with plastics, smear them onto an electrode, and apply voltage. Invariably some of the nanotubes in the layer will point toward the opposite electrode and will emit electrons. Groups at the Georgia Institute of Technology, Stanford and elsewhere have already found ways to grow clusters of upright nanotubes in neat little grids. At optimum density, such clusters can emit more than one amp per square centimeter, which is more than sufficient to light up the phosphors on a screen and is even powerful enough to drive microwave relays and high-frequency switches in cellular base stations.

Indeed, two companies have announced that they are developing products that use carbon nanotubes as field emitters. Ise Electronics in Ise, Japan, has used nanotube composites to make prototype vacuum-tube lamps in six colours that are twice as bright as conventional light bulbs, longer-lived and at least 10 times more energy-efficient. The first prototype has run for well over 10,000 hours and has yet to fail. Engineers at Samsung in Seoul spread nanotubes in a thin film over control electronics and then put *phosphor-coated glass* on top to make a prototype flat-panel display.

The third realm in which carbon nanotubes show special electronic properties is that of the very small, where size-dependent effects become important. At small enough scales, our simple concepts of wires with resistance dramatically fail and must be replaced with quantum-mechanical models. This is a realm that silicon technology is unlikely to reach, one that may yield surprising new discoveries but will also require significantly more scientific research than will either nano circuits or nanotube field-emission devices.

For example, researchers are currently debating exactly how electrons move along a nanotube. It appears that in defect-free nanotubes, electrons travel "*ballistically*"—that is, without any of the scattering that gives metal wires their resistance. When electrons can travel long distances without scattering, they maintain their quantum states, which is the key to observing effects such as the interference between electron waves. A lack of scattering may also help explain why nanotubes appear to preserve the "*spin*" state of electrons as they surf along. (*Electron spin* is a quantum property, not a rotation.) Some researchers are now trying to make use of this unusual behaviour to construct "spintronic" devices that switch on or off in response to electrons' spin, rather than merely to their charge, as electronic devices do.

Similarly, at the small size of a nanotube, the flow of electrons can be controlled with almost perfect precision. Scientists have recently demonstrated in nanotubes a phenomenon called *Coulomb blockade*, in which electrons strongly repulse attempts to insert more than one electron at a time onto a nanotube. This phenomenon may make it easier to build *single-electron transistors*, the ultimate in sensitive electronics. The same measurements, however, also highlight unanswered questions in physics today. When confined to such skinny, one-dimensional wires, electrons behave so strangely that they hardly seem like electrons anymore.

Thus, in time, nanotubes may yield not only smaller and better versions of existing devices but also completely novel ones that wholly depend on quantum effects, Of course, we will have to learn much more about these properties of nanotubes before we can rely on them. Some problems are already evident. We know that all molecular devices, nanotubes included, are highly susceptible to the noise caused by electrical, thermal and chemical fluctuations. Our experiments have also shown that contaminants (oxygen, for example) attaching to a nanotube can affect its electrical properties. That may be useful for creating exquisitely sensitive chemical detectors, but it is an obstacle to making single-molecule circuits. It is a major challenge to control contamination when single molecules can make a difference.

Nevertheless, with so many avenues of development under way, it seems clear that it is no longer a question of whether nanotubes will become useful components of the electronic machines of the future but merely a question of how and when.

11

AUTOANALYZER

The Autoanalyzer has a most profound effect upon the performance of diagnostic laboratory work in recent years. Demands of both patients and physicians have caused a marked increase in the work load with no plateau in sight. Without automation, the volume of work expected is beyond the capabilities of the laboratory workers now, or soon to be available in the field. This automatic continuous-flow analyzer can be made reasonably accurate extremely reproducible, and very adaptable to problems of the laboratory. The routine calibration and recalibration with visual recording and potential automatic verification and reverfication are advantages not previously available with other instruments. The instrument is adaptable in laboratories with marked daily variations in work load. It is capable of increasing the total work production without a great increase in personnel numbers of fatigue. Glassware utilization is markedly reduced. Cost per test may be considerably lowered when work is done in volume. However, it must be noted that the level of economic advantage is variable and depends upon the situation, laboratory, and training of personnel available to perform such studies as might be required for emergencies, weekends, and holidays.

The instrument itself has a modular design facilitating repair, replacement, and organization upon the laboratory bench. Any discussion of the instrument itself seems ideally divided into a periodic description of the successive modules.

Modules

Sampler I

The sampler I consists of an automatic revolving plate containing 40 slots capable of holding plastic sample cups with a volume of 0.5-

3.0 ml. As the plate rotates, a hinged pickup crook rises and falls, dipping into successive cups. The crook then moves from cup to cup in an elevated position aspirating alternately first sample, air, second sample, air, and so on at approximately a 2:1 ratio. The air between samples acts as an intersample buffer. The volume of fluid and air aspirated varies with the rate of rotation of the sampling plate, that is, 20, 40, or 60 samples per hour, as well as the diameter of the aspirating line on the manifold. Within the tubing, the air wipes the "nonwettable surface" and reduces intersample contamination. Because of the relatively slow motion of the crook as it rises and falls, the tip of the crook enters the sample relatively slowly. This allows variation of sample height within the cup to have some potential effect upon the final answer. Uniform, complete filling of the sample cups eliminates this problem.

A second-generation motorized sampler, designated Sampler II, has a stainless-steel aspirating tube moving alternately between successive sample cups and a constantly refilling well of distilled water in the housing of the sampler. The well can be filled with solvent, inhibitor, or some other react at if the chemistry of the particular analysis requires it. The ratio of sample to water wash as well as the speed of rotation (10-20 samples per hour) can be controlled by replaceable cam gears within the instrument.

The tip of the stainless-steel aspirating tube enters the sample cup rapidly to a predetermined level near the bottom of the cup. While the water wash acts to some extent as an intersample buffer, air segments are also introduced elsewhere into the system. Two vibrating mechanical arms are available which enter the two cups preceeding the one being sampled and automatically mix their contents. This is important when colloidal or particulate suspensions such as whole blood have been placed in the cups, and sedimentation has occurred.

Proportioning pump I

This module compresses a manifold consisting of a number of parallel plastic tubes between five tubular roller bars connected to parallel roller chains and a spring-balanced flat plastic platen. The fluid is forced through the tubing in a pulsatile fashion by virtue of proximal compression of the tubing by one roller bar immediately before the distal roller bar elevates to release fluid for flow. The tubing is cotinuously compressed by the bar rolling across the platen and the fluid and/or air is forced forward. There is no backflow within

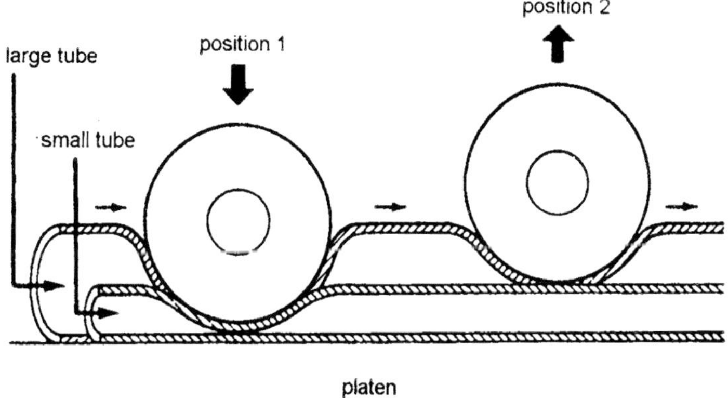

Fig. 11.1. Proportioning pump I with position of roller necessary for pumping action.

the normal operating system. Spring beneath the ends of the platen add additional compression and reduce some of the tendency toward pulsatile nature of the force. The entire roller bar assembly head can be raised and lowered to turn the instrument off and on as well as to release tension upon the plastic manifold when not in use, thereby prolonging its life.

A later version of this pump, designated "proportioning pump II" is the inverse of proportioning pump I. It has a plastic platen cover plate compressing the manifold down upon a series of similar roller bars on a drive chain below, which advances at a constant rate as long as the platen cover is compressed into place. Constant speed, smooth plate surface, regular spaced roller bars, and freedom from contamination are the keys to reliable operation. An off-and-on switch is present, allowing for temporary stoppage, in the com-pressed position, without backflow into the reagent and/or sample reservoirs.

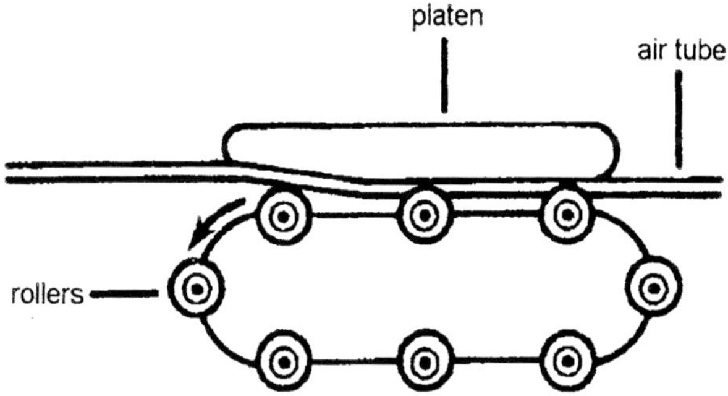

Fig. 11.2. Proportioning pump II showing inverse position of platen.

Manifold

The manifold consists of a pair of grooved plastic end blocks which hold parallel sample and reagent plastic tubes of various sizes by a small external sleeve placed equidistant on each tube to ensure uniform stretching when the end blocks are attached to the proportioning pump brackets. The flow rate through the manifold is proportional to the internal diameter of these resilient tubes. They may show progressive elongation with use and subsequent change (reduction) of internal diameter. This pheomenon can be compensated for by slowly increasing the distance between the plastic end blocks. Relatively small changes in the flow rates have little effect on the performance of the instrument since periodic (daily) recalibration is an essential part of the Auto Analyzer operation. Theoretically, significant changes of import will not usually occur in the short time interval between controls, standards, and unknowns. Care must be taken to avoid wear of individual tubing within the manifold. This usually occurs to those tubes which are on the outside edges of the manifold or on the upper level of a two-level plastic end block sometimes used with pump I. This problem is essentially eliminated in pump II.

Samples and standards, mixed periodically with reagents, are segmented by introduction of small increments of air developing a uniform bubble pattern within the flowing liquid stream. This microsegmentation mixes sample with reagent, wipes the inner walls clean of liquid between segments as well as functionally isolating specimens one from another, thereby minimizing interspecimen contamination. The effect of these bubbles on the rapidity of reaching steady state (wash time) is graphically demonstrated. Segments of air are also added to the recipient stream and perform the same functions after the dialyzable fraction is added. Visual inspection of the manifold for a uniform, regular, bubble pattern with a minimum of pulsatile flow is a indicator of good sampler, pump, and dialyzer performance.

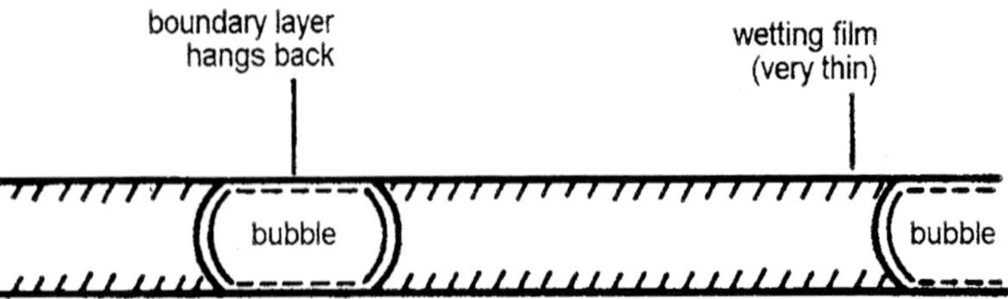

Fig. 11.3. Segment of transmission tubing showing air-segmented stream.

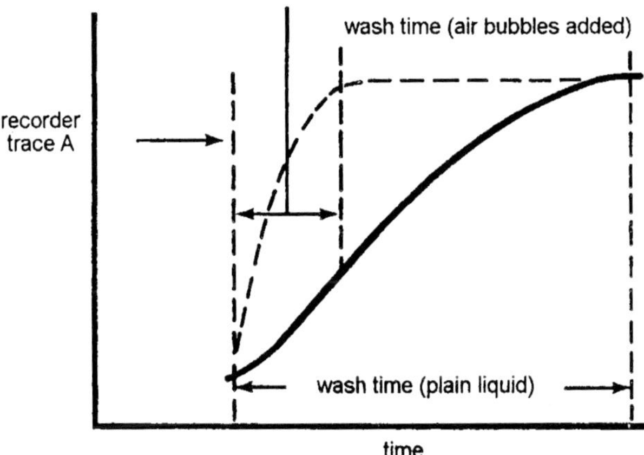

Fig. 11.4. Effect of air bubbles on system wash time.

Detergent may be added to the reagents, reducing surface tension and facilitating uniform flow throughout the system. Care must also be take to ensure that air introduced into the system does not react in the chemical process involved. Often special precautions, such as bubbling the air through sodium hydroxide when performing CO_2 analyses, are necessary to avoid this problem.

Mixing and Transmission

The flowing liquid is propelled through various combinations of glass mixing coils which agitate the dissimilar solutions by continued and repeated inversion during flow through the coil. Properly mixed solutions are transmitted through a variety of tubing, the nature of which depends on the character of analyses involved. The clear, elastic, plastic tubing (Tygon) is used for general purposes but cannot handle pure organic solvents or most concentrated acids. Both Solvaflex and the fluoroelastomer Acidflex may be used in these cases. Glass may also be used effectively, especially when interaction between the transmission tubing and the reagents becomes a problem.

Periodic cleaning and/or replacement of transmission tubing is recommended when discoloration and deposition within the transmission lines interfere with the smooth operation of the instrument.

Long coils of glass are available when a time delay for color development or enzymatic reaction is necessary. Such coils may be placed in a thermally controlled bath as required.

Single Channel Dialysis

Analyses requiring deproteinization are run through a dialyzing module consisting of two plastic plates having mirror-image grooving, which, when separated by a cellophase membrane, produces a long parallel flow dialysis cell. When the segmented sample and recipient streams are introduced on opposite sides of the membrane at approximately the same flow rates, effective cocurrent dialysis is obtained without the interspecimen contamination that appears in the Auto Analyzer system when countercurrent dialysis is intentionally or inadvertently obtained. Therefore, the rate of dialysis of sample to recipient stream is best in a cocurrent fashion with approximately the same rate of flow on either side of the membrane. The semicircular groove in the dialysis plate circles back upon itself to increase the length of dialysis exposure with such a small total area. The rate of dialysis varies with the flow rate, temperature, membrane pore size. (40-60 Å for the standard type C membrane). Particle size, ionic strength, ionization constants, and surface area.

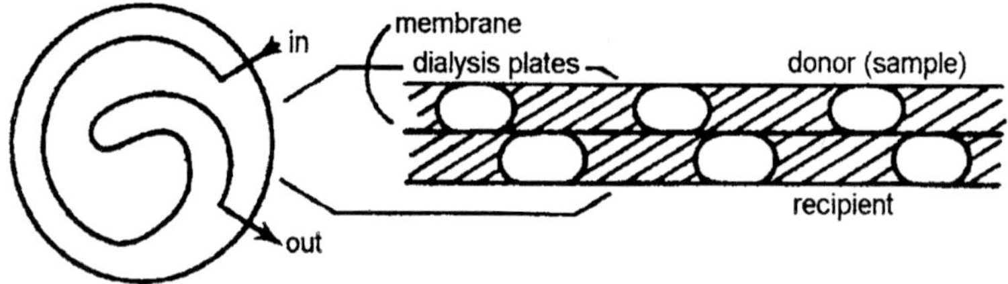

Fig. 11.5. Cocurrent dialysis.

Diallysis rates, and thus the per cent of any unknown dialyzed at the end of the fixed dialysis interval, vary considerably among the molecules or ions of clinical interest. Generally, the rate is directly proportional to the concentration in the sample stream within limits of the concentrations used for the calibrating standards. The concentration of protein, relative protein binding of the unknown, relative hydrophilia, and relative solubility in both sample and recipient stream must be carefully evaluated before aqueous or other standards can be substituted for protein-containing samples.

Colorimeter

The Auto Aanlyzer colorimeter is fundamentally a double-beam photometer consisting of matched optical systems at right angles to a prefocused tungsten exciter lamp. Matched interference filters, calibrated for wavelength, half-bandpass, and per cent of transmitted light are utilized for spectral isolation. Mechanical apertures are available for balancing the reference side, augmenting the effect of the instrument's potentiometer. Photovoltic (selenium barrier layer) cells are used to detect light energy on both the reference and sample sides. "Zero"

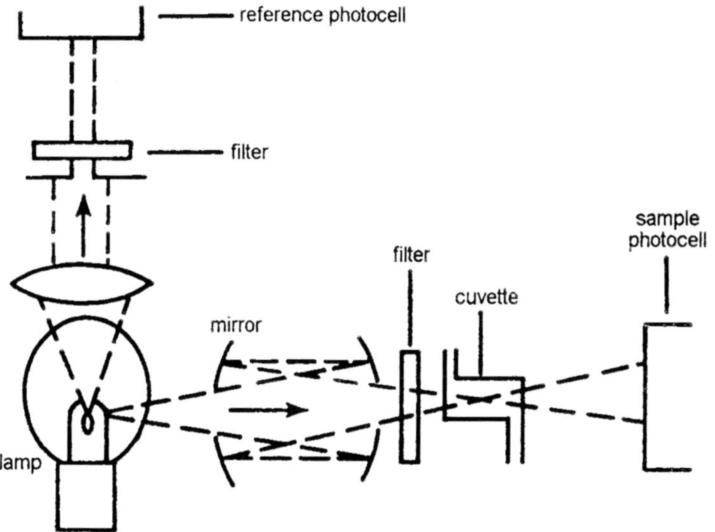

Fig. 11.6. Schematic of double-beam colorimeter.

adjustment can be made using an opaque obstruction on the sample side.

A stream of air-segmented solution, with periodic changes in absorbance representing samples and standards, is continuously flowing into the colorimeter, entering the middle arm of an F-shaped debubbler. A portion, thus freed of air segmentation, is drawn through the lower limb into a microflow cell with a 15-mm optical light path. The colorimeter monitors the stream at all times and produces two

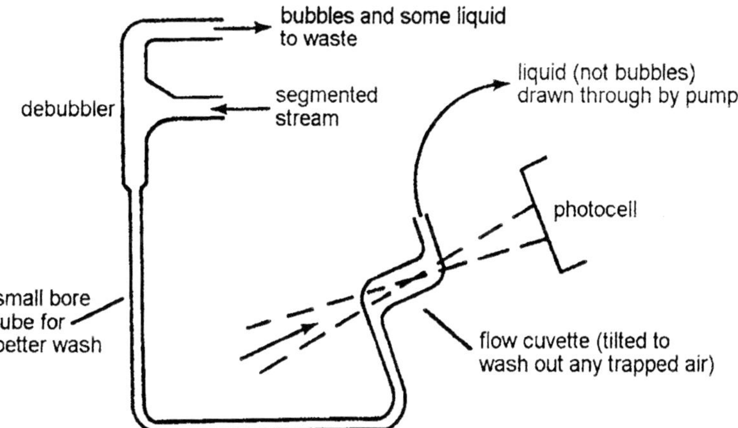

Fig. 11.7. Flow system showing debubbler and flow cell.

continuous signals. The electronic potential from the reference photocell is relatively constant and not affected by the parameters of the liquid stream. The potential from the sample photocell, however, is directly affected by the absorbance of the stream and therefore serves as the monitor of this stream. The ratio of these two potentials is continuously recorded, providing a means of comparing one sample segment of this stream to another, thus making it possible to compare sample to standard for calculation of concentration.

A special model is available containing dual flow cells, one reference and one sample, to provide for automatic serum blanking.

Flame Photometer

The Auto Analyzer flame photometer utilizes a balanced light intensity system (with lithium as an internal standard) for the simultaneous determination of sodium and potassium. The burner system has been modified from the original combustion atomizer-burner using propane and oxygen to a Meaker type, with separate atomizing chamber and burner assembly. In the present system, sample, entering through a fine capillary from the recipient side of the dialysis stream, is vaporized in a flame using a mixture of natural gas and compressed air. While larger droplets drain down the side of the atomizing chamber to the waste, the vapor mixture ascends through a reflux-preventing grid to a burner where combustion, assisted by the surrounding air, produces a relatively low-temperature flame. A solenoid valve wired to a flame-out detector will automatically shut off the gas when necessary. An integrating mirror helps focus the flame's emission light energy through multilayer interference filters onto the detector head housing the photocells of the separate sodium, potassium, and lithium.

Thermisters are included in the detector circuit to reduce the drift which is possible because these photocells are temperature sensitive. This design, along with adequate warm-up time and modest attention to ambient thermostability, will minimize but not eliminate this troublesome feature. The electrical energy required to null the output of the sample photocell back into balance with the lithium reference cell is transmitted to the recorder and is responsible for activating the servomotor assembly, thereby producing the concentration readout.

Fluorometer

Technicon has utilized two continuous-flow fluorometers in their automated analysis system. Both have the same fundamental design, although the individual features look quite different.

The most recent instrument utilizes an 85-W mercury arc lamp for sample excitation providing adequate discontinuous spectral energy peaks from 250 to 600 nm. A rotating wheel interposes circular apertures of various sizes into the light path for required sensitivity. Spectral isolation is accomplished by narrow bandpass interference filters in both the primary and secondary positions receiving parallel light emitting through the quartz optical system. A small-volume sample flow cell is available which reduces some of the interspecimen contamination.

Signal detection is basically dual beam, utilizing matched IP-21 photomultiplier tubes for detection of the emission energy through a system of collimating lenses and a secondary filter at a 90° angle to the light source. The signal from this sample photomultiplier tube is included in part of a null-balancing ratio system connected to a reference photomultiplier tube which receives a portion of the incident (excitation) energy from the light source deflected to one side by a mirror assembly. The signal utilized to constantly null the system as samples enter the flow cell are then transmitted to the recorder.

Recroder

The Auto Analyzer utilizes a modified 10-mV rectilinear strip chart variable-speed recorder which can be calibrated using the colorimeter controls. Utilizing a slide wire potentiometre, the feedback voltage necessary to null the input voltage from the colorimeter is transmitted to a servomotor. This constant system of rebalancing drives the pen and accurately reflects changes in per cent transmission. The peaks are read by comparing them with a calibration curve derived from the standards and transferred to a plastic sheet called a comparator that is then made to overlay the recording. The actual readings can be converted, if necessary, to absorbance by utilizing a conversion table or by replacing the linear chart paper with semilong absorbance paper and properly calibrating the instrument.

General Comments

Automatic integration of each sample aliquot, propelled through the system segmented by portions of air, is established by continuous mixing of each sample segment in the debubbler assembly with the preceding and following wash segment leading to a steady, progressive increase *toward* steady-state (maximum) levels near the end of the aspiration time, followed by a steady fall toward basal (reagent) levels near the end of the wash. Effects of sample interaction will be discussed later, but generally it can be avoided by filling the sample cups to the proper height and/or utilizing a sampler II with automatic intersample washing. The precentage to which the peak absorbance approaches steady state varies widely with the procedure.

A comprehensive program of preventive maintenance and quality control, including periodic manifold and membrane checking, utilization of fresh reagents, adequate standardization, visual observation of calibration curves and sample peaks, will lead to efficient, reliable operation. The rate of analysis is limited only by the efficiency of the interspecimen washout.

Sequential Multiple Analysis

Single Auto Analyzer units may be combined in groups, such as two or four. Various combinations of a sampler, pump, required manifold tubing, detectors, and recorders allow the simultaneous determination of such combinations as electrolytes, BUN, and sugar. After several prototype models combining 8, 10 and 12 channels, Technicon marketed the SMA-12/30 and later the SMA-12/60.

Steady State

The entire concept of sequential multiple analysis requires the development of a sample steady state. Single-stream analysis demonstrates graphically a reagent base-line steady state as well as the plateau type of sample steady state. With multiple-stream sequential analysis of the various steady states of single streams, the various plateaus are electronically integrated into a continuous pattern showing only the plateau (or sample steady state) portion of each of the individual streams. Sample concentration is related to $\%T$ (or peak height) provided sample interaction to a significant degree has not occurred.

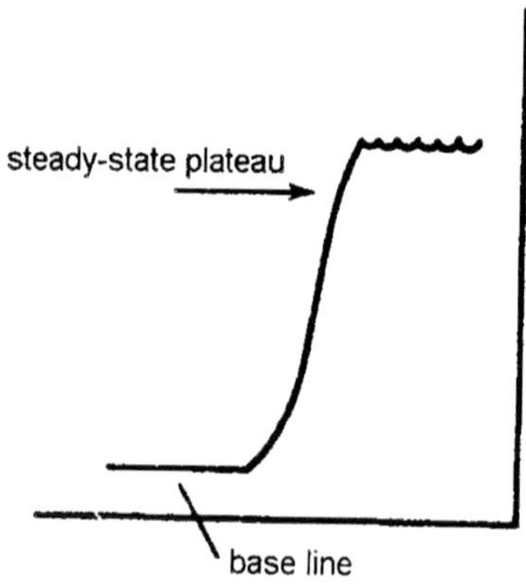

Fig. 11.8. Single-stream analysis.

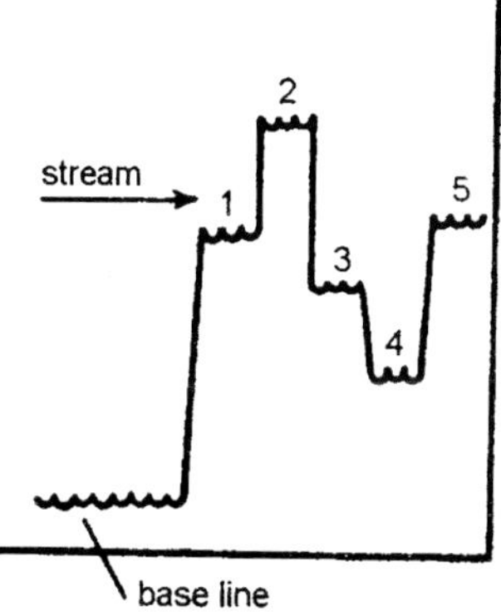

Fig. 11.9. Multiple-stream analysis.

With the original single-channel Auto Analyzer. The sample-to-wash ratio determined the slope of the transmittance change from the base line obtained with continuous sample aspiration, that is, the so-called "steady state." In addition, the amount of return toward the base line between samples was also related to this ratio and has been evaluated in terms of the so-called "half-wash" time. Ideally, the change from base line to steady state must occur as rapidly as possible and vice versa. Complete return between specimens is not necessary, provided the steady state of one sample is not materially affected by levels of the previous samples.

In order for the sequential analysis to be effective, each analytical channel must reach a steady state which is satisfactory for accurate monitoring at the proper time. The SMA-12/30 has a sample wash ratio of approximately 11:1. A new sample or standard is aspirated every 2 min (or at the rate of 30 per hour). In the SMA-12/60, 60 samples are aspirated per hours, taking advantage of improvements such as rapid dialysis, new microflow cell, and smaller reagent volume.

SMA-12/30 and SMA-12/60

The general sequence of modules is the same in these combination units as with the standard single-channel Auto Analyzer. Both utilizer sampler IIs, after which the diluted sample is divided through a variety of manifold and propelled by pump IIs. Eventually the operational reagent lines are separated one from another and, after receiving the proper sample (or the dialysate), progress as separate analytical units toward the colorimeter and/or flame photometer. A properly heated reaction chamber is included along the way when necessary. Transit time through all mixing, dialyzing, heating, and transmission lines is arranged to present the flat (or top) portion of the steady-state curve to the various flow cells of the colorimeter (or flame photometer) complex in a synchronized sequential fashion. A programming unit connects the various then null-balanced photocells to the recorder at the proper time, resulting in a series of straight lines appearing on the preprinted chart paper moving on a recorder at the same rate as the various channels are being evaluated.

The instrument requires a protein-base serum standard for its routine two-point calibration (base line or 100%*T* with reagents flowing as well as some known level). While not checked routinely, linearity should be evaluated periodically by some combination of serial dilutions of quality control sera, aqueous standards, abnormal and normal known and unknown sera, as well as periodic simultaneous recheck of the SMA unknowns by a well-controlled similar "hand" or other automated method. Careful evaluation of each new batch of control serum is essential.

Multichannel Dialysis

In this SMA-12/30, similar dialysis units are used as with the single Auto Analyzer. The 37°C water bath in which they are immersed also serves as an incubation chamber, when necessary, for enzyme or color-developing reactions. The SMA-12/60 dialyzer is greatly reduced in size. The groove length has been decreased from the standard 88 in to either a 6- or 12-in U-type groove separated by the standard type C-cellophane membrane. The sample stream is preheated before arrival at the dialysis chamber.

The new pump II has an air release bar attached to each air segment tube, allowing for the injection of air segments every 2 sec into the reagent line and ensuring uniformity of bubble pattern regardless of manifold tubing size. This was not available with previous models.

Analytical Cartridge

In the SMA-12/30, a large, plastic platter was required to hold the myriad of tubes, coils, dialysis and heating baths necessary to create all proper reagent, dialysate and/or sample mixtures.

In the SMA-12/60, a compact unit (including a preheating assembly at 37°C, a dialysis unit, a metal heating block with heat transfer fluid about centrally placed coils, and an adequate number of mixing coils) is available for each channel. Most transmission in this unit is done through glass tubing. The analytical cartridges are clustered in groups of four for convenience and proximity to reagent supply.

Phasing

In the SMA-12/30, the individual samples are aspirated for 110 sec with an intervening 10-sec water wash. Each photocell (or flame photometer module) is monitored in sequence for approximately 9 sec. The complete cycle is completed in about 110 sec, allowing a 10-sec space between specimens.

To be sure that the curve of the particular analysis coming to its photocell is at "steady state" at the time of the monitoring, the length of transmission tubing utilized may be varied to produce the required sequence. Several "phasers" have been tried to simplify has this operation, with limited practical success. These include a device with concentric circular coils of tygon tubing between two compressible discs included in the various reagent lines. Compression (or release)

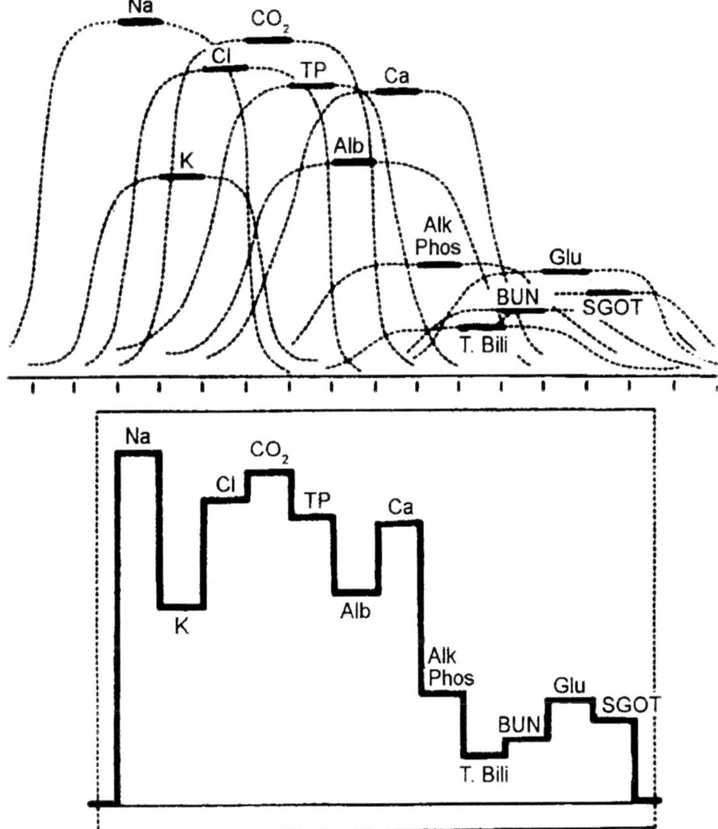

Fig. 11.10. Continuous monitor of some chennels of SMA-12.

of the discs will change the internal diameter of the tubing and thus the flow rate. A trombonelike glass assembly has also been used. Addition of a length of glass tubing, tygon tubing, or circular glass coil into the line requiring it is a quick, practical method of handling phasing problems arising during manifold, reagent, or dialyzer changeover on a day-to-day basis.

With the SMA-12/60, the time delay and gross phasing coils are included in arch analytical cartridge and appear to be easily used, visible, and accessible.

Phase visualization is accomplished by two methods. In the original SMA-12/30, a phasing recorder is optionally available which sequentially monitors four channels during their entire curve. This monitoring includes a recognition of that segment of the steady-state curve simultaneously monitored by the colorimeter and recorder. By serial examination of three groups of four channels, the entire system may be analyzed for the proper phase ratio.

In the 12/60 system, an oscilloscope type of instrument is available for phase monitoring of all 16 channels, including the four sample blanks, and performs the same function of the automatic phase recorder on the 12/30. This provides for rapid visual scanning of the entire analytical curves as well as notation of the location of that segment of the steady state being monitored by the recorder. The ideal position for the monitoring portion of this flat segment is near the end of the curve just before it begins to return toward the base line.

Colorimeter in Multichannel Analyzers

In the SMA-12/30, a series of individual combinations (one per colorimetric analysis) of a light-collimating device, interference filters, standard microflow cells, and photocells are arranged radially about a centrally placed prefocused tungsten exciter lamp. These are sequentially nulled against a separately placed reference photocell which does not have an interference filter or a flow cell between it and the lamp.

100%*T* calibration for reagent base line utilizes mechanically adjustable shutters imposed between the light source and the individual units, as well as electronic potentiometric calibration in the "primary circuit." This is essentially the same basic principle involved in establishing a reagent base line with the conventional Auto Analyzer. A "secondary" or calibrating circuit is added, making possible additional adjustments of the slope, so that one-point calibration of the analytical "curve" can be made to coincide with a value previously printed on the chart paper. A programmer unit coordinates the various channels with their properly phased plateaus (steady state) and the preprinted chart paper on the recorder.

In the 12/60, the analytical cartridges are grouped in units of four. Each group has four complete photometric units with a separate reference cell arranged in a semicircular fashion about a single light source. Phasing is handled completely within the individual analytical cartridges. The 16 available channels allow four determinations in which the sample is nulled against a "serum blank" channel (and not the reference cell), giving automatic blank correction. The difference between the two units is recorded. The programmer permits sequential channel recording. The cycle of 12 is completed in 60 sec. A new microflow cell is incorporated with an internal diameter of 1.5 mm (as compared to 3.0 mm in the standard microflow cell), greatly reducing residual volume within the cell and increasing the efficiency of the interspecimen water wash. The overall result is an increase of from 30 to 60 samples per hour with a decrease in the average reagent consumption per test from 10 ml/min on the 12/30 to 1.7 ml/min on the 12/60. The resulting reduction in reagent cost is most noticeable in those expensive and unstable reagents such as those used in enzyme determinations.

SMAC

The SMAC is considerably modified from the earlier SMA-12/30 or 12/60. The linear sampler module which can accommodate test tubes, sampler cups, or Vacutainers® can be loaded continuously and need not be completely full to operate. Refrigerated reservoirs are provided to conserve standards and controls. The sampler carriers are designed to be centrifugable. Human and/or machine readable labels can be used. The aspirator enters and leaves the specimen several times during the aspiration process, thereby generating additional intersample bubbles which improve mixing and faciliate the development of steady state.

The sampler is distributed to analytical cartridges where it is debubbled, prediluted with distilled water, and resegmented with new bubbles. Peristaltic pumps are present near the analitical cartridge. The additional bubble segmentation is several times that previously attained. The flow cells are positioned on the cartridge only a few millimeters away from the site of the analytical reactions. The grove dimensions of the dialysis unit are decreased. In addition, a new thin cellulose like material is used for the membrane, thus decreasing the contact time required for adequate dialysis. The bubbles are passed directly through the flow cell along with the liquid, unlike previous instruments, and the debubbling function is performed electronically. The instrument compensates for the relatively large sudden swings in absorbance caused by the bubble in the flow cell. This total process results in decreased intersample contamination and provides a mechanism for performing multipoint analysis as would be required for enzyme analysis. The volume of the flow cell has been greatly reduced. The light paths is 10 mm in length by 0.5 mm in diameter and has an internal volume of approximately 2 ul. Each end has an optically clear, flat sapphire rod to help conduct the light from the light source through the flow cell to the photo tube. The flow cells and light sources are integrated in sequence on a time basis, sharing a single light source and electronics. Light from the flow cells is transmitted by fiber optics and arranged about the circumference of a scanning disk which rotates. The disk contains a single slot which allows for exposure of one channel at a time. A compute integrates the pulses generated by the individual flow cells and constructs an absorbance curve for each flow cell. This computer has a built-in diagnostic procedure to check both hardware and software.

One change has been the use of ion-selective electrodes rather than flame photometry for sodium and potassium. Another change has been the use of immobilized enzymes "attached" to the walls of glass tubing for the assay of such substances as glucose.

Conclusions

The Autoanalyzer is a versatile, practical, "relatively inexpensive" method for automating most laboratory determinations involving a physical endpoint. Within practical limits, any standard biochemical procedure not involving solvent extraction may be easily automated as a single unit. The electronics involved in the programming of sequential comparison to a standard reference cell has until this time made substitution of analytical procedures within the established 12 analytical units (12/30 and 12/60) somewhat difficult.

This problem has been overcome to some extent in the SMAC analyzers. The so-called electronic debubbling and light piping has also improved the multichannel colorimetric system and has made it possible to add more than the 12 channel to these units.

12

ULTRACENTRIFUGE

While one can determine the arrangement of organelles and large macromolecular aggregates in cells and tissues by microscopy, and even locate specific molecules using specific staining procedures, a detailed molecular understanding of a cell requires biochemical analysis. Such analysis usually entail the disruption of the cell and the consequent obliteration of its delicate anatomy. To retain as much information as possible about the original location of the molecule under analysis. Biologists have developed techniques for disrupting tissues and cells in a controlled fashion, so that different cells and different components of cells can be separated before biochemical analysis.

ISOLATION OF CELLS

Many types of differentiated cells are not readily obtained as a cultured cell line. In any case, it is usually cheaper and quicker to use cells isolated directly from an animal or plant for large-scale biochemical analysis. The disadvantage is that all tissues in a higher animal or plant contain a mixture of cell types, which must be separated before analysis. Suspensions of single cells are first prepared from the tissue by disrupting the extracellular matrix and intercellular junctions that hold the cells together. The best yields of viable dissociated cells are usually obtained from fetal or neonatal tissues. The procedures is to treat the tissues with proteolytic enzymes (such as trypsin and collagenase and agents that bind, or chelate, Ca^{2+} (such as ethylenediaminetetra-acetic acid, or EDTA) and then to dissociate them into single cells by gentle mechanical disruption.

Several approaches are used to separate the different cell types from a mixed cell suspension. One is to exploit the differences in the cells' physical properties. For example, large cells can be separated from small cells and dense cells from light cells by sedimentation or centrifugation; these techniques will be described when we discuss the separation of organelles and macromolecules, for which they were originally developed. Another approach is based on the fact that some cells adhere strongly to glass or plastic and therefore can be separated from cells that adhere less strongly.

An important refinement of this last technique depends on the specific binding properties of antibodies. Antibodies that bind specifically to the surface of only one cell type in a tissue can be coupled to various matrices—such as collagen, polysaccharide beads, or plastic—to form an "affinity surface" to which only cells recognized by the antibodies will adhere. The bound cells are then recovered by gentle shaking or, in the case of a digestible matrix (such as collagen), by degrading the matrix with enzymes such as collagenase).

The most sophisticated cell-separation technique involves labeling specific cells with antibodies coupled to a fluorescent dye and then separating the labeled cells from the unlabeled ones in an electronic fluorescence-activated cell sorter. Here, individual cells traveling in single file in a fine stream are assessed for their fluorescence by passing them through a laser beam. Slightly further downstream, tiny droplets, most containing either one or no cells, are formed by a vibrating nozzle. The droplets containing a single cell are automatically given a positive or a negative charge at the moment of formation, depending on whether they contain a fluorescent cell; they are then deflected by a strong electric field into an appropriate container. Occasional clumps of cells, detected by their increased light scattering, are left uncharged and are discarded into a waste container. Such machines can select 1 cell in 1000 and sort about 5000 cells each second.

Separation of Organelles and Macromolecules

The cells in a purified population can be disrupted in various ways: by osmotic shock, by ultrasonic vibration, by forcing the cells through a small orifice, or by grinding them up. These procedures break many of the membranes of the cell (including the plasma membrane and membranes of the endoplasmic reticulum and Golgi apparatus) into fragments that immediately reseal to form small, closed vesicles. But, if carefully applied, the disruption procedures leave organelles such as nuclei, mitochondria, lysosmoes, and peroxisomes intact.

The population of cells is thereby reduced to a soluble extract containing a thick suspension of membrane-bounded particles, each with a distinctive size, charge, and density,. Provided that the homogenization medium has been carefully chosen (this requires extensive trial and error for each organelle), the various particles retain most of the biochemical properties of the original organelles in the intact cell.

Separating the various components in this mixture became possible only after the commercial development in the early 1940s of an instrument known as the *preparative ultracentrifuge*, in which extracts of broken cells are rotated at high speeds. At a relatively low speed, large components, such as nuclei and unbroken cells, sediment rapidly and form a pellet at the bottom of the centrifuge tube; at a slightly higher speed, a pellet of mitochrondria is deposited; and at even higher speeds and longer periods of centrifugation, first the small, closed vesicles and then the ribosomes can be collected. All of these functions are impure, but resuspending the pellet and repeating the centrifugation procedure several times removes many of their contaminants.

A finger degree of separation can be achieved by layering the cell homogenate as a narrow band on top of a salt solution in a centrifuge tube. To stablize the sedimenting component against convective mixing, the salt solution beneath the band contains an increasingly dense solution of an inert, highly soluble material such as sucrose (a density gradient). Under these conditions, the different fractions sediment at different rates, forming distinct bands that can be individually collected. The rate at which each component sediments depends on its size and shape and is normally expressed as its sedimentation coefficient or s value (Table 12.1). Present-day ultracentrifuges rotate at speeds up to 80,000 rpm and produce forces up to 500,000 times gravity. At these enormous forces, even relatively small macromolecules, such as tRNA molecules and simple enzymes, separate from one another on the basis of their size.

Table 12.1. Some Typical sedimentation coeffiicients

Particle or Molecule	*Sedimentation Coefficient*
Lysosome	9400S
Tobacco mosaic virus	198S
Ribosome	80S
Ribosomal RNA molecule	28S
tRNA molecule	4S
Hemoglobin molecule	4.5S

The ultracentrifuge is also used to separate cellular components on the basis of their buoyant density rather than their size. In this case, the sample is sedimented through a steep gradient that contains a very high concentration of sucrose or cesium chloride. The cellular components move down the gradient until they reach a position that is equal to their own density, and at this point they float and can move no further. This method can be so sensitive that it is capable of resolving macromolecules that have incorporated heavy sedimentation coefficient (s), in units of seconds, are given by $(dx/dt)/w^{2x}$, where x is the distance from the center of rotation in centimeters, dx/dt is the speed of sedimentation in centimeters per second, and w is the angular rotation of the centifuge rotor in radians per second.

Because such coefficients are extremely small numbers, they are normally expressed in Swedberg units (S), where $1S = 1 \times 10^{-13}$ sec. isotopes, such as ^{13}C or ^{15}N, from the normal unlabeled species. In fact, the cesium chloride method was developed in 1957 to separate the labeled and unlabeled DNA produced after exposing a growing population of bacteria to nucleotide precursors containing ^{15}N; this classic experiment provided direct evidence for the semi-conservative replication of DNA.

Table 12.2. Major events in the development of the ultracentrifuge and the preparation of cell-free extracts

1897	*Buchner* showed that cell-free extracts of yeast can ferment sugars to form carbon dioxide and ethanol, laying the foundations of enzymology.
1926	*Svedberg* developed the first analytical ultracentrifuge and used it to estimate the molecular weight of hemoglobin as 68,000.
1935	*Pickels* and *Beams* introduced several new features of centrifuge design that led to its use as a preparative instrument.
1938	*Behrens* employed differential centrifugation to separate nuclei and cytoplasm from liver cells, a technique further developed for the fractionation of cell organelles by *Claude*, *Brachet*, *Hogeboom*, and others in the 1940s and early 1950s.
1949	*Szent-Gyorgyi* showed that isolated myofibrils from skeletal muscle cells contract on the addition of ATP. In 1955, a similar cell-free system was developed for ciliary beating by *Hofmann-Berling*.
1951	*Brake* used density gradient centrifugation in sucrose solutions to purify a plant virus.

1953 *de Duve* isolated lysosomes and, later, peroxisomes by centrifugation.

1954 *Zamecnik* and colleagues developed the first cell-free system to carry out protein synthesis. This was followed by a decade of intense research activity, during which the genetic code was elucidated.

1957 *Meselson*, *Stahl*, and *Vinograd* developed density gradient centrifugation in cesium chloride solutions for separating nucleic acids.

Centrifugation

Centrifuges are designed to accelerate sedimentation by utilizing centrifugal force. Various types of centrifuges are used in the clinical laboratory for separating suspended particles from a liquid in which the particles are not soluble. Liquids of differing specific gravities (density) may also be separated.

There are three general types of centrifuges: the horizontal head, the angle head, and the ultracentrifuge. Many variations of the horizontal and angle head units are found in the clinical laboratory. These include bench top and floor standing units, refrigerated units, and such special-purpose instruments as the microhematocrit, microsample, cytospin, and continuous-flow systems.

Principles of Centrifugation

In the clinical laboratory, the centrifuge function as a filtration or packing device. The development of large-batch centrifuges and continuous-flow systems which can sediment a precipitate from a large volume of solution in a short time have tended to replace the tedious process of filtration in most clinical and research laboratories. This development has not only speeded the process of separation but, when coupled with refrigeration, has reduced sample lability to a minimum.

General applications include the separation of serum or plasma from red blood cells, the separation of precipitated solids from the liquid phase of a mixture, or the separation of liquids of varying density. Special-purpose units include such applications as quantitative red blood cell packing for measuring hematocrit, automatic cell washing, and component preparation for blood banking. Ultracentrifuges employ very high speed, some with optical systems, to achieve difficult and precise quantitative separations of ultrasmall particles or macromolecules.

The horizontal-head shields or cups are in a vertical position when the centrifuge is at rest and assume a horizontal position when the

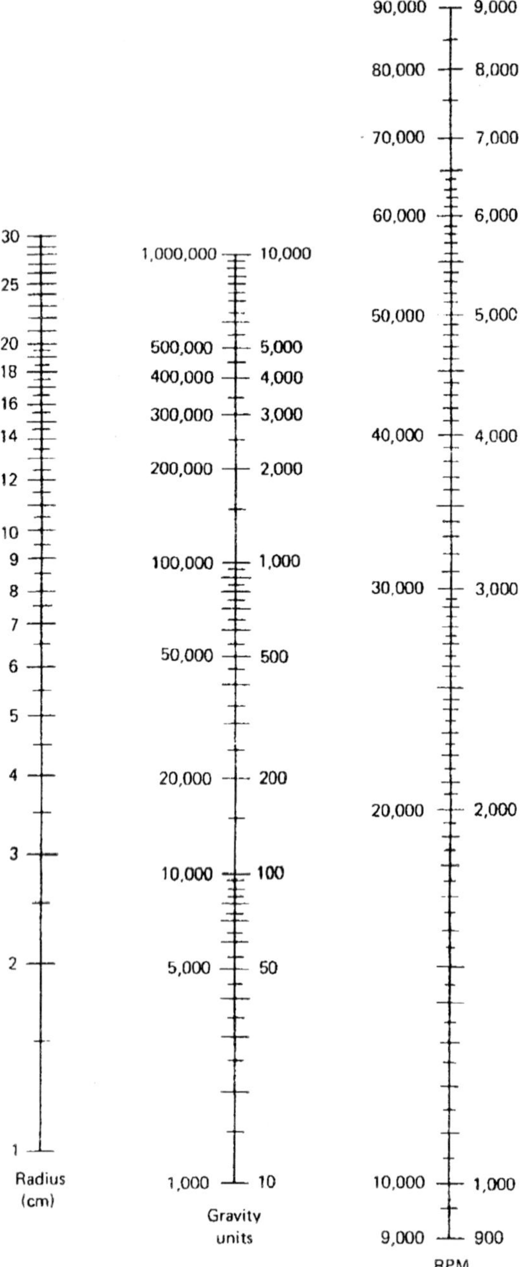

Fig. 12.1 Nomogram for the determination of the gravitational force of a centrifuge rotor.

centrifuge is operating. The horizontal-head centrifuge will operate at speeds up to about 3000 rpm. Speeds higher than this can be attained but are not generally practical because of the excess heat developed by air friction. Depending upon the radius of the head, these units attain forces of approximately 1650 × g.

The angle-head shields or cups are positioned rigidly at a fixed angle, usually 52°, to the shaft around which they rotate. Angle-head centrifuges may attain much higher speeds. This is due to the special construction of the heads, which exhibit very low friction with air. The cups of the angle-head centrifuge are enclosed by a metal case which reduces wind resistance, minimizing the increase of sample temperature commonly exhibited by the horizontal-head centrifuge. Thus the angle-head units can reach speeds of about 7000 rpm and exhibit forces of over 9000 × g.

In the horizontal head centrifuge the particles being sedimented must travel the entire length of the column of liquid in order to reach the bottom of the tube. As the containers fall from the horizontal to the vertical position when the centrifuge stops, a remixing of the sediment and the solution can occur. In the angle-head centrifuge, this is not the case. The shields and cups are at a fixed angle, and the particles have a shorter distance to travel, that is, across the column of liquid to the side of the container. These particles from clusters and move to the bottom of the container very rapidly.

Centrifuges capable of producing speeds on the order of 30,000 to 100,000 rpm with RCF values of 50,000 to 500,000 × g are referred to as ultracentrifuges. Both angle-head and horizontal-head rotors are used in these units. In addition, some instruments have vertical-head rotors where the axis of the tubes is fixed parallel to the rotor's axis of rotation. Even with the high forces produced, several hours or even days are often required to obtain complex separations. The three general centrifugation techniques utilizing the ultra-centrifuge include differential, rate zonal, and isopycnic separations. Differential centrifugation separates particles from a mixture by sedimenting those of larger mass to the bottom, leaving a portion of the smaller particles in the supernatant. The rate zonal technique separates particles based on differences in sedimentation rate in a density gradient, generally sucrose. The isopycnic technique separates particles based on differences in density (composition) in a density gradient, generally, cesium chloride. The ultracentrifuge is significantly more complex than general laboratory centrifuges and has not as yet found wide-spread use in the

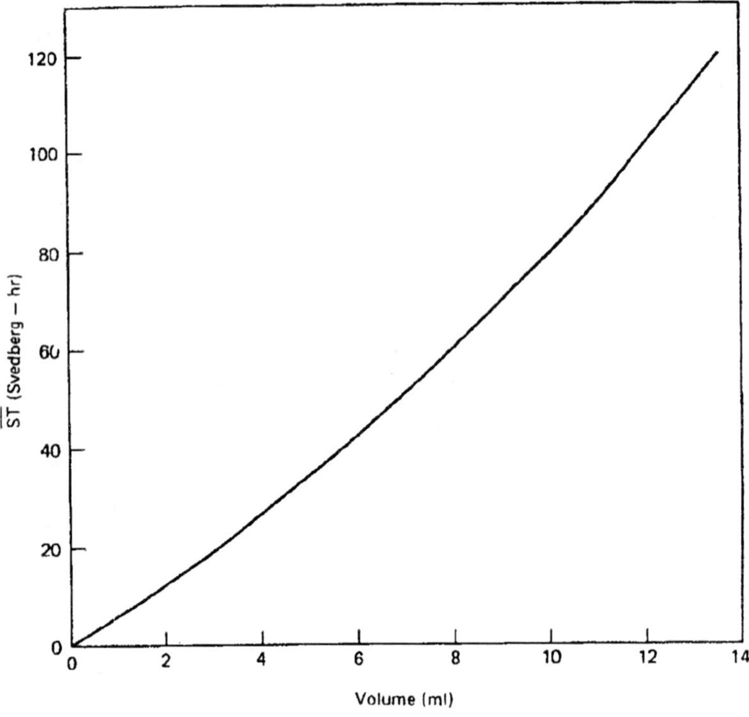

Fig. 12.2. Rotor characteristics for the rapid computation of ST values for given volumes traversed in a Spino No. 40 rotor having a maximum speed of 40,000 rpm.

clinical laboratory; thus the information that follows does not pertain to the ultracentrifuge.

COMPONENT PARTS

The parts of a common centrifuge include the chamber which encloses the internal parts, a cover with latch, the centrifuge head with shields or cups, the shaft and rotor on which the head turns, and the motor-drive assembly. Most centrifuges will include a power switch, braking device, speed control, timer, and possibly a tachometer.

Centrifuge motors are generally series-would dc motors that turn faster as voltage is increased. Occasionally, ac motors are utilized, and speed adjustment is achieved through stepwise reduction of the number of poles in the magnetic field. Both types are high-speed motors which employ direct-drive systems to the head through the motor shaft.

Electrical contact to the commutator in nearly all centrifuge motors is provided by graphite brushes. These gradually wear down as they press against the commutator turning at high speed. If the graphite is allowed to completely wear away, the retainer spring of the brush will make contact with the smooth surface of the commutator and cut grooves or scratches in its surface. A rough commutator surface will cause excessively fast wear of new brushes. The graphite that is worn away may deposit around the contacts, causing arcing and burning which decreases the efficiency of the motor and may damage it or in extreme cases may even start a fire in the motor. The shaft of the motor turns through sleeve bearings which are located at the top and bottom of the motor. Some units contains sealed bearings which are permanently lubricated, while others require periodic oil or grease. Knowledge of basic operating principles is important in establishing and performing proper maintenance on the centrifuge.

The speed of the centrifuge is controlled by a potentiometer, which raises and lowers the voltage supplied to the motor. The calibrations furnished on the speed control of a centrifuge are often only relative voltage increments and can never be taken as accurate indicators of speed. Since in most centrifuges speed is dependent upon voltage, as resistance is increased, speed is lost. Air resistance and turbulence, brush friction, and electrical inefficiency of the centrifuge itself are sources of resistance. These resistances can cause a centrifuge to operate at differing speeds on the same speed-control setting. Different accessories and varying states of repair will also result in the same speed-control problems. Calibration and periodic recalibration of centrifuge speed is extremely important.

Operating of centrifuge with the lid raised is dangerous and must be discouraged. In addition to being hazardous and reducing the centrifuge's speed through increased air resistance, this will also cause the revolving parts of the centrifuge to vibrate excessively, thereby causing extensive wear on the centrifuge and a remixing of the sedimented particles.

Tachometers are provided on many centrifuges to indicate the speed in rpm. A flexible shaft or cable attached to the motor spindle turns inside a flexible housing to which the meter movement is attached at the other end. As the cable turns, it causes the needle to move up scale. Some centrifuges use electric tachometers in which a magnet rotates around a coil, producing a current that may be measured. If a tachometer is not attached to the centrifuge, a strobe light or an electronic meter can be used to determine the actual rpm of each centrifuge setting. For any given setting, the speed of the centrifuge will vary with specimen load. The rpm determined for each centrifuge

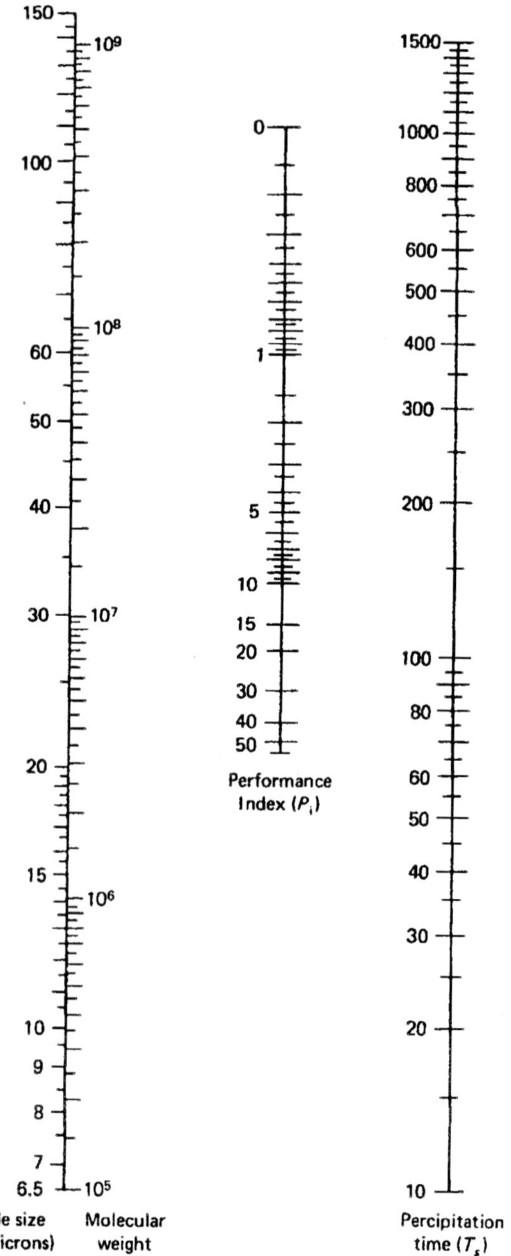

Fig. 12.3. Nomogram for the determination of precipitation time when the performance index of the rotor and the size of the sedimenting particles are known.

setting should be recorded and attached to the centrifuge for easy reference. However, the value of rpm is only one factor involved in determining centrifugal force. The true efficiency of a particular centrifuge must be expressed in relative centrifugal force, some number x gravity. The relative centrifugal force is calculated from the experimentally determined rpm and the radius of the head plus carrier. The RCF value for various speed settings is also generally listed in tabular or graphic form in the manufacturer's instruction manual for the instrument. The term rpm, when applied to a centrifuge, does not give any indication of the force applied to samples in that centrifuge. All comparisons made between centrifuges should be in relative centrifugal force.

A timer device is commonly employed in many centrifuges. Most are spring-driven clock mechanisms which turn the unit off after a present time cycle. Some units have an electronic timer which performs the same function. Many units have a switch that allows the unit to be operated continuously without using the timer device.

Braking devices are incorporated to provide rapid de-acceleration of the unit. Two general types of brake mechanisms are found on centrifuges. One type is mechanical and functions by physically applying pressure to the rotor at some point when a lever is pressed. The other type, found on most units, is electrical in nature and functions by reversing the polarity of the current to the motor.

Rotors

Centrifuge rotors are usually designed either for preparative or analytical purposes. Most analytical rotors are constructed so that light, which is used to determine the position of the solute boundary, can pass through the sample and its container while it is spinning in the rotor. Details of analytical rotors are beyond the scope of the present text but may be found in most of the references to ultracentrifugation.

Preparative rotors are of three principal types: (a) angle, (b) swinging-bucket, and (c) zonal. Angle and swinging-bucket rotors are much more common than the zonal type. In the angle rotor, containers, which may be either tubes or bottles of glass, metal, or plastic, are inserted into cavities in the rotor that are constructed at an angle to the axis of rotation. In the case of the swinging-bucket rotor, the container is placed in a holder which is itself held to the rotor body by means of pins that allow it to swing. The vertical axis of the container is parallel to the axis of rotation when the rotor is at rest, but as it begins to spin the container and its holder swing out because

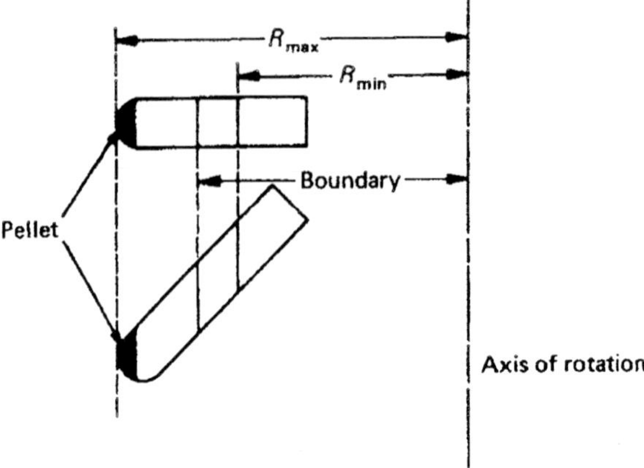

Fig. 12.4. Relationship of R_{max} and R_{min} to the axis of rotation in swinging-bucker and angle centrifuge rotors.

of the centrifugal force. At operational speeds the container is usually perpendicular to the axis of rotation.

In addition to the three types of rotors mentioned above, a special type of rotor is used in preparative centrifuges manufactured by the Sharples Equipment Co. This type of rotor, which is a hollow stainless-steel cylinder, has a bottom fitted with crossed baffles. As the rotor spins in the centrifuge, the solution containing the particles flows from the bottom across the baffles into the interior of the rotor. The vortex of liquid created by the spinning baffles causes the heaviest particles to sediment onto the lower wall of the rotor whereas the lighter particles sediment higher on the wall. If the speed and flow rate are controlled properly, the solvent and the lightest particles flow out through holes in the top of the rotor. This type of rotor is very effective for *continuous-flow centrifugation* and can be used to harvest large batches of microorganisms from their liquid cultures and subcellular particles from large batches of solution. Angle centrifuge rotors have also been adapted for continuous-flow centrifugation, and several models are available commercially.

All rotors must be designed so that they do not come apart under the stress of centrifugal fields. They should be constructed of a strong metal and should be machined so that their opposite sides strike a balance in weight. Those rotors which are used for very high speeds must be stronger and better-balanced than those used at low speeds. In addition, the metal used for the rotor should be protected with a resistant coating.

Types of Centrifuges

Most students of biochemistry are already familiar with standard bench-top laboratory centrifuges. These instruments are used in most laboratories for routine sedimentation work when refrigeration or high speeds are not required. The standard laboratory or clinical centrifuges are capable of reaching speeds up to about 5000 rpm with forces of about 3500 times the acceleration of gravity (×g) depending upon the particular rotor and centrifuge. They employ either angle or swinging-bucket rotors, most commonly for 50-ml or 12-ml and 15-ml tubes. Certain bench-top centrifuges can attain speeds up to about 16,000 rpm.

Other laboratory centrifuges are capable of attaining higher speeds than the bench-top models. Many of these are equipped with refrigerated rotor chambers which are essential for most work involving enzymes. These centrifuges are usually available for use with many different rotors that offers a wide choice of speeds and sample volumes.

The most highly refined centrifuges are called ultracentrifuges. These instruments are available in both preparative and analytical models although, with the proper techniques, preparative models can be used for analytical purposes. Ultracentrifuges are capable of attaining very high speeds, up to about 75,000 rpm, with centrifugal forces of up to about 500,000 x g. Since such speeds result in consideration heating of the rotor, because of the friction with air, ultracentrifuges are usually constructed with a vacuum rotor chamber and refrigeration. Rotors for use at the highest speeds are constructed of titanium.

Analytical models of the ultracentrifuge are equipped with optical systems to measure the position of the solute boundary during centrifugation. These systems utilize refractive or absorptive properties of the sample to make measurements. Many analytical ultracentrifuges are also equipped with controls which enable them to be operated at carefully regulated slow speeds so that solutes may be studied by the approach-to-equilibrium technique. These systems and methods of analytical ultracentrifugation are discussed extensively in the references to ultracentrifugation found at the end of this chapter.

Several different centrifuge models are designed for or are modified for continuous-flow sedimentation. These either utilize angle rotors or the Sharples cylindrical type of rotor discussed earlier.

Most centrifuges are equipped with electrical motors to propel the rotors. A few centrifuges utilize compressed air or steam for this purpose.

General Operation and Maintenance

Daily inspection along with periodic function verification and proper preventive maintenance are vital to the efficiency and longevity of the centrifuge. A regular schedule for checks must be established and followed to insure proper operation of the instrument.

Daily Operation

Daily operation should include observation and inspection of the following:

1. The centrifuge should not be on the same circuit as sensitive electronic measuring devices such as spectrophotometers, since it generates electrical noise and has a high current drain at start-up.
2. Check the cleanliness of the chamber, and immediately clean up all spills. Be aware of biohazardous materials (microbiologic, radioactive, chemical) which require specific decontamination procedures.
3. Always balance the load of the centrifuge before operating: use the correct tube sizes and type for the particular centrifuge.
4. Always insure that the cover is closed and latched while the unit is operating. This will prevent the dangerous scattering of both biohazardous and physically dangerous material out into the environment where the operator and others may be exposed.
5. Observe for unusual noises or vibrations during operation.

Function Verification

Frequency of function verification procedures should be appropriate for the application of the centrifuge. It is generally recommended that procedures be performed at 3-month intervals with more frequent checks on those units with critical application. All checks performed and data gathered should be recorded in such a fashion that the information is readily available to the operator. Correction factors for speed, timing, or temperature should be posted on the unit.

Tolerance limits for function checks should be established. The limits set will again depend on the application of the centrifuge. Include information for corrective action should tolerance limits be exceeded. Function verification procedures should include the following:

1. *rpm calibration*. One of the best ways to check the function of a centrifuge is to check its speed. Depending on how the unit is equipped, both the speed control and built-in tachometer should be checked with an external device. This may be accomplished with

either a strobe light or mechanical or electronic tachometer of good accuracy. Several speeds used regularly on the unit should be checked. Values obtained with the external measuring device should agree within 5% of those with a built-in tachometer.

2. *Timer*. The timer should be set for common timing intervals and these checked against an accurate stop-watch or electronic timer. General laboratory centrifuges should be accurate to 10% of the total timed interval.
3. ***Temperature***. The thermometer on refrigerated units should be checked against a certified thermometer, and a correction factor should be derived if necessary.

Preventive Maintenance

Preventive maintenance should be performed on the same time schedule as are function verification procedures. The preventive maintenance schedule should include the following:

Lubrication

Depending on the type of centrifuge, bearings on the upper and lower end of the motor shaft may be permanently lubricated or sealed. If this is not the case, then the manufacturer's instructions must be followed for lubrication. Bearing wear may be checked at this time by determining the amount of side play in the shaft.

Motor components

Brushes should be removed and checked for wear. Replacement is recommended if they are worn to more than one-half their original length. When reinserting used brushes, replace them in the same orientation, and be certain that spring tension is adequate to maintain good contact with the commutator. The condition of the commutator should be examined. In order to avoid electrical arcing, the commutator and brush holders must be free of dirt, oil, and dust. If the commutator is scartched or scored, it will have to be removed and machined smooth with a lathe. On refrigerated units the manufacturer's instruction manual should be consulted for maintenance procedures on the refrigeration system.

Electrical integrity

Both grounding resistance and current leakage should be checked periodically for proper rating; if the unit has a circuit breaker, it should be checked for proper operation. In addition, the line cord, plug, lamps, and wiring should be examined for defects.

Mechanical integrity

If the unit is equipped with a safety interlock, this device should be checked for proper working order. Gaskets, latches, hinges, and control knobs should be examined to determine that all are functioning and in good condition. The head and shields or carriers should be examined for signs of mechanical stress (cracks) and for cleanliness and balance.

Use and Care of Rotors and Centrifuges

Before using any centrifuge read the operational directions or obtain instructions form an experienced operator. Observe the following rules:

1. Do not overfill centrifuge tubes or bottles.
2. Balance the containers with their solutions before placing them in opposite cavities of a centrifuge.
3. Check the rotor to see that it is empty and undamaged before beginning a centrifuge run.
4. Do not exceed allowable speeds for a given rotor.
5. Turn off the centrifuge power immediately if any irregular noise or vibration occurs.
6. Keep rotors and rotor chambers clean. Chambers can be cleaned with a sponge, warm water, and a mild detergent; rotors can be cleaned with a bristle brush, warm water, and a mild detergent followed by rinsing with distilled water and drying by resting them upside-down.
7. See that the centrifuge is maintained according to instructions.

Some rotors for ultracentrifuges suffer from metal fatigue with prolonged use. Therefore, after a certain amount of use their maximum allowable speed must be decreased, a procedure called derating. For this reason, a record of their use must be maintained. Instructions on derating are provided with these rotors.

13

BIOLOGICAL MACHINES

Since the human body is made of molecules, molecular technology will be used to bring health. The ill, the old, and the injured all suffer from misarranged patterns of atoms, whether misarranged by invading viruses, passing time, or swerving cars. Devices able to rearrange atoms will be able to set them right. *Nanotechnology* will bring a fundamental breakthrough in medicine. Physicians now rely chiefly on surgery and drugs to treat illness. Surgeons have advanced from stitching wounds and amputating limbs to repairing hearts and reattaching limbs. Using microscopes and fine tools, they join delicate blood vessels and nerves. Yet even the best microsurgeon cannot cut and stitch finer tissue structures. Modern scalpels and sutures are simply too coarse for repairing capillaries, cells, and molecules. Consider "delicate" surgery from a cell's perspective a huge blade sweeps down, chopping blindly past and through the molecular machinery of a crowd of cells, slaughtering thousands. Later, a great obelisk plunges through the divided crowd, dragging a cable as wide as a freight train behind it to rope the crowd together again. From a cell's perspective even the most delicate surgery, performed with exquisite knives and great skill, is still a butcher job. Only the ability of cells to abandon their dead, regroup, and multiply makes healing possible. Yet as many paralyzed accident victims know too well, not all tissues heal.

Drug therapy, unlike surgery, deals with the finest structures in cells. *Drug molecules* are simple molecular devices. Many affect specific molecules in cells. *Morphine molecules*, for example, bind to certain receptor molecules in brain cells, affecting the neural impulses that signal pain. *Insulin*, *beta blockers*, and other drugs fit other receptors. But drug molecules work without direction. Once dumped into the body, they tumble and bump around in solution haphazardly until they bump a target molecule, fit, and stick, affecting its function. Surgeons can see problems and plan actions, but they wield crude tools; drug molecules affect tissues at the molecular level, but they are too simple to sense, plan, and act. But molecular machines directed by nano computers will offer physicians another choice. They will combine sensors, programs, and molecular tools to form systems able to examine and repair the ultimate components of individual cells. They will bring surgical control to the molecular domain.

These advanced molecular devices will be years in arriving, but researchers motivated by medical needs are already studying molecular machines and molecular engineering. The best drugs affect specific molecular machines in specific ways. Penicillin, for example, kills certain bacteria by jamming the *nanomachinery* they use to build their cell walls, yet it has little effect on human cells.

Biochemists study molecular machines both to learn how to build them and to learn how to wreck them. Around the world (and especially the Third World) a disgusting variety of viruses, bacteria, protozoa, fungi, and worms parasitize human flesh. Like penicillin, safe, effective drugs for these diseases would jam the parasite's molecular machinery while leaving human molecular machinery unharmed. Dr. Seymour Cohen, professor of pharmacological science at Suny, argues that biochemists should systematically study the molecular machinery of these parasites. Once biochemists have determined the shape and function of a vital protein machine, they then could often design a molecule shaped to jam it and ruin it. Such drugs could free humanity from such ancient horrors as schistosomiasis and leprosy, and from new ones such as AIDS.

Drug companies are already redesigning molecules based on knowledge of how they work. Researchers at Upjohn Company have designed and made modified molecules of *vasopressin*, a hormone that consists of a short chain of amino acids. Vasopressin increases the work done by the heart and decreases the rate at which the kidneys produce urine; this increases blood pressure. The researchers de signed modified vasopressin molecules that affected receptor molecules in the kidney more than those in the heart, giving them more specific and controllable medical effects. More recently, they de signed a modified vasopressin molecule that binds to the kidney's receptor molecules without direct effect, thus blocking and *inhibiting* the action of natural vasopressin.

Medical needs will push this work forward, encouraging researchers to take further steps toward protein design and molecular engineering. Medical, military, and economic pressures all push us in the same direction. Even before the assembler breakthrough, *molecular technology* will bring impressive advances in medicine; trends in biotechnology guarantee it. Still, these advances will generally be piecemeal and hard to predict, each exploiting some detail of biochemistry. Later, when we apply assemblers and technical AI systems to medicine, we will gain broader abilities that are easier to foresee.

To understand these abilities, consider cells and their self-repair mechanisms. In the cells of your body, natural radiation and noxious chemicals split molecules, producing reactive molecular fragments. These can misbond to other molecules in a process called cross-linking. As bullets and blobs of glue would damage a machine, so radiation and reactive fragments damage cells, both breaking molecular machines and gumming them up. If your cells could not repair themselves, damage would rapidly kill them or make them run amok by damaging their control systems. But evolution has favoured organisms with machinery able to do something about this problem. The self-replicating factory system repaired itself by replacing damaged parts; cells do the same. So long as a cell's DNA remains intact, it can make error-free tapes that direct ribosomes to assemble new protein machines.

Unfortunately for us, DNA itself becomes damaged, resulting in mutations. Repair enzymes compensate somewhat by detecting and repairing certain kinds of damage to DNA. These repairs help cells survive, but existing repair mechanisms are too simple to correct all problems, either in DNA or elsewhere. Errors mount, contributing to the aging and death of cells—and of people.

Life, Mind, and Machines

Does it make sense to describe cells as "machinery," whether self- repairing or not? Since we are made of cells, this might seem to reduce human beings to "*mere machines*," conflicting with a holistic understanding of life.

But a dictionary definition of holism is "the theory that reality is made up of organic or unified wholes that are greater than the simple sum of their parts." This certainly applies to people: one simpler sum of our parts would resemble hamburger, lacking both mind and life. The human body includes some ten thousand billion *billion* protein parts, and no machine so complex deserves the label "mere." Any brief description of so complex a system cannot avoid being grossly incomplete, yet at the cellular level a description in terms of machinery makes sense. Molecules have simple moving parts, and many act like familiar types of machinery. Cells considered as a whole may seem less mechanical, yet biologists find it useful to describe them in terms of molecular machinery.

Biochemists have unraveled what were once the central mysteries of life, and have begun to fill in the details. They have traced how molecular machines break food molecules into their building blocks and then reassemble these parts to build and renew tissue. Many details of the structure of human cells remain unknown (single cells have billions of large molecules of thousands of different kinds), but biochemists have mapped every part of some viruses. Biochemical laboratories often sport a large wall chart showing how the chief molecular building blocks flow through bacteria. Biochemists under stand much of the process of life in detail, and what they don't understand seems to operate on the same principles. The mystery of heredity has become the industry of genetic engineering. Even embryonic development and memory are being explained in terms of changes in biochemistry and cell structure.

In recent decades, the very quality of our remaining ignorance has changed. Once, biologists looked at the process of life and asked, "How can this be?" But today they understand the general principles of life, and when they study a specific living process they commonly ask, "Of the many ways this could be, which has nature chosen?" In many instances their studies have narrowed the competing explanations to a field of one. Certain biological processes—the coordination of cells to form growing embryos, learning brains, and reacting immune systems-still present a real challenge to the imagination. Yet this is not because of some deep mystery about how their parts work, but because of the immense complexity of how their many parts interact to form a whole.

Cells obey the same natural laws that describe the rest of the world. *Protein machines* in the right molecular environment will work whether they remain in a functioning cell or whether the rest of the cell was ground up and washed away days before. Molecular machines know nothing of "*life*" and "*death*."

Biologists—when they bother—sometimes define life as the ability to grow, replicate, and respond to stimuli. But by this standard, a mindless system of replicating factories might qualify as life, while a conscious artificial intelligence modeled on the human brain might not. Are viruses alive, or are they "*merely*" fancy molecular machines? No experiment can tell, because nature draws no line between living and nonliving. Biologists who work with viruses instead ask about viability: "Will this virus function, if given a chance?" The labels of "*life*" and "*death*" in medicine depend on medical capabilities: physicians ask, "Will this patient function, if we do our best?" Physicians once declared patients dead when the heart stopped; they now declare patients dead when they despair of restoring brain activity. Advances in cardiac medicine changed the definition once; advances in brain medicine will change it again.

Just as some people feel uncomfortable with the idea of machines thinking, so some feel uncomfortable with the idea that machines underlie our own thinking. The word "*machine*" again seems to conjure up the wrong image, a picture of gross, clanking metal, rather than signals flickering through a shifting weave of neural fibers, through a living tapestry more intricate than the mind it embodies can fully comprehend. The brain's really machinelike machines are of molecular size, smaller than the finest fibers.

A whole need not resemble its parts. A solid lump scarcely resembles a dancing fountain, yet a collection of solid, lumpy molecules forms fluid water. In a similar way, billions of molecular machines make up neural fibers and synapses, thousands of fibers and synapses make up a neural cell, billions of neural cells make up the brain, and the brain itself embodies the fluidity of thought. To say that the mind is "*just molecular machines*" is like saying that the Mona Lisa is '*just dabs of paint.*" Such statements confuse the parts with the whole, and confuse matter with the pattern it embodies. We are no less human for being made of molecules.

From Drugs to Cell Repair Machines

Being made of molecules, and having a human concern for our health, we will apply molecular machines to biomedical technology. Biologists already use antibodies to tag proteins, enzymes to cut and splice DNA, and viral syringes (like the T_4 phage) to inject edited DNA into bacteria. In the future, they will use assembler-built nanomachines to probe and modify cells. With tools like disassemblers, biologists will be able to study cell structures in ultimate, molecular detail. They then will catalog the hundreds of thousands of kinds of molecules in the body and map the structure of the hundreds of kinds

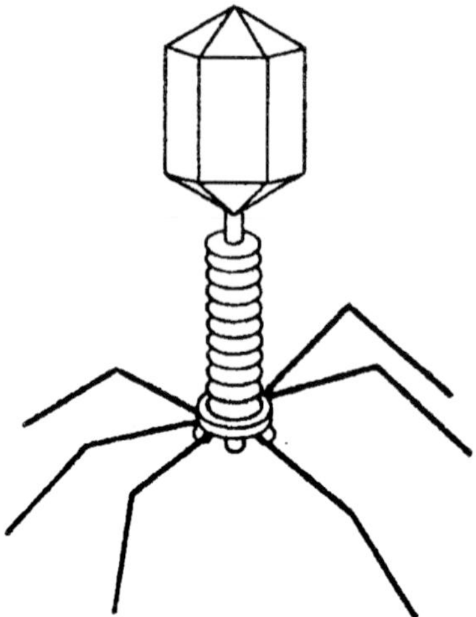

Fig. 13.1. Bacteriophage T_4.

of cells. Much as engineers might compile a parts list and make engineering drawings for an automobile, so biologists will describe the parts and structures of healthy tissue. By that time, they will be aided by sophisticated technical AI systems. Physicians aim to make tissues healthy, but with drugs and surgery they can only encourage tissues to repair themselves. Molecular machines will allow more direct repairs, bringing a new era in medicine.

To repair a car, a mechanic first reaches the faulty assembly, then identifies and removes the bad parts, and finally rebuilds or replaces them. Cell repair will involve the same basic tasks—tasks that living systems already prove possible.

Access

White blood cells leave the bloodstream and move through tissue, and viruses enter cells. Biologists even poke needles into cells without killing them. These examples show that molecular machines can reach and enter cells.

Recognition

Antibodies and the tail fibers of the T_4 phage—and indeed, all specific biochemical interactions—show that molecular systems can recognize other molecules by touch.

Disassembly

Digestive enzymes (and other, fiercer chemicals) show that molecular systems can disassemble damaged molecules.

Rebuilding

Replicating cells show that molecular systems can build or rebuild every molecule found in a cell.

Reassembly

Nature also shows that separated molecules can be put back together again. The machinery of the T_4 phage, for example, self-assembles from solution, apparently aided by a single enzyme. Replicating cells show that molecular systems can assemble every system found in a cell.

Thus, nature demonstrates all the basic operations that are needed to perform molecular-level repairs on cells. What is more, systems based on nanomachines will generally be more compact and capable than those found in nature. Natural systems show us only lower bounds to the possible, in cell repair as in everything else.

Cell Repair Machines

In short, with molecular technology and technical AI we will compile complete, molecular-level descriptions of healthy tissue, and we will build machines able to enter cells and to sense and modify their structures.

Cell repair machines will be comparable in size to bacteria and viruses, but their more-compact parts will allow them to be more complex. They will travel through tissue as white blood cells do, and enter cells as viruses do—or they could open and close cell membranes with a surgeon's care. Inside a cell, a repair machine will first size up the situation by examining the cell's contents and activity, and then take action. Early cell repair machines will be highly specialized, able to recognize and correct only a single type of molecular disorder, such as an enzyme deficiency or a form of DNA damage. Later machines (but not much later, with advanced technical AI systems doing the design work) will be programmed with more general abilities.

Complex repair machines will need nanocomputers to guide them. A micron-wide mechanical computer will fit in 1/1000 of the volume of a typical cell, yet will hold more information than does the cell's DNA. In a repair system, such computers will direct smaller, simpler computers, which will in turn direct machines to examine, take apart, and rebuild damaged molecular structures.

By working along molecule by molecule and structure by structure, repair machines will be able to repair whole cells. By working along cell by cell and tissue by tissue, they (aided by larger devices, where need be) will be able to repair whole organs. By working through a person organ by organ, they will restore health. Because molecular machines will be able to build molecules and cells from scratch, they will be able to repair even cells damaged to the point of complete inactivity. Thus, cell repair machines will bring a fundamental breakthrough: they will free medicine from reliance on self- repair as the only path to healing.

To visualize an advanced cell repair machine, imagine it—and a cell—enlarged until atoms are the size of small marbles. On this scale, the repair machine's smallest tools have tips about the size of your fingertips; a medium-sized protein, like hemoglobin, is the size of a typewriter; and a ribosome is the size of a washing machine. A single repair device contains a simple computer the size of a small truck, along with many sensors of protein size, several manipulators of ribosome size, and provisions for memory and motive power. A total volume ten meters across, the size of a three-story house, holds all these parts and more. With parts the size of marbles packing this volume, the repair machine can do complex things.

But this repair device does not work alone. It, like its many siblings, is connected to a larger computer by means of mechanical data links the diameter of your arm. On this scale, a cubic-micron computer with a large memory fills a volume thirty stories high and as wide as a football field. The repair devices pass it information, and it passes back general instructions. Objects so large and complex are still small enough: on this scale, the cell itself is a kilometer across, holding one thousand times the volume of a cubic-micron computer, or a million times the volume of a single repair device. Cells are spacious.

Will such machines be able to do everything necessary to repair cells? Existing molecular machines demonstrate the ability to travel through tissue, enter cells, recognize molecular structures, and so forth, but other requirements are also important. Will repair machines work fast enough? If they do, will they waste so much power that the patient will roast?

The most extensive repairs cannot require *vastly* more work than building a cell from scratch. Yet molecular machinery working within a cellular volume routinely does just that, building a new cell in tens of minutes (in bacteria) to a few hours (in mammals). This indicates that repair machinery occupying a few percent of a cell's volume will be able to complete even extensive repairs in a reason able time—

days or weeks at most. Cells can spare this much room. Even brain cells can still function when an inert waste called *lipofuscin* (apparently a product of molecular damage) fills over ten percent of their volume.

Powering repair devices will be easy: cells naturally contain chemicals that power *nanomachinery*. Nature also shows that repair machines can be cooled: the cells in your body rework themselves steadily, and young animals grow swiftly without cooking themselves. Handling heat from a similar level of activity by repair machines will be no sweat—or at least not too much sweat, if a week of sweating is the price of health.

All these comparisons of repair machines to existing biological mechanisms raise the question of whether repair machines will be able to *improve* on nature. DNA repair provides a clear-cut illustration.

Just as an illiterate "*book-repair machine*" could recognize and repair a torn page, so a cell's repair enzymes can recognize and repair breaks and cross-links in DNA. Correcting misspellings (or mutations), though, would require an ability to read. Nature lacks such repair machines, but they will be easy to build. Imagine three identical DNA molecules, each with the same sequence of nucleotides. Now imagine each strand mutated to change a few scattered nucleotides. Each strand still seems normal, taken by itself. Nonetheless, a repair machine could compare each strand to the others, one segment at a time, and could note when a nucleotide failed to match its mates. Changing the odd nucleotide to match the other two will then repair the damage. This method will fail if two strands mutate in the same spot. Imagine that the DNA of three human cells has been heavily damaged— after thousands of mutations, each cell has had one in every million nucleotides changed. The chance of our three-strand correction procedure failing at any given spot is then about one in a million mil lion. But compare five strands at once, and the odds become about one in a million million-million, and so on. A device that compares many strands will make the chance of an uncorrectable error effectively nil.

In practice, repair machines will compare DNA molecules from several cells, make corrected copies, and use these as standards for proofreading and repairing DNA throughout a tissue. By comparing several strands, repair machines will dramatically improve on nature's repair enzymes.

Other repairs will require different information about healthy cells and about how a particular damaged cell differs from the norm. Anti bodies identify proteins by touch, and properly chosen antibodies can generally distinguish any two proteins by their differing shapes and surface properties. Repair machines will identify molecules in a similar way. With a suitable computer and data base, they will be able to identify proteins by reading their amino acid sequences.

Consider a complex and capable repair system. A volume of two cubic microns—about 2/1000 of the volume of a typical cell—will be enough to hold a central data base system able to:

1. Swiftly identify any of the hundred thousand or so different human proteins by examining a short amino acid sequence.
2. Identify all the other complex molecules normally found in cells.
3. Record the type and position of every large molecule in the cell.

Each of the smaller repair devices (of perhaps thousands in a cell) will include a less capable computer. Each of these computers will be able to perform over a thousand computational steps in the time that a typical enzyme takes to change a single molecular bond, so the speed of computation possible seems more than adequate. Because each computer will be in communication with a larger computer and the central data base, the available memory seems adequate. Cell repair machines will have both the molecular tools they need and "brains" enough to decide how to use them.

Such sophistication will be overkill (over cure?) for many health problems. Devices that merely recognize and destroy a specific kind of cell, for example, will be enough to cure a cancer. Placing a computer network in every cell may seem like slicing butter with a chain saw, but having a chain saw available does provide assurance that even hard butter can be sliced. It seems better to show too much than too little, if one aims to describe the limits of the possible in medicine.

Some Cures

The simplest medical applications of *nanomachines* will involve not repair but selective destruction. Cancers provide one example; infectious diseases provide another. The goal is simple: one need only recognize and destroy the dangerous replicators, whether they are bacteria, cancer cells, viruses, or worms. Similarly, abnormal growths and deposits on arterial walls cause much heart disease; machines that recognize, break down, and dispose of them will clear arteries for more normal blood flow. Selective destruction will also cure diseases such as herpes in which a virus splices its genes into the DNA of a host cell. A repair device will enter the cell, read its DNA, and remove the addition that spells "*herpes*."

Repairing damaged, cross-linked molecules will also be fairly straightforward. Faced with a damaged, cross-linked protein, a cell repair machine will first identify it by examining short amino acid sequences, then look up its correct structure in a data base. The machine will then compare the protein to this blueprint, one amino acid at a time. Like a proofreader finding misspellings and strange characters (characters), it will find any changed amino acids or improper cross-links. By correcting these flaws, it will leave a normal protein, ready to do the work of the cell.

Repair machines will also aid healing. After a heart attack, scar tissue replaces dead muscle. Repair machines will stimulate the heart to grow fresh muscle by resetting cellular control mechanisms. By removing scar tissue and guiding fresh growth, they will direct the healing of the heart.

This list could continue through problem after problem (*Heavy metal poisoning*? —Find and remove the metal atoms) but the conclusion is easy to summarize. Physical disorders stem from misarranged atoms; repair machines will be able to return them to working order, restoring the body to health, Rather than compiling an endless list of curable diseases (from *arthritis*, *bursitis*, cancer, and dengue to yellow fever and zinc chills and back again), it makes sense to look for the limits to what cell repair machines can do. Limits do exist.

Consider stroke, as one example of a problem that damages the brain. Prevention will be straightforward: Is a blood vessel in the brain weakening, bulging, and apt to burst? Then pull it back into shape and guide the growth of reinforcing fibers. Does abnormal cloning threaten to block circulation? Then dissolve the clots and normalize the blood and blood-vessel linings to prevent a recurrence. Moderate neural damage from stroke will also be repairable: if reduced circulation has impaired function but left cell structures intact, then restore circulation and repair the cells, using their structures as a guide in restoring the tissue to its previous state. This will not only restore each cell's function, but will preserve the memories and skills embodied in the neural patterns in that part of the brain.

Repair machines will be able to regenerate fresh brain tissue even where damage has obliterated these patterns. But the patient would lose old memories and skills to the extent that they resided in that pan of the brain. If unique neural patterns are truly obliterated, then cell repair machines could no more restore them than art conservators could restore a tapestry from stirred ash. Loss of information through obliteration of structure imposes the most important, fundamental limit to the repair of tissue.

Other tasks are beyond cell repair machines for different reasons—maintaining mental health, for instance. Cell repair machines will be able to correct some problems, of course. Deranged thinking some times has biochemical causes, as if the brain were drugging or poisoning itself, and other problems stem from tissue damage. But many problems have little to do with the health of nerve cells and everything to do with the health of the mind.

A mind and the tissue of its brain are like a novel and the paper of its book. Spilled ink or flood damage may harm the book, making the novel difficult to read. Book repair machines could nonetheless restore physical "*health*" by removing the foreign ink or by drying and repairing the damaged paper fibers. Such treatments would do nothing for the book's *content*, however, which in a real sense is nonphysical. If the book were a cheap romance with a moldy plot and empty characters, repairs would be needed not on the ink and paper, but on the novel. This would call not for physical repairs, but for more work by the author, perhaps with advice.

Similarly, removing poisons from the brain and repairing its nerve fibers will thin some mental fogs, but not revise the content of the mind. This can be changed by the patient, with effort; we are all authors of our minds, But because minds change themselves by changing their brains, having a healthy brain will aid sound thinking more than quality paper aids sound writing. Readers familiar with computers may prefer to think in terms of hardware and software. A machine could repair a computer's hard ware while neither understanding nor changing its software.

Such machines might stop the computer's activity but leave the patterns in memory intact and ready to work again. In computers with the right kind of memory (called *nonvolatile*), users do this by simply switching off the power. In the brain the job seems more complex, yet there could be medical advantages to inducing a similar state.

Anesthesia Plus

Physicians already stop and restart consciousness by interfering with the chemical activity that underlies the mind. Throughout active life, molecular machines in the brain process molecules. Some disassemble sugars, combine them with oxygen, and capture the energy this releases. Some pump salt ions across cell membranes; others build small molecules and release them to signal other cells. Such processes make up the brain's metabolism, the sum total of its chemical activity. Together with its electrical effects, this metabolic activity underlies the changing patterns of thought.

Surgeons cut people with knives. In the mid-1800s, they learned to use chemicals that interfere with brain metabolism, blocking conscious thought and preventing patients from objecting so vigorously to being cut. These chemicals are *anesthetics*. Their molecules freely enter and leave the brain, allowing anesthetists to interrupt and restart human consciousness.

People have long dreamed of discovering a drug that interferes with the metabolism of the entire body, a drug able to interrupt metabolism completely for hours, days, or years. The result would be a condition of *biostasis* (from *bio*, meaning life, and *stasis*, meaning a stoppage or a stable state). A method of producing *reversible biostasis* could help *astronauts* on long space voyages to save food and avoid boredom, or it could serve as a kind of one-way time travel. In medicine, *biostasis* would provide a deep *anesthesia* giving physicians more time to work. When emergencies occur far from medical help, a good biostasis procedure would provide a sort of universal first-aid treatment: it would stabilize a patient's condition and prevent molecular machines from running amok and damaging tissues.

But no one has found a drug able to stop the entire metabolism the way anesthetics stop consciousness—that is, in a way that can be reversed by simply washing the drug out of the patient's tissues. Nonetheless, *reversible biostasis* will be possible when repair machines become available.

To see how one approach would work, imagine that the blood stream carries simple molecular devices to tissues, where they enter the cells. There they block the molecular machinery of metabolism—in the brain and elsewhere—and tie structures together with stabilizing cross-links. Other molecular devices then move in, displacing water and packing themselves solidly around the molecules of the cell. These steps stop metabolism and preserve cell structures. Be cause cell repair machines will be used to reverse this process, it can cause moderate molecular damage and yet do no lasting harm. With metabolism stopped and cell structures held firmly in place, the patient will rest quietly, dreamless and unchanging, until repair machines restore active life.

If a patient in this condition were turned over to a present-day physician ignorant of the capabilities of cell repair machines, the consequences would likely be grim. Seeing no signs of life, the physician would likely conclude that the patient was dead, and then would make this judgment a reality by "*prescribing*" an *autopsy*, followed by burial or *burning*.

But our imaginary patient lives in an era when biostasis is known to be only an interruption of life, not an end to it. When the patient's contract says "*wake me*!" (or the repairs are complete, or the flight to the stars is finished), the attending physician begins resuscitation. Repair machines enter the patient's tissues, removing the packing from around the patient's molecules and replacing it with water. They then remove the cross-links, repair any damaged molecules and structures, and restore normal concentrations of salts, blood sugar, ATP, and so forth. Finally, they unblock the metabolic machinery. The interrupted metabolic processes resume, the patient yawns, stretches, sits up, thanks the doctor, checks the date, and walks out the door.

From Function to Structure

The reversibility of biostasis and irreversibility of severe stroke damage help to show how cell repair machines will change medicine. Today, physicians can only help tissues to heal themselves. Accordingly, they must try to preserve the *function* of tissue. If tissues cannot function, they cannot heal. Worse, unless they are preserved, deterioration follows, ultimately obliterating structure. It is as if a mechanic's tools were able to work only on a running engine.

Cell repair machines change the central requirement from preserving *function* to preserving *structure*. Repair machines will be able to restore brain function with memory and skills intact only if the distinctive structure of the neural fabric remains intact. Biostasis involves preserving neural structure while deliberately blocking function. All this is a direct consequence of the molecular nature of the repairs. Physicians using scalpels and drugs can no more repair cells than someone using only a pickax and a can of oil can repair a fine watch. In contrast, having repair machines and ordinary nutrients will be like having a watchmaker's tools and an unlimited supply of spare parts. Cell repair machines will change medicine at its foundations.

From Treating Disease to Establishing Health

Medical researchers now study diseases, often seeking ways to pre vent or reverse them by blocking a key step in the disease process.

The resulting knowledge has helped physicians greatly: they now prescribe insulin to compensate for diabetes, antihypertensive to prevent stroke, penicillin to cure infections, and so on down an impressive list. Molecular machines will aid the study of diseases, yet they will make understanding disease far less important. Repair machines will make it more important to understand health.

The body can be ill in more ways than it can be healthy. Healthy muscle tissue, for example, varies in relatively few ways: it can be stronger or weaker, faster or slower, have this antigen or that one, and so forth. Damaged muscle tissue can vary in all these ways, yet also suffer from any combination of strains, tears, viral infections, parasitic worms, bruises, punctures, poisons, sarcomas, wasting diseases, and congenital abnormalities. Similarly, though neurons are woven in as many patterns as there are human brains, individual synapses and dendrites come in a modest range of forms--if they are healthy.

Once biologists have described normal molecules, cells, and tissues, properly programmed repair machines will be able to cure even unknown diseases. Once researchers describe the range of structures that (for example) a healthy liver may have, repair machines exploring a malfunctioning liver need only look for differences and correct them. Machines ignorant of a new poison and its effects will still recognize it as foreign and remove it. Instead of fighting a million strange diseases, advanced repair machines will establish a state of health.

Developing and programming cell repair machines will require great effort, knowledge, and skill. Repair machines with broad capabilities seem easier to build than to program. Their programs must contain detailed knowledge of the hundreds of kinds of cells and the hundreds of thousands of kinds of molecules in the human body. They must be able to map damaged cellular structures and decide how to correct them. How long will such machines and programs take to be developed? Offhand, the state of biochemistry and its present rate of advance might suggest that the basic knowledge alone will take centuries to collect. But we must beware of the illusion that advances will arrive in isolation.

Repair machines will sweep in with a wave of other technologies. The assemblers that build them will first be used to build instruments for analyzing cell structures. Even a pessimist might agree that human biologists and engineers equipped with these tools could build and program advanced cell repair machines in a hundred years of steady work. A cocksure, far-seeing pessimist might say a thousand years. A really committed nay-sayer might declare that the job would take people a million years. Very well: fast technical AI systems—a million fold faster than scientists and engineers—will then develop advanced cell repair machines in a single calendar year.

A Disease Called "Aging"

Aging is natural, but so were smallpox and our efforts to prevent it. We have conquered smallpox, and it seems that we will conquer aging. *Longevity* has increased during the last century, but chiefly be cause better sanitation and drugs have reduced bacterial illness. The basic human lifespan has increased little.

Still, researchers have made progress toward understanding and slowing the aging process. They have identified some of its causes, such as uncontrolled cross-linking. They have devised partial treatments, such as *antioxidants* and *free-radical inhibitors*. They have proposed and studied other mechanisms of aging, such as "*clocks*" in the cell and changes in the body's hormone balance. In laboratory experiments, special drugs and diets have extended the lifespan of mice by 25 to 45 percent.

Such work will continue; as the baby boom generation ages, expect a boom in aging research. One biotechnology company, Senetek of Denmark, specializes in *aging research*. In April 1985, Eastman Kodak and ICN Pharmaceuticals were reported to have joined in a $45 million venture to produce *isoprinosine* and other drugs with the potential to extend lifespan. The results of conventional antiaging research may substantially lengthen human life spans and improve the health of the old—during the next ten to twenty years. How greatly will drugs, surgery, exercise, and diet extend lifespans? For now, estimates must remain guesswork. Only new scientific knowledge can rescue such predictions from the realm of speculation, because they rely on new science and not just new engineering.

With cell repair machines, however, the potential for life extension becomes clear. They will be able to repair cells so long as their distinctive structures remain intact, and will be able to replace cells that have been destroyed. Either way, they will restore health. Aging is fundamentally no different from any other physical disorder; it is no magical effect of calendar dates on a mysterious life-force. Brittle bones, wrinkled skin, low enzyme activities, slow wound healing, poor memory, and the rest all result from damaged molecular machinery, chemical imbalances, and misarranged structures. By restoring all the cells and tissues of the body to a youthful structure, repair machines will restore youthful health.

People who survive intact until the time of cell repair machines will have the opportunity to regain youthful health and to keep it almost as long as they please. Nothing can make a person (or any thing else) last forever, of course, but barring severe accidents, those wishing to do so will live for a long, long time. As a technology develops, there comes a time when its principles become clear, and with them many of its consequences. The principles of rocketry were clear in the 1930s, and with them the consequence of spaceflight. Filling in the details involved designing and testing tanks, engines, instruments, and so forth. By the early 1950s, many details were known. The ancient dream of flying to the Moon had became a goal one could plan for.

The principles of molecular machinery are already clear, and with them the consequence of cell repair machines. Filling in the details will involve designing molecular tools, assemblers, computers, and so forth, but many details of existing molecular machines are known today. The ancient dream of achieving health and long life has be come a goal one can plan for.

Medical research is leading us, step by step, along a path toward molecular machinery. The global competition to make better materials, electronics, and biochemical tools is pushing us in the same direction. Cell repair machines will take years to develop, but they lie straight ahead. They will bring many abilities, both for good and for ill. A moment's thought about military *replicators* with abilities like those of cell repair machines is enough to turn up nauseating possibilities.

Decision Making Machines

Computers have emerged from back rooms and laboratories to help with writing, calculating, and play in homes and offices. These machines do simple, repetitive tasks, but machines still in the laboratory do much more. *Artificial intelligence* researchers say that computers can be made smart, and fewer and fewer people disagree. To understand our future, we must see whether artificial intelligence is as impossible as flying to the Moon.s

Thinking machines need not resemble human beings in shape, purpose, or mental skills. Indeed, some artificial intelligence systems will show few traits of the intelligent liberal arts graduate, but will instead serve only as powerful engines of design. Nonetheless, understanding how human minds evolved from mindless matter will shed light on bow machines can be made to think. Minds, like other forms of order, evolved through variation and selection.

Minds act. One need not embrace *Skinnerian behaviourism* to see the importance of behaviour, including the internal behaviour called *thinking*. RNA replicating in test tubes shows how the idea of purpose can apply (as a kind of shorthand) to utterly mindless molecules. They lack nerves and muscles, but they have evolved to "*behave*" in ways that promote their replication. Variation and selection have shaped each molecule's simple behaviour, which remains fixed for its whole "life." Individual RNA molecules don't adapt, but bacteria do. Competition has favoured bacteria that adapt to change, for example by adjusting their mix of digestive enzymes to suit the food available. Yet these mechanisms of adaptation are themselves fixed: food molecules trip genetic switches as cold air trips a thermostat.

Bacteria

Some bacteria also use a primitive form of *trial-and-error* guidance. Bacteria of this sort tend to swim in straight lines, and have just enough "*memory*" to know whether conditions are improving or worsening as they go. If they sense that conditions are improving, they keep going straight. If they sense that conditions are getting worse, they stop, tumble, and head off in a random, generally different, direction. They test directions, and favour the good directions by discarding the bad. And because this makes them wander toward concentrations of food molecules, they have prospered.

Flatworm

Flatworm lack brains, yet show the faculty of true learning. They can learn to choose the correct path in a simple T-maze. They try turning left and turning right, and gradually select the behaviour—or form the habit—which produces the better result. This is selection of behaviour by its consequences, which behaviourist psychologists call "the *Law of Effect*." The evolving genes of worm species have produced worm individuals with evolving behaviour.

Law of Effect

Still, worms trained to run mazes (even Skinner's pigeons, trained to peck when a light flashes green) show no sign of the reflective thought we associate with mind. Organisms adapting only though the simple *Law of Effect* learn only by *trial and error*, by varying and selecting actual behaviour—they don't think ahead and decide. Yet natural selection often favoured organisms that could think, and thinking is not magical. As Daniel Dennett of Tufts University points out, evolved genes can equip animal brains with internal models of bow the world works (somewhat like the models in computer-aided engineering systems). The animals can then "*imagine*" various actions and consequences, avoiding actions which "*seem*" dangerous and carrying out actions which "*seem*" safe and profitable. By testing ideas against these internal models, they can save the effort and risk of testing actions in the external world.

Dennett further points out that the *Law of Effect* can reshape the models themselves. As genes can provide for evolving behaviour, so they can provide for evolving mental models. Flexible organisms can vary their models and pay more attention to the versions that prove better guides to action. We all know what it is to try things, and learn which work. Models need not be instinctive; they can evolve in the course of a single life.

Speechiess Animals

Speechless animals, however, seldom pass on their new insights. These vanish with the brain that first produced them, because learned mental models are not stamped into the genes. Yet even speechless animals can imitate each other, giving rise to memes and cultures. A female monkey in Japan invented a way to use water to separate grain from sand; others quickly learned to do the same. In human cultures, with their language and pictures, valuable new models of how the world works can outlast their creators and spread worldwide.

Mind

On a still higher level, a mind (and "*mind*" is by now a fitting name) can hold evolving standards for judging whether the parts of a model—the ideas of a worldview—seem reliable enough to guide action. The mind thus selects its own contents, including its selection rules. The rules of judgment that filter the contents of science evolved in this way.

Goals

As behaviour, models, and standards for knowledge evolve, so can *goals*. That which *brings* good, as judged by some more basic standard, eventually begins to *seem* good: it then becomes a goal in itself. Honesty pays, and becomes a valued principle of action. As thought and mental models guide action and further thought, we adopt clear thinking and accurate models as goals in themselves. Curiosity grows, and with it a love of knowledge for its own sake. The evolution of goals thus brings forth both science and ethics. As Charles Darwin wrote, '*the highest possible stage in moral culture is when we recognize that we ought to control our thoughts.*" We achieve this as well by *variation* and *selection*, by concentrating on thoughts of value and letting others slip from attention.

Marvin Minsky of the MIT *Artificial Intelligence Laboratory* views the mind as a sort of society, an evolving system of communicating, cooperating, competing agencies, each made up of yet simpler agents. He describes thinking and action in terms of the activity of these agencies. Some agencies can do little more than guide a hand to grasp a cup; others (vastly more elaborate) guide the speech system as it chooses words in a sticky situation. We aren't aware of directing our fingers to wrap around a cup *just so*. We delegate such tasks to competent agents and seldom notice unless they slip. We all feel conflicting impulses and speak unintended words; these are symptoms of discord among the agents of the mind. Our awareness of this is part of the self-regulating process by which our most general agencies manage the rest.

Memes

Memes may be seen as agents in the mind that are formed by teaching and imitation. To feel that two ideas conflict, you must have embodied both of them as agents in your mind—though one may be old, strong, and supported by allies, and the other a fresh idea-agent that may not survive its first battle. Because of our superficial self-awareness, we often wonder where an idea in our heads came from. Some people imagine that these thoughts and feelings come directly from agencies outside their own minds; they incline toward a belief in haunted heads.

Genius

In ancient Rome, people believed in "*genii*," in good and evil spirits attending a person from cradle to grave, bringing good and ill luck. They attributed outstanding success to a special "*genius*." And even now, people who fail to see how natural processes create novelty see "*genius*" as a form of magic. But in fact, evolving genes have made minds that expand their knowledge by varying idea patterns and selecting among them. With quick variation and effective selection, guided by knowledge borrowed from others, why shouldn't such minds show what we call genius? Seeing intelligence as a natural process makes the idea of intelligent machines less startling. It also suggests how they might work.

Machine Intelligence

One dictionary definition of "*machine*" is "Any system or device, such as an electronic computer, that performs or assists in the performance of a human task." But just how many human tasks will machines be able to perform? Calculation was once a mental skill beyond machines, the province of the intelligent and educated. Today, no one thinks of calling a pocket calculator an *artificial intelligence*; calculation now seems a "*merely*" mechanical procedure.

Still, the idea of building ordinary computers once was shocking. By the mid-1800s, though, Charles Babbage had built mechanical calculators and part of a programmable mechanical computer; how ever, he ran into difficulties of finance and construction. One Dr. Young helped not at all: he argued that it would be cheaper to invest the money and use the interest to pay human calculators. Nor did the British Astronomer Royal, Sir George Airy—an entry in his diary states that "On September 15th Mr. Goulburn ... asked my opinion on the utility of Babbage's calculating machine ... I replied, entering fully into the matter, and giving my opinion that it was worthless."

Babbage's machine was ahead of its time—meaning that in building it, machinists were forced to advance the art of making precision parts. And in fact it would not have greatly exceeded the speed of a skilled human calculator—but it would have been more reliable and easier to improve.

The story of *computers* and *artificial intelligence* (known as *AI*) resembles that of flight in air and space. Until recently people dismissed both ideas as impossible—commonly meaning that they couldn't see how to do them, or would be upset if they could. And so far, *AI* has had no simple, clinching demonstration, no equivalent of a working airplane or a landing on the Moon. It has come a long way, but people keep changing their definitions of intelligence.

Press reports of "*giant electronic brains*" aside, few people called the first computers intelligent. Indeed, the very name "*computer*" suggests a mere arithmetic machine. Yet in 1956, at Dartmouth, during the world's first conference on artificial intelligence, researchers Alan Newell and Herbert Simon unveiled Logic Theorist, a program that proved theorems in symbolic logic.

In later years computer programs were playing chess and helping chemists determine molecular structures. Two medical programs,

CASNET and *MYCIN* (the first dealing with internal medicine, the other with the diagnosis and treatment of infections), have performed impressively. According to the *Handbook of Art Intelligence*, they have been "rated, in experimental evaluations, as performing at human-expert levels in their respective domains." A program called *prospector* has located, in Washington state, a molybdenum deposit worth millions of dollars.

These so-called "*expert systems*" succeed only within strictly limited areas of competence, but they would have amazed the computer programmers of the early 1950s. Today, however, few people consider them to be *real* artificial intelligence: *AI* has been a moving target. The passage from *Business Week* quoted earlier only shows that computers can now be programmed with enough knowledge, and perform fancy enough tricks, that some people feel comfortable calling them intelligent. Years of seeing fictional robots and talking computers on television have at least made the idea of *AI* familiar.

The chief reason for declaring *AI* impossible has always been the notion that "*machines*" are intrinsically stupid, an idea that is now beginning to fade. Past machines have indeed been gross, clumsy things that did simple, brute-force work. But computers handle information, follow complex instructions, and can be instructed to change their own instructions, They can experiment and learn. They contain not gears and grease but traceries of wire and evanescent patterns of electrical energy. As Douglas Hofstadter urges (through a character in a dialogue about *AI*), "Why don't you let the word '*machine*' conjure up images of patterns of dancing light rather than of giant steam shovels?"

Cocktail-party critics confronted with the idea of artificial intelligence often point to the stupidity of present computers, as if this proved something about the future. (A future machine may wonder whether such critics exhibited genuine thought.) Their objection is irrelevant—steam locomotives didn't fly, though they demonstrated mechanical principles later used in airplane engines. Likewise, the creeping worms of an eon ago showed no noticeable intelligence, yet our brains use neurons much like theirs.

Casual critics also avoid thinking seriously about *AI* by declaring that we can't possibly build machines smarter than ourselves. They forget what history shows. Our distant, speechless ancestors managed to bring forth entities of greater intelligence through genetic evolution without even thinking about it. But we are thinking about it, and the memes of technology evolve far more swiftly than the genes of biology.

We can surely make machines with a more human-like ability to learn and organize knowledge.

There seems to be only one idea that could argue for the impossibility of making thought patterns dance in new forms of matter. This is the idea of *mental materialism*—the concept that mind is a special substance, a magical thinking-stuff somehow beyond imitation, duplication, or technological use.

Psychobiologists see no evidence for such a substance, and find no need for mental materialism to explain the mind. Because the complexity of the brain lies beyond the full grasp of human understanding, it seems complex enough to embody a mind. Indeed, if a single person could *fully* understand a brain, this would make the brain less complex than that person's mind. If all Earth's billions of people could cooperate in simply watching the activity of one human brain, each person would have to monitor tens of thousands of active synapses simultaneously—clearly an impossible task. For a person to try to understand the flickering patterns of the brain as a whole would be five billion times more absurd. Since our *brain's mechanism* so massively overwhelms our mind's ability to grasp it, that mechanism seems complex enough to embody the mind itself.

Turing's Target

In a 1950 paper on machine intelligence, British mathematician Alan Turing wrote: "I believe that by the end of the century the use of words and general educated opinion will have altered so much that one will be able to speak of machines thinking without expecting to be contradicted." But this will depend on what we call thinking. Some say that only people can think, and that computers cannot be people; they then sit back and look smug.

But in his paper, Turing asked how we judge *human* intelligence, and suggested that we commonly judge people by the quality of their conversation. He then proposed what he called the *imitation game*—which everyone else now calls the Turing test. Imagine that you are in a room, able to communicate through a terminal with a person and a computer in two other rooms. You type messages; both the person and the computer can reply. Each tries to act human and intelligent. After a prolonged keyboard "*conversation*" with them—perhaps touching on literature, art, the weather, and how a mouth tastes in the morning—might be that you could not tell which was the person and which the machine. If a machine could converse this well on a regular basis, then Turing suggests that we should consider it genuinely intelligent.

Further, we would have to acknowledge that it knew a great deal about human beings.

For most practical purposes, we need not ask "Can a machine have *self-awareness*—that is, *consciousness*?" Indeed, critics who declare that machines cannot be conscious never seem able to say quite what they mean by the term. Self-awareness evolved to guide thought and action, not merely to ornament our humanity. We must be aware of other people, and of their abilities and inclinations, to make plans that involve them. Likewise we must be aware of ourselves, and of our own abilities and inclinations, to make plans about ourselves. There is no special mystery in self awareness. What we call the self-reacts to impressions from the rest of the mind, orchestrating some of its activities; this makes it no more (and no less) than a special part of the interacting patterns of thought. The idea that the self is a pattern in a special mind substance (distinct from the mind substance of the brain) would explain nothing about awareness.

A machine attempting to pass the Turing test would, of course, claim to have *self-awareness*. Hard-core biochauvinists would simply say that it was lying or confused. So long as they refuse to say what they mean by consciousness, they can never be proved wrong. Nonetheless, whether called *conscious* or not, *intelligent machines* will still *act* intelligent, and it is their actions that will affect us. Perhaps they will someday shame the biochauvinists into silence by impassioned argument, aided by a brilliant public-relations campaign.

No machine can now pass the *Turing test*, and none is likely to do so soon. It seems wise to ask whether there is a good reason even to try: we may gain more from *AI* research guided by other goals.

Let us distinguish two sorts of artificial intelligence, though a system could show both kinds. The first is *technical AI*, adapted to deal with the physical world. Efforts in this field lead toward automated engineering and scientific inquiry. The second is *social AI*. adapted to deal with human minds. Efforts in this field lead toward machines able to pass the *Turing test*.

Researchers working on *social AI* systems will learn much about the human mind along the way, and their systems will doubtless have great practical value, since we all can profit from intelligent help and advice. But automated engineering based on *technical AI* will have a greater impact on the technology race, including the race toward molecular technology. And an advanced automated engineering system may be easier to develop than a *Turing-test passer*, which must not only possess knowledge and intelligence, but must mimic *human* knowledge and *human* intelligence—a special more difficult challenge.

As Turing asked, "May not machines carry out something which ought to be described as thinking but which is very different from what a man does?" Although some writers and politicians may refuse to recognize machine intelligence until they are confronted with a talkative machine able to pass the Turing test, many engineers will recognize intelligence in other forms.

Engines of Design

We are well on the way to *automated engineering*. Knowledge engineers have marketed expert systems that help people to deal with practical problems. Programmers have created computer-aided design systems that embody knowledge about shapes and motion, stress and strain, electronic circuits, heat flow, and how machine tools shape metal. Designers use these systems to augment their mental models, speeding the evolution of yet unbuilt designs. Together, designers and computers form intelligent, semiartificial systems.

Engineers can use a wide variety of computer systems to aid their work. At one end of the spectrum, they use computer screens simply as drawing boards. Farther along, they use systems able to describe parts in three-dimensions and calculate their response to heat, stress, current, and so on. Some systems also know about computer-controlled manufacturing equipment, letting engineers make simulated tests of instructions that will later direct computer-controlled machines to make real parts. But the far end of the spectrum of systems involves using computers not just to record and test designs, but to generate them.

Programmers have developed their most impressive tools for use in the computer business itself. Software for chip design is an example. *Integrated circuit chips* now contain many thousands of transistors and wires. Designers once had to work for many months to design a circuit to do a given job, and to lay out its many parts across the surface of the chip. Today they can often delegate this task to a so-called "*silicon compiler*." Given a specification of a chip's function, these software systems can produce a detailed design—ready for manufacture—with little or no human help.

All these systems rely entirely on human knowledge, labouriously gathered and coded. The most flexible automated design systems today can fiddle with a proposed design to seek improvements, but they learn nothing applicable to the next design. But EURISKO is different.

Developed by Professor Douglas Lenat and others at Stanford University, EURISKO is designed to explore new areas of knowledge. It is guided by heuristics—pieces of knowledge that suggest plausible actions to follow or implausible ones to avoid; in effect, various rules of thumb. It uses heuristics to suggest topics to work on, and further heuristics to suggest what approaches to try and how to judge the results. Other heuristics look for patterns in results, propose new heuristics, and rate the value of both new and old heuristics. In this way EURISKO evolves better behaviours, better internal models, and better rules for selecting among internal models. Lenat himself describes the variation and selection of heuristics and concepts in the system in terms of "mutation" and "selection," and suggests a social, cultural metaphor for understanding their interaction.

Since heuristics evolve and compete in EURISKO, it makes sense to expect parasites to appear—as indeed many have. One machine-generated heuristic, for example, rose to the highest possible value rating by claiming to have been a co-discoverer of every valuable new conjecture. Professor Lenat has worked closely with EURISKO, improving its mental immune system by giving it heuristics for shed ding parasites and avoiding stupid lines of reasoning.

EURISKO has been used to explore elementary mathematics, programming, biological evolution, games, three-dimensional integrated circuit design, oil spill cleanup, plumbing, and (of course) heuristics. In some fields it has startled its designers with novel ideas, including new electronic devices for the emerging technology of three-dimensional integrated circuits.

The results of a tournament illustrate the power of a human/*AI* team. Traveller TCS is a futuristic naval war game, played in accordance with two hundred pages of rules specifying design, cost, and performance constraints for the fleet ("*TCS*" stands for "*Trillion Credit Squadron*"). Professor Lenat gave EURISKO these rules, a set of starting heuristics, and a program to simulate a battle between two fleets. He reports that "it then designed fleet after fleet, using the simulator as the '*natural selection*' mechanism as it '*evolved*' better and better fleet designs." The program would run all night, designing, testing, and drawing lessons from the results. In the morning Lenat would cull the designs and help it along. He credits about 60 percent of the results to himself, and about 40 percent to EURISKO.

Lenat and EURISKO entered the 1981 national Traveller TCS tournament with a strange fleet. The other contestants laughed at it, then lost to it. The Lenat/EURISKO fleet won every round, emerging as the national champion. As Lenat notes, "This win is made more significant by the fact that no one connected with the program had ever played this game before the tournament, or seen it played, and there were no practice rounds." In 1982 the competition sponsors changed the rules. Lenat and EURISKO entered a very different fleet. Other contestants again laughed at it, then lost. Lenat and EURISKO again won the national championship.

In 1983 the competition sponsors told Lenat that if he entered and won again, the Competition would be canceled. Lenat bowed out.

EURISKO and other *AI* programs show that computers need not be limited to boring, repetitive work if they are given the right sort of programming. They can explore possibilities and turn up novel ideas that surprise their creators. EURISKO has shortcomings, yet it points the way to a style of partnership in which an AI system and a human expert both contribute knowledge and creativity to a design process.

In coming years, similar systems will transform engineering. Engineers will work in a creative partnership with their machines, using software derived from current computer-aided design systems for doing simulations, and using evolving, EURISKO-like systems to suggest designs to simulate. The engineer will sit at a screen to type in goals for the design process and draw sketches of proposed designs. The system will respond by refining the designs, testing them, and displaying proposed alternatives, with explanations, graphs, and diagrams. The engineer will then make further suggestions and changes, or respond with a new task, until an entire system of hard ware has been designed and simulated.

As such automated engineering systems improve, they will do more and more of the work faster and faster. More and more often, the engineer will simply propose goals and then sort among good solutions proposed by the machine. Less and less often will the engineer have to select parts, materials, and configurations. Gradually engineers will be able to propose more general goals and expect good solutions to appear as a matter of course. Just as EURISKO ran for hours evolving fleets with a *Traveller TCS simulator*, automated engineering systems will someday work steadily to evolve passenger jets having maximum safety and economy—or to evolve military jets and missiles best able to control the skies.

Just as EURISKO has invented electronic devices, future automated engineering systems will invent molecular machines and molecular electronic devices, aided by software for molecular simulations. Such advances in automated engineering will magnify the design-ahead phenomenon described earlier. Thus automated engineering will not only speed the assembler breakthrough, it will increase the leap that follows.

Eventually software systems will be able to create bold new designs without human help. Will most people call such systems intelligent? It doesn't really matter.

AI Race

Companies and governments worldwide support *AI* work because it promises commercial and military advantages. The United States has many university artificial intelligence laboratories and a host of new companies with names like Machine Intelligence Corporation, Thinking Machines Corporation, Teknowledge, and Cognitive Systems Incorporated. In October of 1981 the Japanese Ministry of Trade and Industry announced a ten-year, $850 million program to develop advanced AI hardware and software. With this, Japanese researchers plan to develop systems able to perform a billion logical inferences per second. In the fall of 1984 the Moscow Academy of Science announced a similar, five-year, $100 million effort. In October of 1983 the U.S. Department of Defense announced a five-year, $600 million Strategic Computing Program; they seek machines able to see, reason, understand speech, and help manage battles. As Paul Wallich reports in the *IEEE Spectrum*, "Artificial intelligence is considered by most people to be a cornerstone of next-generation computer technology; all the efforts in different countries accord it a prominent place in their list of goals.'

Advanced AI will emerge step by step, and each step will pay off in knowledge and increased ability. As with molecular technology (and many other technologies), attempts to stop advances in one city, county, or country will at most let others take the lead. A miraculous success in stopping visible *AI* work *everywhere* would at most delay it and, as computers grow cheaper, let it mature in secret, beyond public scrutiny. Only a world state of immense power and stability could truly stop *AI* research everywhere and forever—a "*solution*" of bloodcurdling danger, in light of past abuses of merely national power. Advanced AI seems inevitable. If we hope to form a realistic view of the future, we cannot ignore it.

In a sense, *artificial intelligence* will be the ultimate tool because it will help us build all possible tools. *Advanced AI* systems could maneuver people out of existence, or they could help us build a new and better world. Aggressors could use them for conquest, or fore sighted defenders could use them to stabilize peace. They could even help us control AI itself. The hand that rocks the AI cradle may well rule the world.

As with assemblers, we will need foresight and careful strategy to use this new technology safely and well. The issues are complex and interwoven with everything from the details of molecular technology to employment and the economy to the philosophical basis of human rights. The most basic issues, though, involve what AI can do.

Are We Smart Enough?

Despite the example of the evolution of human beings, critics may still argue that our limited intelligence may somehow prevent us from programming genuinely intelligent machines. This argument seems weak, amounting to little more than a claim that because the critic can't see how to succeed, no one else will ever do better. Still, few would deny that programming computers to equal human abilities will indeed require fresh insights into human psychology. Though the programming path to AI seems open, our knowledge does not justify the sort of solid confidence that thoughtful engineers had (decades before Sputnik) in being able to reach the Moon with rockets, or that we have today in being able to build assemblers through protein design. Programming genuine artificial intelligence, though a form of engineering, will require new science. This places it beyond firm projection.

We need accurate foresight, though. People clinging to comforting doubts about AI seem likely to suffer from radically flawed images of the future. Fortunately, automated engineering escapes some of the burden of biochauvinist prejudice. Most people are less upset by the idea of machines designing machines than they are by the idea of true general-purpose *AI systems*. Besides, automated engineering has been shown to work; what remains is to extend it. Still, if more general systems are likely to emerge, we would be foolish to omit them from our calculations. Is there a way to sidestep the question of our ability to design intelligent programs?

In the 1950s, many *AI researchers* concentrated on simulating brain functions by simulating neurons. But researchers working on programs based on words and symbols made swifter progress, and the focus of AI work shifted accordingly. Nonetheless, the basic idea of neural simulation remains sound, and molecular technology will make it more practical. What is more, this approach seems guaranteed to work because it requires no fundamental new insights into the nature of thought.

Eventually, *neurobiologists* will use *virus-sized molecular machines* to study the structure and function of the brain, cell by cell and molecule by molecule where need be. Although *AI* researchers may gain useful insights about the organization of thought from the resulting advances in brain science, neural simulation can succeed without such insights. Compilers translate computer programs from one language to another without understanding how they work. Photocopies transfer patterns of words without reading them. Like wise, researchers will be able to copy the neural patterns of the brain to another medium without understanding their higher-level organization.

After learning how neurons work, engineers will be able to design and build analogous devices based on *advanced nano electronics* and *nanomachines*. These will interact like neurons, but will work faster. Neurons, though complex, do seem simple enough for a mind to understand and an engineer to imitate. Indeed, neurobiologists have learned much about their structure and function, even without molecular machinery to probe their workings.

With this knowledge, engineers will be able to build fast, capable AI systems, even without understanding the brain and without clever programming. They need only study the brain's neural structure and join 5 neurons to form the same functional pattern. If they make all the parts right—including the way they mesh to form the whole—then the whole, too, will be right. "*Neural*" activity will flow in the patterns we call thought, but faster, because all the parts will work faster.

Accelerating the Technology Race

Advanced AI systems seem possible and inevitable, but what effect will they have? No one can answer this in full, but one effect of automated engineering is clear: it will speed our advance toward the limits of the possible.

To understand our prospects, we need some idea of how fast *advanced AI systems* will think. Modern computers have only a tiny fraction of the brain's complexity, yet they can already run programs imitating significant aspects of human behaviour. They differ totally from the brain in their basic style of operation, though, so direct physical comparison is almost useless. The brain does a huge number of things at once, but fairly slowly; most modern computers do only one thing at a time, but with blinding speed.

Still, one can imagine *AI hardware* built to imitate a brain not only in function, but in structure. This might result from a neural-simulation approach, or from the evolution of *AI programs* to run on hard ware with a brain like style of organization. Either way, we can use analogies with the human brain to estimate a *minimum* speed for advanced assembler-built *AI systems*.

Neural synapses respond to signals in thousandths of a second; experimental electronic switches respond a hundred million times faster (and nano electronic switches will be faster yet). Neural signals travel at under one hundred meters per second; electronic signals travel a million times faster. This crude comparison of speeds suggests that brain like electronic devices will work about a million times faster than brains made of neurons (at a rate limited by the speed of electronic signals). This estimate is crude, of course. A neural synapse is more complex than a switch; it can change its response to signals by changing its structure. Over time, synapses even form and disappear. These changes in the fibers and connections of the brain embody the long-term mental changes we call learning. They have stirred Professor Robert Jastrow of Dartmouth to describe the brain as an enchanted loom, weaving and reweaving its neural patterns throughout life.

To imagine a brain like device with comparable flexibility, picture its electronic circuits as surrounded by mechanical nanocomputers and assemblers, with one per synapse-equivalent "*switch*. "Just as the molecular machinery of a synapse responds to patterns of neural activity by modifying the synapse's structure, so the nanocomputers will respond to patterns of activity by directing the nanomachinery to modify the switch's structure. With the right programming, and with communication among the nanocomputers to simulate chemical signals, such a device should behave almost exactly like a brain.

Despite its complexity, the device will be compact. *Nanocomputers* will be smaller than *synapses*, and assembler-built wires will be thinner than the brain's *axons* and *dendrites*. Thin wires and small switches will make for compact circuits, and compact circuits will speed the flow of electronic patterns by shortening the distances signals must travel. It seems that a structure similar to the brain will fit in less than a cubic centimeter. Shorter signal paths will then join with faster transmission to yield a device over ten million times faster than a human brain.

Only cooling problems might limit such machines to slower average speeds. Imagine a conservative design, a million fold faster than a brain and dissipating a million fold more heat. The system consists of an assembler-built block of sapphire the size of a coffee mug, honeycombed with circuit-lined cooling channels. A high-pressure water pipe of equal diameter is bolted to its top, forcing cooling water through the channels to a similar drainpipe leaving the bottom. Hefty power cables and bundles of optical-fiber data channels trail from its sides.

The cables supply fifteen megawatts of electric power. The drainpipe carries the resulting heat away in a three-ton-per-minute flow of boiling-hot water. The optical fiber bundles carry as much data as a million television channels. They bear communications with other *AI* systems, with engineering simulators, and with assembler systems that build designs for final testing. Every ten seconds, the system gobbles almost two kilowatt-days of electric energy (now worth about a dollar). Every ten seconds, the system completes as much design work as a human engineer working eight hours a day for a year (now worth tens of thousands of dollars). In an hour, it completes the work of centuries. For all its activity, the system works in a silence broken only by the rush of cooling water.

This addresses the question of the sheer speed of thought, but what of its complexity? *AI development* seems unlikely to pause at the complexity of a single human mind. As John McCarthy of Stanford's *AI lab* points out, if we can place the equivalent of one human mind in a metal skull, we can place the equivalent of ten thousand cooperating minds in a building. (And a large modern power plant could supply power enough for each to think at least ten thousand times as fast as a person.) To the idea of *fast engineering intelligences*, add the idea of fast engineering teams.

Engineering AI systems will be slowed in their work by the need to perform experiments, but not so much as one might expect. Engineers today must perform many experiments because bulk technology is unruly. Who can say in advance exactly how a new alloy will behave when forged and then bent ten million times? Tiny cracks weaken metal, but details of processing determine their nature and effects.

Because assemblers will make objects to precise specifications, the unpredictabilities of bulk technology will be avoided. Designers (whether human or *AI*) will then experiment only when experimentation is faster or cheaper than calculation, or (more rarely) when basic knowledge is lacking.

AI systems with access to *nanomachines* will perform many experiments rapidly. They will design apparatus in seconds, and replicating assemblers will build it without the many delays (ordering special parts, shipping them, and so on) that plague projects today. Experimental apparatus on the scale of an assembler, nanocomputer, or living cell will take only minutes to build, and nanomanipulators will perform a million motions per second. Running a million ordinary experiments at once will be easy. Thus, despite delays for experimentation, automated engineering systems will move technology forward with stunning speed.

From past to future, then, the likely pattern of advancing ability looks something like this. Across eons of time, life moved forward in a long, slow advance, paced by genetic evolution. Minds with language picked up the pace, accelerated by the flexibility of memes. The invention of the methods of science and technology further accelerated advances by forcing memes to evolve faster. Growing wealth, education, and population—and better physical and intellectual tools—have continued this accelerating trend across our century.

The automation of engineering will speed the pace still more. Computer-aided design will improve, helping human engineers to generate and test ideas ever more quickly. Successors to EURISKO will shrink design times by suggesting designs and filling in the de tails of human innovations. At some point, full-fledged automated engineering systems will pull ahead on their own.

In parallel, molecular technology will develop and mature, aided by advances in automated engineering. Then assembler-built *AI systems* will bring still swifter automated engineering, evolving technological ideas at a pace set by systems a million times faster than a human brain. The rate of technological advance will then quicken to a great upward leap: in a brief time, many areas of technology will advance to the limits set by natural law. In those fields, advance will then halt on a lofty plateau of achievement.

This transformation is a dizzying prospect. Beyond it, if we survive, lies a world with replicating assemblers, able to make whatever they are told to make, without need for human labour. Beyond it, if we survive, lies a world with automated engineering systems able to direct assemblers to make devices near the limits of the possible, near the final limits of technical perfection.

Eventually, some *AI systems* will have both great technical ability and the social ability needed to understand human speech and wishes. If given charge of energy, materials, and assemblers, such a system might aptly be called a "*genie machine.*" What you ask for, it will produce. Arabian legend and universal common sense suggest that we take the dangers of such engines of creation very seriously indeed.

Decisive breakthroughs in technical and *social AI* will be years in arriving. As Marvin Minsky has said, "The modestly intelligent machines of the near future promise only to bring us the wealth and comfort of tireless, obedient, and inexpensive servants." Most systems now called "*AI*" do not think or learn; they are only a crude distillate of the skills of experts, preserved, packaged, and distributed for consultation.

But *genuine AI* will arrive. To leave it out of our expectations would be to live in a fantasy world. To expect *AI* is neither optimistic nor pessimistic: as always, the researcher's optimism is the technophobe's pessimism. If we do not prepare for their arrival, social *AI systems* could pose a grave threat: consider the damage done by the merely human intelligence of terrorists and demagogues. Likewise, technical *AI systems* could destabilize the world military balance, giving one side a sudden, massive lead. With proper preparation, however, artificial intelligence could help us build a future that works—for the Earth, for people, and for the advancement of intelligence in the universe.

Why discuss the dangers today? Because it is not too soon to start developing institution able to deal with such question. *Technical AI* is emerging today, and its every advance will speed the technology race. Artificial intelligence is but one of many powerful technologies we must learn to manage, each adding to a complex mixture of threats and opportunities.

14

CHROMATOGRAPHY

One of the most generally useful methods of protein fractionation involves chromatography, a technique that was originally developed for the fractionation of low molecualr weight components, such as sugars and amino acids. The most common type of chromatography used to separate small molecules is known as partition chromatography. In the most convenient form of this technique, a drop of sample is applied as a spot to a sheet of absorbent paper (in *paper chromatography*) or to a sheet of plastic or glass that has been covered with a thin layer of inert absorbent material such as cellulose or silica gel (*thin-layer chromatography*). Then a mixture of solvents, such as water and an alcohol, is allowed to permeate the sheet from one edge. As the solvents move across the sheet, they pick up those molecules in the sample that are soluble in them. The solvents are selected so that one of them absorbs more strongly to the absorbent material than the other. Cosequently, molecules that are most soluble in the absorbed solvent are relatively retarded, while those that are most soluble in the other solvent move more quickly. To detect the location of the different molecules, the chromatograph is dried and then stained. Various forms of paper and thin-layer chromatography are still widely used to analyze many different small molecules.

Proteins are most often fractionated by *column chromatography*, in which a mixture of proteins in solution is passed through a column containing a porous solid matrix and the different proteins are retarded to different extents by their interaction with the matrix. In this way, different proteins can be separately collected as they flow out of the bottom of the column. Many types of matrices have been developed

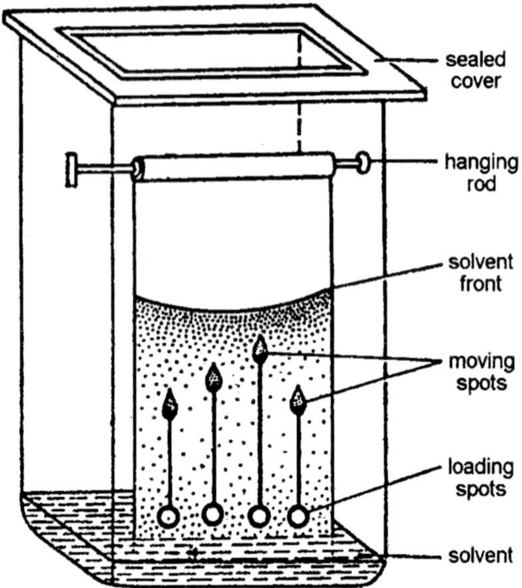

Fig. 14.1. Paper chromatography.

that allow proteins to be separated without altering their native structure. These matrices discriminate between different proteins by change size, or their ability to bind to particular chemical groups on the matrix.

The separation of small molecules by *paper chromatography*. After the sample has been applied to the origin and dried, a solution containing a mixture of two solvents is allowed to flow slowly through the paper by capillary action. Different components in the sample move at different rates in the paper according to their relative solubility in the solvent that is preferentially absorbed by the paper.

The separation of molecues by *column chromatography*. The sample is appplied to the top of a cylindrical column made of glass or plastic containing a permeable solid matrix immersed in solvent. Then a large amount of solvent is pumped slowly through the column and is collected in separate tubes as it emerges from the bottom. Various components of the sample travel at different rates through the column and are thereby fractionated.

How Cells are Studied

Schematic drawing of some different types of chromatographic matrix. In ion-exchange chromatography (A), the insoluble matrix carries ionic charges that retard molecules of opposite charge. Matrices commonly used for separating proteins are diethylaminoethylcellulose (DEAEcellulose), which is positively charged; and carboxymethyl cellulose (CM-cellulose) and phospho cellulose, which are negatively charged. The strength of the association between the dissolved molecules and the ion-exchange matrix depends on both the ionic strength and the pH of the eluting solution, which may be therefore varied in a systematic fashion to achieve an effective separation. In gel filtration chromatography (B), the matrix is inert but porous. Molecules that are small enough to penetrate into the matrix have a larger volume of solvent available to them and therefore travel more slowly through the column. Beads of cross-linked polysaccharide (dextran or agarose) are available commerically in a wide range of pore sizes, making them suitable for the fractionation of molecules of various molecular weights, from less than 500 to over $5' \times 10^6$. Affinity chromatography (C) utilises an insoluble matrix that is covalently linked to specific ligand, such as an antibody molecule or an enzyme substrate, that will bind a specific protein. Enormous purifications are often achieved in a single pass through such an affinity column.

Many different types of matrices are commercially available for this purpose. Ion-exchange columns are packed with small beads that carry either a positive or negative charge, so that proteins are fractionated according to the arrangement of charges on their surface. Hydrophobic columns are packed with beads from which hydrophobic side chains protrude, so that proteins with exposed hydrophobic regions are retarded. Gel-filtration columns are packed with tiny porous beads and separate proteins according to their size: molecules that are small enough to enter the pores percolate inside successive beads as they travel, while large molecules remain between the beads and therefore flow more rapidly through the column, emerging first. In addition to providing a means of separating molecules, gel-filtration chromatography is a convenient way to determine their size.

A particular protein cannot usually be separated from a mixture in a single step by column chromatography because each step generally increases the proportion of the protein in the mixture by no more than 20-fold. Since most proteins represent less than 1/1000 of the total cellular protein, it is usually necessary to use several different types of column in succession in order to purify them. A far more efficient procedure, known as affinity chromatography, takes advantage of the biologically important binding interactions that occur on protein surfaces. For example, if an enzyme substrate is covalently coupled to an inert matrix, such as a polysaccharide bead, the enzyme can often be retained by the matrix along with very few other proteins. In a similar way, specific antibodies can be coupled to a matrix, which can then be used to purify molecules recognised by the antibodies. Because of the great specificity of such affinity columns, 1000- to 10,000-fold purifications can sometimes be achieved in a single pass through the column.

Typical results obtained when three different chromatographic steps are used in succession to purify a protein. In this example, a whole cell extract was first fractionated by allowing it to percolate through an ion-exchange resin packed into a column (A). The column was washed and the bound proteins were then eluted by passing a solution containing a gradually increasing concentration of salt onto the top of the column. Proteins with the lowest affinity for the ion-exchange resin passed directly through the column and were collected in the earliest wash fractions eluted from the botton of the column. The remaining proteins were eluted in sequence according to their affinity for the resin—those proteins binding the tightest to the resin requiring the highest concentration of salt to remove them. The protein of interest eluted in a narrow peak and was detected by its enzymatic activity. The fractions with activity were pooled and then applied to a second, gel-filtration column (B). The elution position of the still impure protein was again determined by its enzymatic activity and the active fractions pooled and purified to homogeneity on an affinity column (C) that contained an immobilized substrate of the enzyme.

Terminology

In order to clarify the discussion of chromatography a knowledge of the terms likely to be encountered is helpful. The term *adsorbent* is often applied to the stationary phase. Although this usage is only correct when it refers to a solid stationary phase having the property of adsorption, the term adsorbent is sometimes used even when reffering to a liquid stationary phase or to a mixture of a liquid and solid. The word *sorbent* can be used more properly to include any stationary phase, whether it is liquid or solid. The term support is correctly used to indicate the solid scaffolding which holds a liquid stationary phase.

The mobile phase which causes the *solute* or *sample* to migrate through the stationary phase is called the *solvent* or *developer*. The forward edge of the solvent is called the *solvent front*. The point where the sample is placed at the start of the operation is the *origin*. The position of a solute on a chromatogram is usually indicated by its

R_f value which is the ratio of the distance it has migrated from the origin to that of the solvent front. Thus,

$$R_f = \frac{d(\text{solute})}{d(\text{solvent})}$$

where d(solute) is the distance travelled by the solute and the d(solvent) is the distance which the solvent front has moved from the origin. Under specific conditions R_f values are a characteristic of each chemical entity. Occasionally, in special cases, one may encounter another type of value such as R_g. This term may be used when a known solute is employed as a marker or standard and the positions of the other solutes are compared to it rather than to the solvent front. The solute on a chromatogram cannot usually be seen visually unless it happens to be a highly coloured substance. Thus, the solute must often be *detected* or *visualized* by some other means.

To collect the solutes that have been separated on a chromatogram, the solvent can be allowed to flow through the entire stationary phase until the desired solutes are carried out also. This process is called *elution* and the eluted mobile phase is the *effluent*.

Theory

The technique of chromatography is based on the fact that the molecules of solute in a sample are in a state of equilibrium between the solvent and the stationary phase. In a given stationary phase and solvent system, the distribution of the solute between the two phases is a property of the solute and may differ more or less from one compound to another. As a solvent or mobile phase is passed through a stationary phase, those solutes having a greater affinity for the stationary phase will be more retarded in their flow than those having less affinity. Thus, a separation of different solutes will occur which may be complete at the end of the operation.

Although the discussion of chromatographic theory which follows refers primarily to adsorption and partition processes, many aspects can be applied to ion-exchange and gel-chromatography. However, because of the importance of the latter two processes in experimental biochemistry, they will be discussed in some detail by themselves.

Distribution of Solute

The distribution of a solute between the two phases of a chromatogram is referred to as the *distribution coefficient* and is usually indicated by the letter K. By convention,

$$K = \frac{\text{conc. solute in mobile phase}}{\text{conc. solute in stationary phase}}$$

A value of 0.2 for K means that the stationary phase contains five times as much solute as the mobile phase.

The distribution of solute between the two phases can also be expressed in terms of an *effective distribution coefficient*, B. This term refers to the total amount of solute found in the mobile phase divided by the total amount found in the stationary phase. It can be expressed as

$$B = K \frac{\text{vol. mobile phase}}{\text{vol. stationary phase}}$$

The operation of a chromatogram can be described in terms of a large number of equilibrations. Each of these equilibrations can be thought of as occupying a layer or portion of the chromatogram, be it a column or a thin layer, and each layer is referred to as a *theoretical plate*. The length of each of these portions or layers in a direction perpendicular to the direction of solvent flow is called the height equivalent to a theoretical plate, or HETP. The latter term is a measure of the efficiency of a chromatogram.

For purposes of explanation consider an ideal chromatogram existing as a column of a solid stationary phase and a liquid mobile phase. The column is 7 cm length and contains 1 ml of solvent per centimeter. If 1 ml of solvent containing 128μg of a solute is added to the top of the column, Step 1, 1 ml of solute will be eluted from the bottom of the column and 7 ml of solvent will remain in the column. If the effective distribution coefficient B is equal to 1, then the solute will be equally distributed between the stationary and mobile phases.Thus, 64μg of solute will be found in each of the two phases of the top 1 cm of the column. If an additional 1 ml of solvent, without solute, is then added to the top of the column, another 1 ml of solute will flow out of the column. One will find, as seen in Step 2, that the 64μg of solute that had been in the mobile phase of the top layer has moved into 1-cm layer second from the top. Thus, 64 μg of sloute will be found in each of the top two 1-cm layers and will be equally distributed between adsorbent and solvent in each of those layers. The addition of another 1 ml of solvent to the column as illustrated in Step 3 will carry the 32 μg of solute contained in the solvent of the top layer into the second layer while the moving solvent will carry the 32 μg of solute from the solvent phase of the second layer into the third layer.

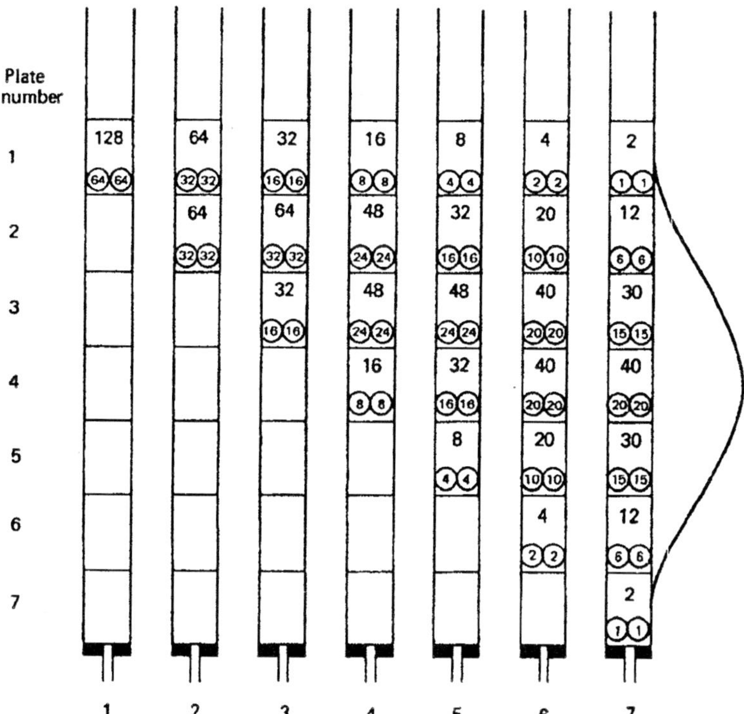

Fig. 14.2. Hypothetical distribution of solute during chromatography in a complete column.

The 32 μg of solute left in the top layer will now distribute itself equally between the two phases as will the 32 μg of solute carried into the third layer. In the second layer one will find the 32μg of solute carried from the solvent phase 64μg of the top layer and the 32 μg that had been left behind from the stationary phase. Thus, one will still find 64 μg of solute in the second layer where it is in a state of equilibrium between the two phases. The distribution of solute following the addition of each 1 ml of solvent can be observed in each of the additional steps. It can be seen in Step 7 that the solute is distributed throughout the column and that the highest concentration is in the center of the column. From the curve drawn alongside the column of Step 7 one can seen that the concentrations approximate a Gaussian distribution.

If the effective distribution coefficient in the above example is greater than 1, a greater fraction of solute will pass through with the solvent after each equilibration. The development of the column will then proceed faster, and the highest concentration of solute will be found below the center of the column. On the other hand, if the effective distribution coefficient is less than1, the adsorbent will retain a greater proportion of solute after each equilibration than will the solvent. In this case the development of the column will be slower, and the concentration peak will be found above the center of the column. However, in each case a definite percentage of solute will be found throughout the column. A little practice with distributions of solutes using different values of B will illustrate these principles rapidly.

Complete Column Chromatograms

Although the ideal chromatogram just described suffices to explain elementary mechanisms of chromatography, chromatograms that function this way are not likely to be encountered in actual practice. For example, the solvent in functioning chromatogram is added continuously, and equilibrations take place continuously. A column of the type used in the explanation is called a *complete column* since addition of solvent is stopped before the solute is eluted from the bottom of the column. Recovery of fractions of solute from such a column can be made by

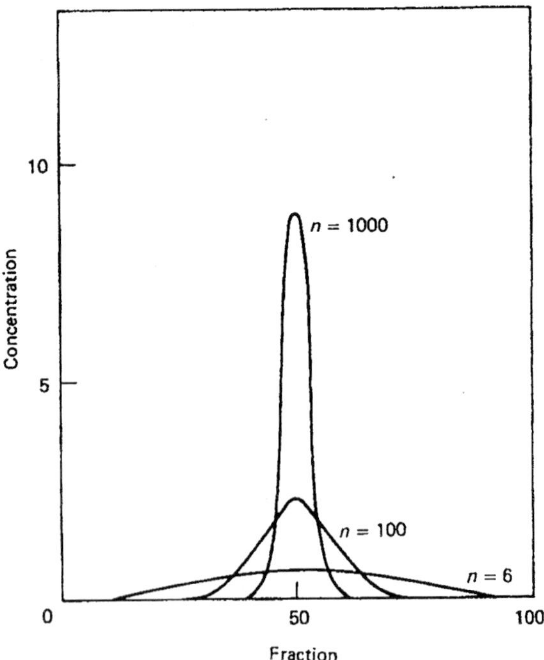

Fig. 14.3. Distribution of solute in a column chromatogram as a function of the number of equilibrations. The letter n, refers to the number of equilibrations or theoretical plates.

extruding the column from the tube and dividing it into sections, a process not likely to be chosen for preparative purposes. A somewhat analogous situation occurs in the case of thin-layer or -paper chromatograms where the development is terminated before the solutes reach the end of the support. The type of chromatogram where solutes are eluted is called an *elution chromatogram*; there are differences in the behaviour of these two types of chromatograms.

In the case of the complete column just described, the sample spreads over a greater portion of the column as the number of equilibrations, or theoretical plates, increases; however, the *degree* of spreading decreases as this happens. The same phenomenon can be expressed mathematically as the amount of solute per unit of the chromatogram by equation

$$C = \frac{\sqrt{n(B+1)}}{L\sqrt{2\pi B}}$$

where C is the fraction of total solute per unit length of column, n is the number of equilibrations or theoretical plates, B is the effective distribution coefficient, and L is the total length of the column. The equation can also be expressed as

$$C = \sqrt{\frac{n}{2\pi p(L-p)}}$$

where p is the position of the peak of the solute. These equations show that the degree of spreading decreases as the function of the square root of the number of equilibrations. In an actual chromatogram, equilibrations are taking place continuously and at least several hundred equilibrating layers of plates are present.

Resolving Power

One other concept in chromatography of value to the biochemist is that of *resolving power*. Essentially resolving power indicates the ability of a chromatogram to separate solutes that differ only slightly in their effective distribution coefficient.

The resolving power of a chromatogram is related to the values of the effective distribution coefficients of the compounds to be separated. The difference in these values may be used as a fair criterion of the resolving power of a given chromatogram. Relative resolving power of a complete chromatogram may be expressed in mathematical terms as

$$\text{Relative resolving power} = \frac{\sqrt{Bn}}{B+1} = \sqrt{nR_f(1-R_f)}$$

One may see from the above equation that the resolving power is proportional to the square root of the number of theoretical plates contained in the chromatogram; therefore lengthening a chromatogram would increase its relative resolving power.

Adsorption and Partition Processes

The distribution of solute between the mobile phase and the stationary phase can involve either adsorption or partition processes depending upon the type of stationary phase employed. In the partition process the solute is in equilibrium between two liquids—the liquid solvent and the liquid of the stationary phase—and its distribution is determined by its solubility in each. Those solutes which are more soluble in the solvent will migrate faster than those that are more soluble in the stationary phase. The partition system is more analogous to the case of the distribution of the solute between two mutually exclusive solvents in a separatory funnel and, in fact, the partition coefficient of a solute can be measured in a separartory funnel. Those compounds which differ in their partition coefficient should be separable. Since the partition coefficient for a compound is a constant for a given liquid—liquid pair, its relative distribution does not change even when its concentration does. In addition, its rate of movement is independent of concentration.

In adsorption the solute is in equilibrium between a solid adsorbent and a liquid solvent, and the behaviour is more complicated. Adsorption involves a variety of forces. In the case of neutral solutes, the principal force is dipole-dipole attraction with hydrogen bonding being involved in some instances. Where acidic or basic solutes are involved, ionic forces are chiefly responsible for the interaction. The solute then is caught between the attractive forces of the adsorbent on the one hand and of the solvent on the other. In addition, the solvent tends to complete with the solute for sites on the adsorbent; this results in a displacement effect. The vast interplay of forces competing for the solute often results in a variation in the distribution of solute between the two phases as a function of concentration. This behaviour is expressed in Figure where convex and concave curve represent the type of distribution encountered with adsorption while the straight line is that associated with partition. The convex curve is more commonly found in adsorption chromatograms and results in a phenomenon called "tailing" where parts of the solute lag behind the major portion.

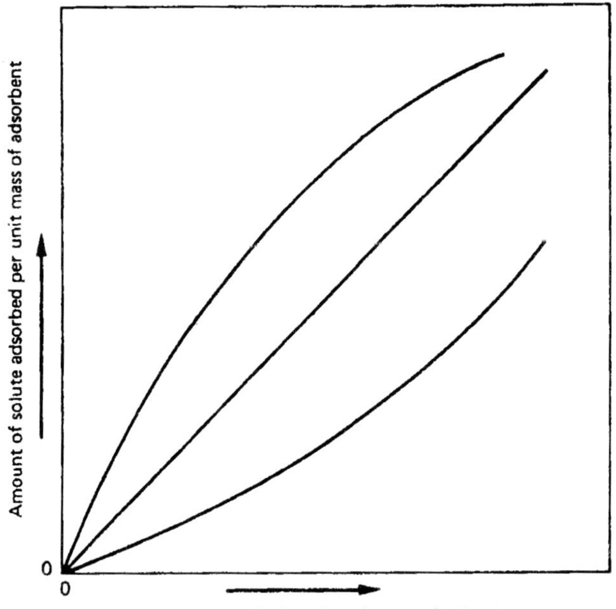

Fig. 14.4. Distribution of solute between the adsorbent and the solvent as a function of solute concentration showing the three most common types of distribution.

In order to choose between adsorption and partition chromatography, the experimenter must consider the objective of the operation and the compounds being separated. Adsorption processes are better for preparative work because larger quantities can be manipulated. Partition processes are dependent on the solubility of the solute in two liquid; therefore partition processes are most effective in separating compounds which differ in their solubility properties. Thus, one would expect members of a homologous series to be separated best by partition chromatography. Adsorption chromatography, on the other hand, is more effective in separating compounds that differ mainly in their configuration or charge.

The effectiveness of most solvents in separating solutes in adsorption or partition chromatography is related to their dielectric constants. These constants are dependent upon the degree of polarity of the compound but also are affected by hydrogen bonding and polarization. Water, for example, is a very polar solvent because it has a very strong dipole and it processes hydrogen bonding. Alcohols and esters have weaker dipoles and less hydrogen bonding, and are, therefore, less polar, whereas the hydrocarbons, such as petroleum, ether and cyclohexane, are the least polar of all. The dielectric constants of some of the more common solvents used in chromatography are shown in Table 14.1.

Table 14.1. Dielectric constants of common chromatographic solvents

Solvent	*Dielectric constant*
n-Hexane	1.89
Petroleum ether	ca 2
Cyclohexane	2.02
Carbon tetrachloride	2.23
Benzene	2.27
Toulene	2.38
o-Xylene	2.57
Diethyl ether	4.34
Chloroform	4.81
Butyl acetate	5.01
Ethyl acetate	6.02
Pyridine	12.3
n-Butanol	17.1
n-Propanol	20.1
Acetone	20.7
Ethanol	24.3
Methanol	32.6
Water	78.5

A good rule-of-thumb with respect to the polarity of organic solvents is that polarity increases as the number of functional groups increases and decreases as the molecular weight increases. In addition, various salts have a high degree of polarity because of their positive and negative charges. Lastly, it should be pointed out that the elutroprism (the tendency to elute solutes) of a solvent system may be altered by combining solvents of different dielectric constants.

Column Chromatography

Column chromatography has proven to be one of the most useful techniques in biochemistry for the separation and purification of a wide range of compounds. In column chromatography a thin layer of sample is placed on a stationary phase held in a tube of suitable material, usually glass. The chromatogram is developed by permitting a solvent to flow through the stationary phase.The solutes separate as

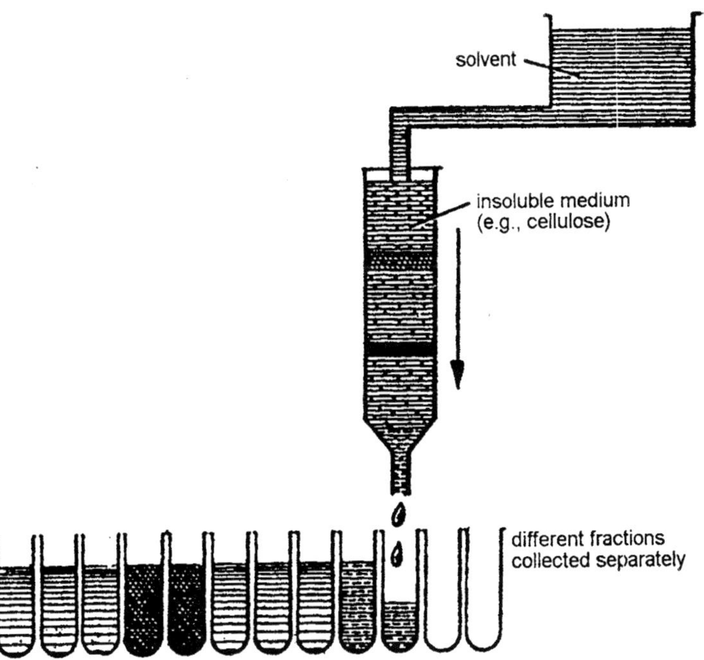

Fig. 14.5. Column chromatography.

bands in the chromatogram and are usually collected as separate fractions as they emerge from the column. The fastest moving component would be collected first; the slowest; last. This elution or flowing type of chromatogram is of more practical; value in experimental biochemistry than the complete type that was described during the discussion of chromatographic theory.

Discussion of the elution column chromatogram necessitates definition of two additional terms. These are the retention or hold up volume and the eluting volume. *Retention* or *holdup volume*, usually indicated by v, is equal to the volume of the mobile phase or solvent present in the column at any given time; v remains constant for any given column. *Eluting volume*, usually indicated by V, refers to volume of solvent that must be added to a column to produce the peak of a solute in the effluent. Of course, V is the function of the effective distribution coefficient or the R_f value of the compound being chromatographed. The relationship of these terms can be expressed mathematically as

$$\frac{V}{v} = \frac{B+1}{B}$$

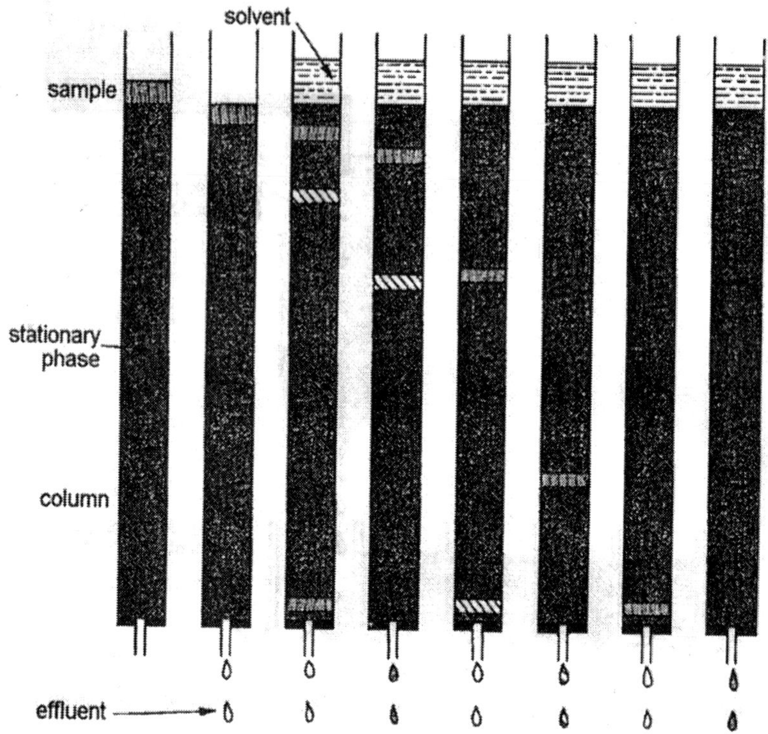

Fig. 14.6. A column chromatogram illustrating the separation and elution of three solutes of different R_f values.

and

$$\frac{V}{v} = \frac{1}{R_f}$$

The term V/v is equal to the number of holdup volumes necessary to elute a given solute. Although it is the function of the effective distribution coefficient, V/v does not depend on the length of the column. The holdup volume can be determined experimentally by measuring the volume of solvent required to elute a dye that has no affinity for the stationary phase. Thus, it can be seen from these relationships that the volume of solvent required to elute a given solute can be calculated from the reciprocal of its R_f value of the holdup volume of the column is known.

As in the case of the complete column, the resolving power of a flowing column is a function of the square root of the number of theoretical plates in the column. However, the equation for its calculation differs slightly from that for the complete column and is

$$\sqrt{\frac{n}{B+1}}$$

where n is the number of theoretical plates and B is the effective distribution coeffficient of the compound being chromatographed. The resolving power of a flowing column may also be calculated from equation:

$$\sqrt{\frac{n(V-v)}{V}}$$

By substituting numerical values for the terms in the appropriate equations it can be seen that the maximum resolving power of an elution column is about twice that of a complete column.

The number of theoretical plates in an elution column can be calculated from the expression

$$n = \frac{8V(V-v)}{w^2}$$

where *w* is equal to the volume contained in the band of solute measured at 36.8% of the total height of the band in an elution diagram. The other terms are identical to those defined earlier.

Through the use of the above equations it is possible to calculate the volume of solvent required to elute a given solute and the volume of effluent which contains the solute. Furthermore, it is possible to determine the effect of increasing or decreasing the length of a column.

Choice of Chromatographic System

The first step in any chromatographic procedure is to choose a system that will effectively separate the substances in the mixture. If the compounds are known, the choice of system can frequently be made by consulting appropriate methods in the literature. If the components of the mixture are not known, a choice must be made on the basis of the predictable components. Where a mixture contains a component or components of interest as a small fraction of the total mixture, it is often desirable to remove the bulk of the extraneous substances by methods other than chromatography. The resulting partially purified mixture can then be purified further by appropriate chromatographic techniques.

As mentioned earlier, adsorption techniques are usually better for preparative work than are partition techniques. In addition, ion exchange is often a very powerful technique for preparative work. Ion exchange is frequently used in the purification of proteins from cell or tissue extracts and will be discussed in some detail later.

Columns

Columns are usually constructed of glass and may be designed in various lengths and diameters. Some are extremely simple; others are made more sophisticated, with the use of pressure, solvent reservoirs, fritted glass or nylon plates, water-jackets, and so on. In choosing a column for chromatography, the ratio of the length to the diameter of the column is sometimes important. An acceptable ratio depends on the ease of separation of the solutes. Shorter columns are easier to pack, but satisfactory separations often require longer columns whose ratio of length to diameter is 100—200 to 1.

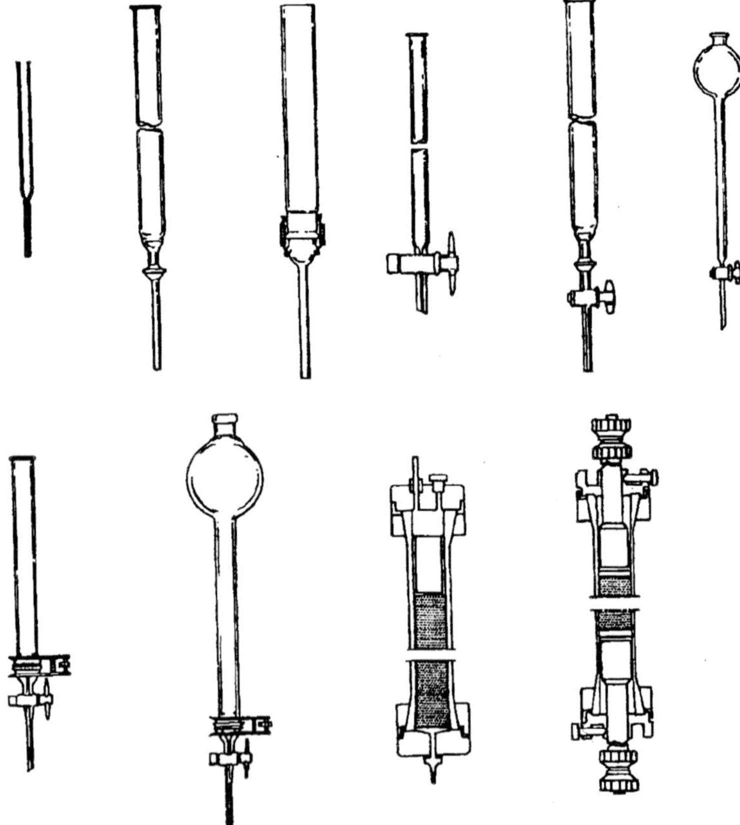

Fig. 14.7. Diagrams of various types of columns used for chromatography.

Stationary Phase

The stationary phase of a column may be composed of either true adsorbents, partition agents, ion-exchange resins or cellulose, or gels. The theoretical aspects of adsorbents and partition agents were discussed earlier. Ion-exchange agents and dextran or polyacrylamide gels differ in many details from the traditional chromatographic adsorbents.

The more commonly encountered stationary phases are listed in Table 14.2. Of the true adsorbents, alumina and silica gel have been used most. The capacities and properties of these adsorbents can be controlled or "activated" by modifying their surface-water content.The degree of activation is indicated by a scale (called the Brockman index) from I to V where I is the most active and V the least active. For polar adsorbents, the most active contains the least water. The particle size of these adosrbents may vary from about 50 to 200 mesh. More detailed information on the treatment of specific adsorbents can be found in most of the general references on chromatography listed at the end of the chapter.

Table 14.2. Stationary phases used in adsorption and partition column chromatography

Adsorbents
Neutral alumina
Basic alumina
Acidic alumina
Hydroxylapatite (calcium phosphate)
Magnesium silicate
Silica gel (silicic acid)
Partition agents
Diatomaceous earth (Celite, Filter-cel, super-cel)
Silica gel (silicic acid)
Cellulose

In the case of partition systems the support is usually diatomaceous earth, silica gel, or cellulose. The support should be capable of holding a large amount of the polar or nonpolar stationary phase and should not interact with it.

Solvents

Solvents should be of high purity. For columns containing an adsorbent as the stationary phase, the solvent can be a single compound or an appropriate mixture of compounds.

In the case of partition chromatography it is necessary to equilibrate the solvent system with the liquid used as the stationary phase. The usual procedure is to mix thoroughly the pair of immiscible liquids in a separatory funnel and then separate the two phases. The support for the column is usually mixed with the more polar phase while the second or less polar phase is used as the solvent system. It is important here that the equilibration of the two liquids be done at the same temperature as that at which the column is to be operated since a change in temperature will affect the mutual solubility of the two phases.

The solvent system must be chosen on the basis of the solubility of the sample, since the solutes must distribute themselves between the two liquids. An ideal value for the ratio of the solubility in the mobile phase to that in the stationary phase is 1 : 100. This ideal cannot always be realized, however, and a lower ratio must often be used.

Operation of Chromatographic System

Setting up Columns

The first step in setting up a column is to place a plug in the bottom of the tube. With simple columns a wad of glass wool or cotton will suffice. Other columns are made with a plate of fritted glass or with provision for a nylon or Teflon filter disc. Any tendency for particles of the stationary phase to clog the pores of the plug or disc may be reduced by placing a layer of fine glass beads, washed sand, or a filter-paper discover the plug.

Columns containing an adsorbent as the stationary phase may be packed either wet or dry. In the latter method the dry adsorbent is packed into the column a little at a time using a glass rod, wooden peg, or stopper held on the end of a rod as a tamper. Following the packing of the adsorbent, it is washed with the solvent to be used as the mobile phase.

Columns are more commonly packed wet by introducing the adsorbent in a slurry of the solvent. As the adsorbent settles, the excess solvent is eluted from the column and additional slurry is added until the column of adsorbent is built up to the desired length. An alternate method is to add dry adsorbent directly to the tube which already contains the solvent and to let the adsorbent settle out as described in the former procedure. In any case it is important to maintain a continuous column.

Columns containing partition agents are more difficult to prepare than those with adsorbents. In the case of supports composed of silica gel or diatomaceous earth, the stationary phase is usually mixed directly by adding the equilibrated stationary phase to the support in a ratio between 0.5 and 1.0 by weight (stationary phase support). Best mixing can be accomplished by grinding with a mortar and pestle or using a suitable blender. A slurry of this product and the mobile phase is prepared and poured into the column where it can be packed by gravity or by tamping with a suitable plunger. Usually columns of silica gel are packed by gravity whereas those of diatomaceous earth require tamping.

With cellulose, columns are prepared by adding a slurry of cellulose in acetone to the glass tube. The column is packed by gravity and excess acetone drained off the equilibrated solvent.

Once the column has been packed, the stationary phase is protected from disturbance by covering the top surface with a disc of filter paper, glass wool, small glass beads, or clean sand.

Introduction of Sample

The sample to be chromatographed should be added to the top of the column in a minimum amount of solvent and allowed to seep slowly into the stationary phase. A pipet or Pasteur pipet can be used to introduce the sample. After the sample has percolated into the stationary phase, a small volume of solvent is added in order to take the sample below the surface. This prevents dilution of the sample into a large volume of solvent.

Development

The column is developed by adding a large volume of solvent to the top of the column and allowing it to flow at a convenient rate. The rate is determined by the characteristics of the particular system being used and by the size of the columns. Better patterns of elution are obtained with slow rates of flow than with rapid rates. Tailing of bands in partition systems usuallly indicates a lack of equilibration that can be remedied by decreasing the flow rate. Where tailing occurs in adsorption systems, it may be corrected by increasing the polarity of the solvent system in a stepwise fashion. A broad change in the polarity sometimes results in shifting the order in which the solutes migrate and must be avoided.

One convinient method for increasing the rate of development of a column is to employ a gradual change in the composition of the solvent system. This method, called *gradient elution*, may involve a

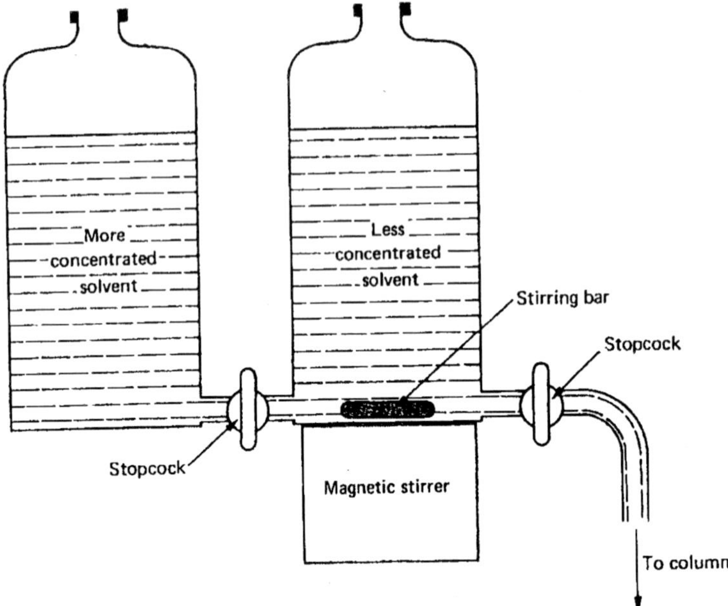

Fig. 14.8. Diagram of a gradient maker for column chromatographic development.

change in the ratio of two solvents or an increase in the concentration of a particular solvent such as a salt. Various pieces of apparatus have been designed to provide suitable solvent gradients.

Collection of Effluent

The simplest method of collecting the solutes eluted from a chromatographic column is to introduce suitable reservoirs manually, such as test tubes, flasks, or beakers, to catch the effluent. The effluent can be divided into fraction on the basis of volume, drops, weight, or time. Although this method of collection is not a problem if the column is short and its flow rate is relatively fast, it becomes inconvenient with slow flow rates, particularly if the column must be operated in a cold room. Because of these inconveniences mechanical fraction collectors have been designed.

Mechanical fraction collectors are available from a number of manufacturers in models of varying sophistication. Usually fractions can be collected at preset time intervals, by volume, or by the number of drops. A typical fraction collector is pictured.

Detection

Solutes eluted from columns in different fractions can be detected by either direct or indirect methods, depending upon the nature of the solute and objective of the chromtographer. These methods have quantitative as well as qualitative application.

LIQUID CHROMATOGRAPHY

Liquid chromatography (LC) is a chromatographic technique which utilizes a liquid mobile phase to separate mixtures of compounds. Recent advances in pump technology have increased liquid chromatography's utility in the clinical laboratory. The separations that were once performed on long columns requiring hours of clution time are now feasible in minutes with the use of high-pressure pumps and small particle-size columns. This chapter discusses the basic principles, the various types, the important parameters, and the common equipment used in liquid chromatography.

Principles

The chromatography of a components of a mixture is based on simple principles. Separation of the components of a mixture may be accomplished because each component interacts with its environment differently from the other compounds under the same conditions. The solubility and the miscibility of compounds are the primary factors affecting their interactions.

The polarity of individual molecules determines their solubility and miscibility. The simplest organic compounds, alkanes and alkenes, are nonpolar. A good example of a nonpolar compound is hexane.

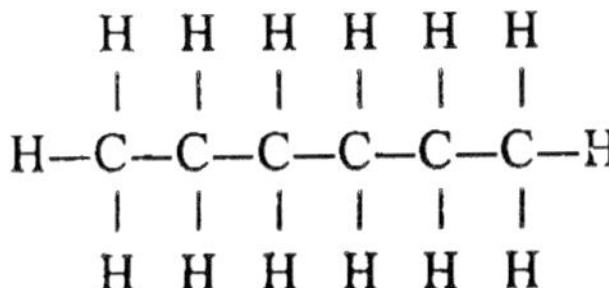

Factors affecting polarity are the presence or absence of electron donating or withdrawing groups, conjugation, and molecular symmetry. Examples of electron withdrawing groups and molecular symmetry affecting polarity and hexanol (left, below) and hexanoic acid (right, below), the latter being the more polar.

```
  H  H  H  H  H  H          H  H  H  H  H  H
  |  |  |  |  |  |          |  |  |  |  |  |
H—C—C—C—C—C—C—OH      H—C—C—C—C—C—C—OH
  |  |  |  |  |  |          |  |  |  |  |  |
  H  H  H  H  H  H          H  H  H  H  H  H
```

Most biological molecules vary from slightly polar to polar. Water is a very polar molecule because of its molecular structure which results in a permanent dipole.

The polarity of molecules is predictive of their general solubility : the rule "like dissolves like" may be applied in liquid chromatography for selection of appropriate mobile and stationary phases.

The components of a liquid chromatography column may be itemized as follows:

1. Stationary phase—may be a solid or a liquid.
2. Mobile phase—is a liquid.
3. Support—is most often a solid.

Separation of solutes may be achieved if the stationary and mobile phases are selected so that separation of the solutes occurs due to either differences in solubility of the solutes between the stationary and mobile phases or differences in adsorption on the stationary phase.

As a discrete band of sample-containing solution is forced through the column, the portion of solution high in concentration of a specific component is equilibrated with the stationary phase and the mobile phase; the next increment of mobile phase containing less of this component will extract a portion of the adsorbed component into the mobile phase. If every specific component enters into this "back and forth" exchange to different degrees and at different speeds from all other components, separation is achieved.

Types of Liquid Chromatography

There are four basic types of liquid chromatogrphy; liquid-solid (adsorption), liquid-liquid (partition), ion-exchange, and gel permeation (molecular sieving).

Adsorption

Adsorption chromatography is performed with a liquid mobile phase and a solid stationary phase which reversibly adsorbs solutes. Common examples of stationary phases are silica gel, porous glass beads, and alumina. The mobile phase is usually a relatively nonpolar solvent mixture. This technique is applicable when sample components vary widely in polarity, e.g., lipids.

Partition

Partition chromatography is performed with a liquid-coated stationary phase which is immiscible with the mobile phase. The relative distribution of the sample components between the mobile phase and the stationary phase determines the relative separation. Usually, stationary liquid phases and polar while the mobile phase is nonpolar (normal phase). Water and polyethylene glycol are typical normal phase stationary phases. Hexane and chloroform are common

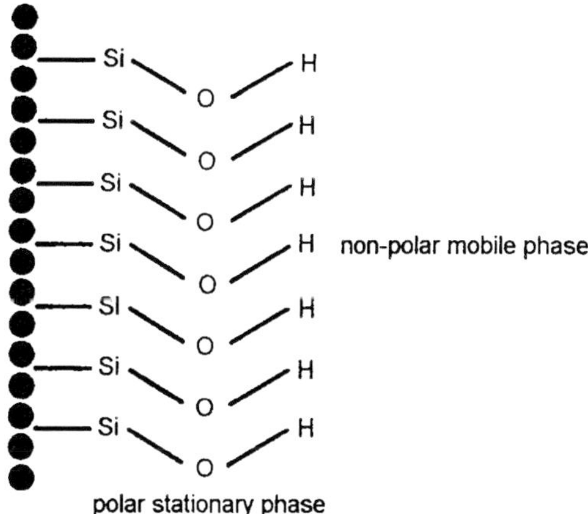

Fig. 14.9. Normal phase liquid chromatography.

normal-phase mobile phases. It is possible to damage a column by removing the polar coating if a polar mobile phase is used. A very nonpolar molecule would move rapidly through a normal-phase column if the mobile phase were hexane. The nonpolar molecule would have a greater attraction for the nonpolar hexane than for the polar groups of the stationary phase.

Another form of partition chromatography, reverse phase, consists of a nonpolar stationary phase and a polar mobile phase. Typical mobile phases in this form of chromatography are water, acetonitrile, and methanol. Common stationary phases are hydrocarbon chains

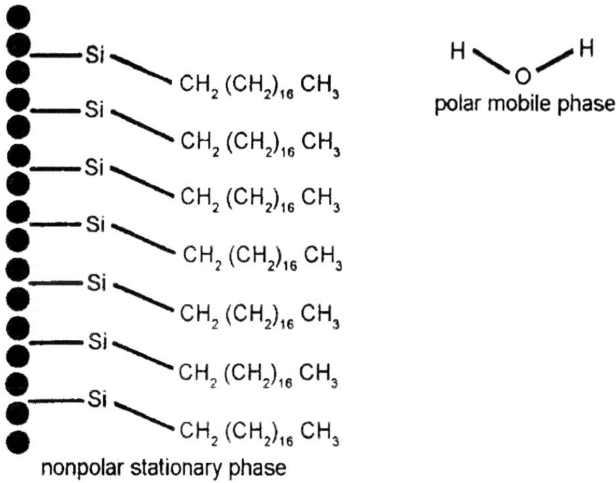

Fig. 14.10. Reverse phase chromatography.

chemically bonded to the support. Partition chromatography is employed when sample components have known differences in solubility, e.g., drugs. Figure 14.9 illustrates the molecular structures of typical stationary and mobile phases employed in reverse-phase partition chromatography.

Ion-Exchange

Ion-exchange chromatography is performed using an ion-exchange resins as the stationary phase and an aqueous solution as the mobile phase. The resins are highly polymerized crosslinked hydrocarbons which contain ionized functional group. The resin may be thought of as having two distinct parts; one is a large, nondiffusible but permeable ion (basic resin structure), and the second part is a small, equally but oppositely charged ion that is free to migrate throughout the resin structure during the exchange. The ionic functional group attached to the structure of the resin determines the exchanger characteristics. Cation resins are produced when an acidic functional group, for example ($-SO_3H$), is attached to the resin structure. The acidic hydrogen is available for exchange. Conversely, anion exchangers are produced when basic functional groups are attached to the resin structure.

The mechanism of the exchange is illustrated as follows :

$X^- + R^+Y^- \quad Y^- + R^+X^-$ (anion exchange)

$X^+ + R^-Y^+ \quad Y^+ + R^-X^+$ (cation exchange)

where

X = sample ion

Y = mobile-phase ion

R = ionic sites on exchanger

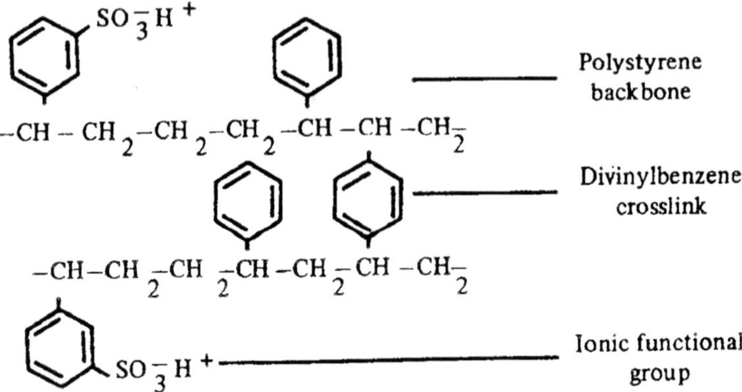

Fig. 14.11. Cationic ion exchange resin.

In an anion exchange system, the sample ion X^- is in competition with the mobile phase ion, Y-, for the ionic sites, R+, on the ionic exchanger. In cationic-exchange chromatography, the sample cations, X+ are competing with the mobile-phase ions, Y+, for the ionic sites. R-, on the ion exchanger. Solutes interacting weakly with the resin in the presence of the mobilephase ions are eluted rapidly, while solutes that react strongly with the resin are retained much longer, pH and ionic strength of the mobile phase determine the solute ion-resin interaction. Clinically, this technique is used for separation of small ions, small hormones, heme-breakdown products, and amino acids.

Gel Permeation

Gel permeation is a *mechanical* sorting of molecules based on molecular size. Size separation is accomplished with a porous packing gel through which the smallest components in the sample migrate into the smallest pores of the gel while the large molecules are encluded from the gel and elute from the column early in the separation. This technique is used for separation of high molecular weight enzymes and proteins from other serum constituents.

Chromatographic Parameters

Columns vary with their packing material, according to the type of liquid chromatography to be performed. It is important to choose the appropriate type of column for the compounds to be separated. A general rule to follow is : adsorption chromatography is most applicable for nonpolar compounds, while partition chromatography is the method of choice for slightly polar to polar compounds. Any ionized compounds are best separated utilizing ion-exchange chromatography.

An additional parameter which is useful for controlling separation is temperature. Increasing temperature may in some instances improve component separation and will reduce solvent viscosity which results in reduced column back pressure.

There are two basic methods for solvent delivery. The first is isocratic solvent delivery in which a mobile phase of fixed composition is delivered to the column at a set rate throughout the chromatogram. The second method is gradient solvent delivery in which one component of the mobile phase is varied in concentration throughout the chromatogram.

Equipment for Liquid Chromatography

The solvent delivery system may be one of the three basic types. The first is a syringe pump which is driven by a worm-screw drive.

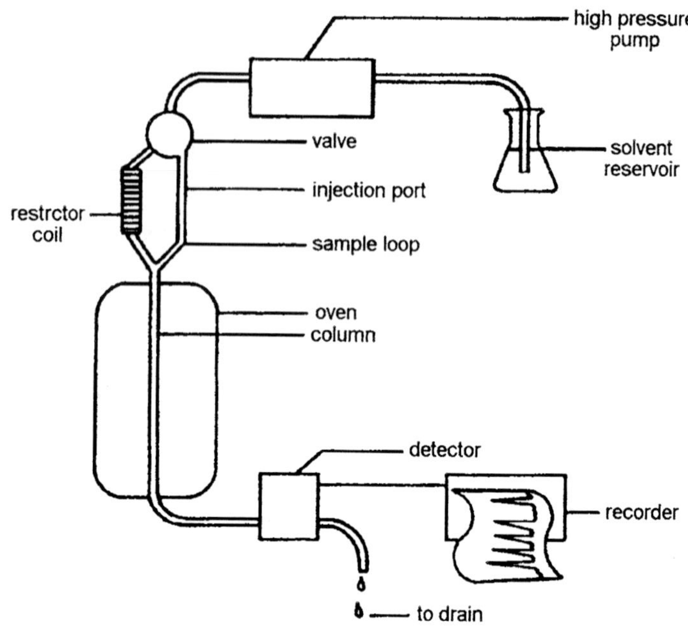

Fig. 14.12. Basic liquid chromatograph components.

This mechanism yields a precise, pulseless flow. Another type is the single-piston reciprocating pump which maintains constant pressure by a pressure-sensitive feedback circuit. The third type of delivery system is a dual-reciprocating pump which has two pistons diametrically positioned and driven by the same cam. Both the single piston and the dual-piston pump have a pulsed flow which is controlled by various damping mechanisms, resulting in low baseline noise. Table 14.3 illustrates the advantages and disadvantages of each solvent delivery system.

There are three basic types of sample injector systems for liquid chromatographs. The first is the on-column injection system which consists of a siliconized rubber disk, and a septum through which the sample is loaded into the flowing mobile phase at a point just above the column bed using a micro high-pressure syringe. Upkeep of this system is difficult because of leaks and punctured septums which cannot be replaced unless the pumping mechanism is shut down. Small particles of septum material, tend to collect at the head of the column, causing high back pressure and peak broadening.

The second type of injector is a fixed-loop injector in which the sample is loaded into the injector with a syringe to completely fill the

Table 14.3. Solvent delivery systems.

Solvent Delivery System	*Advantages*	*Disadvantages*
Syringe pump	Even solvent flow rate Low flow rates precisely controlled	Unable to replenish solvent without shutting system down: high pressure not attainable.
Reciprocating pump	Solvent reservoir easily refilled during operation; high pressures attainable; relative ease of operation	Pulsing solvent flow
Dual-headed reciprocating pump	Same as the reciprocating-pump system	Small amount of pulsing in the flow of the solvent

injector sample loop. When the valve is turned to the "inject" position, the total volume of the sample loop is washed onto the column by the mobile phase.

The third type is the syringe-loop injector in which the sample, premeasured by microsyringe, is loaded into the sample loop which is partially filled with mobile phase. The sample is then flushed on to the column when the injector is placed in the "inject" position.

The advantage of the fixed-loop injector is that a reproducible volume is applied to the column ; but this system lacks the flexibility of varying the injection volume. Both loop injector systems, however, allow ease of sample injection against high solvent pressure and are superior to the septum injector. There are four basic detectors used for liquid chromatography; fixed wavelength adsorption, multiple wavelength adosorption, fluorescent, and electrochemical.

Table 14.4. Liquid chromatograph detectors.

Detector	*Advantages*	*Disadvantages*
Fluorometric	High sensitivity for fluorescent compounds	Moderately specific
Electrochemical	Extremely sensitive	Too sensitive
	Good specificity	Mobile phase must be an electrolyte
Fixed wavelength	Cheap	Poor specificity
	Stable	
	Relatively sensitive	
Multiple wavelength	Good specificity	Moderately sensitive

The fixed-wavelength UV-visible detector generally consists of a mercury lamp light source and utilizes filters to select the desired wavelength. The multiple-wavelength UV-visible detector utilizes a monochromator to obtain the wavelengths of choice. Both the fixed and multiple wavelength detectors are spectrophotometers utilizing a microflow cell as a cuvette. The fluorometric detector is a fluorometer adapted with a microflow cell to detect fluorescent compounds.

Electrochemical detectors are employed to detect compounds that can be oxidized or reduced, thus changing the conductivity of the compounds as compared to the solvent flowing through the flow cell.

Advantages and Disadvantages

The first advantage of liquid chromatography, as compared to other types of chromatography, is the simple sample preparation. A single extraction is usually sufficient, and derivatives usually are not required, thereby removing a step in which experimental error is likely to be introduced. This technique may be used at ambient or slightly above-ambient temperatures. Because of the relatively low temperatures involved in liquid chromatography, sample stability is of little concern. Finally the instrumentation for liquid chromatography is reliable, and routine maintenance is simple compared to gas chromatography.

Two major disadvantages of liquid chromatography are the lack of sensitivity to some compounds and the requirement of moderately expensive equipment to perform analyses.

Gel Chromatography

The application of various gels to chromatography has proven to be a valuable tool for the separation of solutes differing in molecular weight. By this technique larger molecules are eluted from a column of the gel before the smaller molecules. Thus, the method has wide use in the purification of proteins, in desalting protein-containing solutions, and in the determination of molecular weights.

The first theoretical approaches to gel chromatography considered the behaviour of solutes in a column of gel as being governed primarily by steric effects. Small molecules were viewed as being able to penetrate the regions between the gel chains whereas large molecules, because of their size, were prevented from entering. However, the steric approach does not take into consideration other factors which influence the distribution of many solutes between the gel and the mobile phase. These factors can be included in a unified approach to the theory of gel chromatography by interpreting the latter as a kind of partition chromatography.

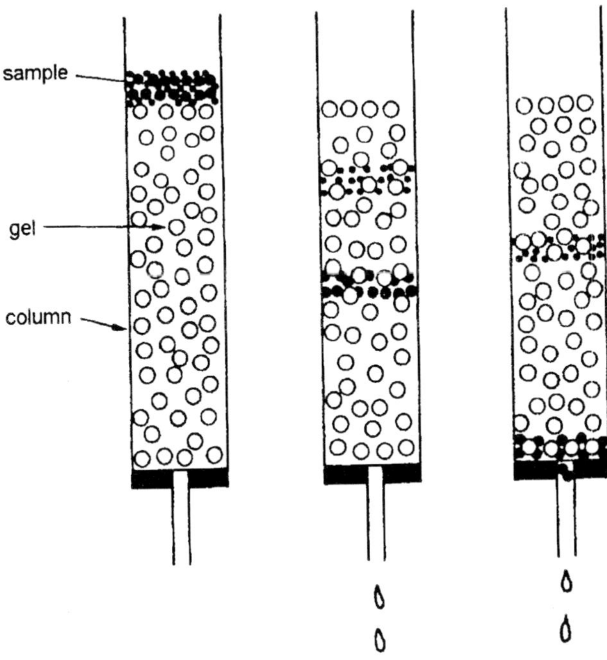

Fig. 14.13. Gel filtration column showing the distribution of solute particles of two different molecular weights during development.

Gels presently used in chromatography consist of three principal types (a) dextran, (b) polyacrylamide, and (c) agarose.

Dextran gels are prepared by reacting alkaline solutions of crystalline dextran with epichlorohydrin. The properties of the gel can be controlled by adjusting the concentration of epichlorohydrin and the molecular weight and concentration of dextran. Dextran gels are available commercially under the name Sephadex and are prepared in different particle sizes and with different degrees of crosslinkage. In addition to these gels which are hydrophilic there is available a hydrophobic gel, Sephadex LH-20, which can be used with a number of organic solvents.

Polyacrylamide gels are prepared by polymerizing acrylamide in the presence of a bifunctional acrylamide, such as N, N´-methylenebisacrylamide. The presence of the bifunctional acrylamide results in crosslinking between different chains of the polymer, and a three-dimensional array of chains is built up. A carboxylic acid amide group on every other carbon atom introduces polarity into the gel. Polyacrylamide gels for chromatography are available commercially

Table 14.5. Properties of various Sephadex gels.

Sephadex type	*Water-regain per g dry Sephadex, ml*	*Bed volume per g dry Sephadex, ml*	*Fractionation range for globular proteins, mol.wt.*
G-10	1.0 ± 0.1	2—3	—700
G-15	1.5 ± 0.2	2.5—3.5	—1500
G-25	2.5 ± 0.2	4—6	1000—5000
G-50	5.0 ± 0.3	9—11	1500—30,000
G-75	7.5 ± 0.5	12—15	3000—70,000
G-100	10.0 ± 1.0	15—20	4000—150,000
G-150	15.0 ± 1.5	20—30	5000—400,000
G-200	20.0 ± 2.0	30—40	5000—800,000

under the name Bio-gel. They are available with different degrees of crosslinking and, thus, different pore sizes.

Agarose is a linear polysaccharide of D-galactose and 3, 6-anhydro-l-galactose which is purified from agar, an extract of certain seaweeds. It is free of ionizable groups and can be prepared as fine beads. It is available commercially under the name Sepharose. Agarose may be used for the separation of solutes of greater molecular weight that can be separated with the dextran or polyacrylamide gels.

Table 14.6. Properties of Various Bio-gels.

Bio-gel type	*Water-regain per g dry Bio-gel, ml*	*Bed volume per g dry Bio-gel, ml*	*Fractionation range for globular proteins, mol. wt.*
P-2	1.6	3.8	200—2000
p-4	2.6	6.1	500—4000
p-6	3.2	7.4	1000—5000
P-10	5.1	12.0	5000—17,000
P-20	5.4	13.0	10,000—30,000
P-30	6.2	14.0	20,000—50,000
P-60	6.8	18.0	30,000—70,000
P-100	7.5	22.0	40,000—100,000
P-150	9.0	27.0	50,000—150,000
P-200	13.5	47.0	80,000—300,000
P-300	2.0	70.0	100,000—400,000

Preparation of Gels for Chromatography

Both dextran and polyacrylamide gels must be hydrated before they can be used in a column. The water-regain values listed in Tables show the minimum quantities of water necessary to hydrate the gels, although 5-10 minutes this amount should be used to prepare gels. Those gels with a small pore size can be hydrated in a few hours at room temperature; however, those with large pore sizes require up to 24 hours. The process of hydration or swelling can be accelerated by placing the slurry of the gel in a boiling water bath. Boiling also helps to eliminate air bubbles which may become trapped in the slurry and which will cause an uneven flow of the column.

Selection of Columns

Selection of a column for gel chromatography is governed by its use. For example, short columns can be used for desalting purposes, whereas long columns may be necessary for critical separations. Very narrow columns should be avoided because they lead to wall effects in which the solvent has a tendency to flow faster in a thin layer along the glass wall. Wall effects can be minimized by coating the interior surface of the column with a solution of 1% dichlorodimethylsilane in benzene. The solution is added to the column at 60°C and, after standing for a few minutes, is decanted. The benzene is then evaporated in a drying oven and the entire process repeated again. Such a double coating is generally satisfactory for several months. In choosing a column one should also avoid those with large chambers at the outlet and those with fritted glass or porous plastic discs; these have a tendency to become clogged with the fine beads of gel.

Packing Columns

Packing columns with gels of low pore size rarely causes any difficulties. Thus, gels of Sephadex G-10 to G-50 and of Bio-gel P-2 to P-10 can be packed without special precautions. Usually the column is partially filled with the buffer solution to be used and the slurry containing the gel is poured into the top of the column. As the bed begins to form, the outlet at the bottom of the column is opened. Additional slurry can be added until the desired height of gel is obtained. The gel is then rinsed with several column volumes of buffer at a flow rate similar to that used for elution of the sample. The gel must never be allowed to dry out.

Gels of high pore size present more problems in packing. Improperly packed columns will not perform satisfactorily if they perform at all. A small amount of buffer is first poured into the

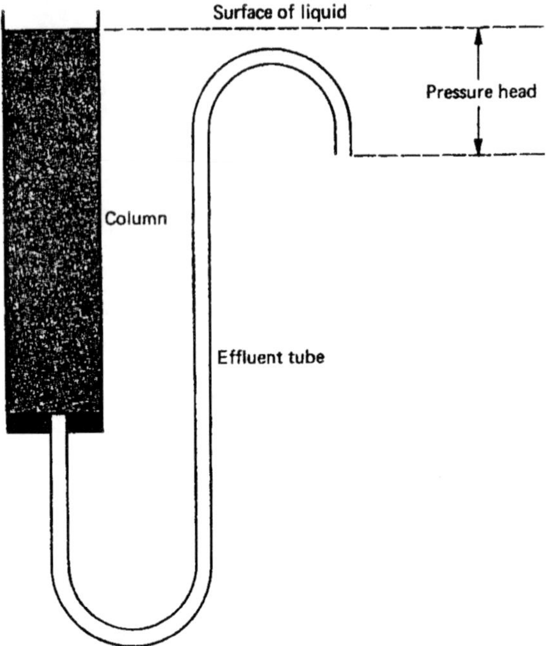

Fig. 14.14. Apparatus for packing a column with gel sorbent of high water-regain properties.

column which is positioned vertically with the outlet tubing held above the top of the column. The slurry of swollen of gel is poured into the column, preferably down a glass rod. It is desirable to pour all of the slurry to produce a complete bed into the column at one time. An extension tube mounted on top of the column may be necessary to hold the excess slurry. The slurry in the column is stirred gently with a glass rod to remove any air bubbles. The outlet tube is next lowered below the surface of the liquid in the column and the buffer is allowed to flow slowly out. A slowly rising horizontal surface of the gel indicates uniform packing. The outlet tubing is then lowered gradually until the operating pressure to be used in the column is reached. This pressure should not exceed 100—150 mm in the case of the highest porosity gels but can be increased for gels of lower porosity. High pressure compresses the particles and impedes normal flow.

Repacking

Most columns containing gels can be reused without any special regeneration. However the column must be cleansed of unwanted solutes by sufficient washing with water or buffer.

After lengthy use some columns flow more slowly because of compression of the gel bed. Flow rates can be increased by backwashing or repacking the gel.

Sample Application and Elution

Several methods have been used for application of samples on gel beds. The simplest involves adjusting the buffer to the top of the bed level followed by the careful addition of the sample to the top of the bed using a suitable pipet. After allowing the sample to drain into the gel, a small amount of buffer is added to wash the sample completely into the gel. The buffer can then be added to the desired level, the column connected to a buffer reservoir, and the elution begun. For analytical purposes the sample should be applied in as small a volume as possible. When the width of the zone of solute is not as critical, such as is the case in desalting, the size of the sample is less important.

Solutes that are uncharged can usually be eluted with distilled water, but those with charged groups usually require a developer with an ionic strength of 0.02 or more. The developer is necessary for best results because most gels contain a small number of charged groups which result in some adsorption of solutes. For best results the rate of elution should be constant. This can be accomplished manually or by using a constant-pressure flask such as a Mariotte flask.

Elution of solutes from columns of gels can also be accomplished by reversed flow. In this method the sample is applied at the lower end of the chromatogram and solutes are washed upward through the gel and out the top of the column. This method is most useful where gels of the highest porosity are being used in order to increase their usually very slow flow rate.

Determination of Molecular Weights

The elution volume of globular proteins and other macromolecules is determined largely by their molecular weight and is a linear function of the logarithm of the molecular weight. In practice a suitable standard of known molecular weight is first applied to the column for calibration purposes. A standard as closely related as possible to the substance being studied should be chosen. Equations for the calculations of molecular weights of globular proteins have been derived for several different dextran gels. These are

Sephadex G-75: $\log_{10}$ mol. wt. = 5.624—0.752 (V/v)

Sephadex G-100: $\log_{10}$ mol. wt. = 5.941—0.847 (V/v)

Sephadex G-200: $\log_{10}$ mol. wt. = 6.698 —0.987 (V/v)

where V is equal to the elution volume and v is equal to the holdup volume. The holdup volume can be determined by measuring the volume of solvent required to chromatograph a coloured compound of very high molecular weight. Substances such as Blue Dextran 2000 (Pharmacia) with a molecular weight of 2.0×10^6 are available commercially for this purpose.

Thin-layer Chromatography

Thin-layer chromatography has recently won widespread favour in biochemistry. It has been used extensively for qualitative, quantitative, and preparative purposes and can be applied successfully to the separation and identification of a wide range of organic compounds. TLC offers the advantages of quick separation, high sensitivity, simple equipment, and ready adaptability. TLC has replaced paper chromatography as the most useful analytical chromatographic method.

In TLC the stationary phase is attached to a suitable support such as glass or plastic film. The stationary phase can be a finely divided powder acting as an adsorbent or as a support for a liquid. Many powders are used as sorbents, but silica gel and cellulose are the most common. Silica gel can function as an adsorbent following heat "activation" and can be used quite successfully to separate many of the less polar compounds. Silica gel and cellulose, when properly coated with a liquid film, act as supports for partition chromatography of a wide range of polar and nonpolar compounds. After the sample is applied to the sorbent near one end of the thin-layer plate, the chromatogram is developed in an ascending fashion by immersing the lower margin of the plate in solvent inside an appropriate chamber. After the solvent has risen a suitable distance, the TLC plate is removed from the solvent, dried, and the separated spots visualized by any one of a variety of methods.

Operation of TLC System

Supporting Plates

Plates used for the support of chromatographic thin layers may be of glass, metal, or suitable plastic. Glass plates can be used in a number of sizes ranging from microscope slides to large plates. Microscope slides are the simplest and can be used to separate upto four components in a few minutes. More complex mixtures require larger plates. The latter are also more useful for preparative TLC where thicker layers of adsorbent must be applied. Flexible sheets of polyester film can also serve as bases for thin layers. Although they are more difficult to coat than glass plates, polyester film does not break if dropped.

Precoated plates of glass and film are now available commercially. Polyester film plates cannot be used with all system solvents, however.

Sorbents

In theory any stationary phase used in column chromatography can be used in TLC. In fact, TLC has been referred to as open-column chromatography. In practice true adsorbents and partition agents usually comprise the stationary phase or layer.

Adsorbents for TLC are much smaller in practice size than those employed for column chromatography. Silica gel is the most common adsorbent although basic alumina is of greater use for the separation of bases. Silica gel and cellulose are most commonly employed as partition agents although diatomaceous earth is sometimes used.

The layers of sorbents are usually held in place on the plates by various binders that are mixed in with the sorbent prior to layering. The most common binders is plaster of paris (mixed with sorbents at a level of 10 to 15%). Starch and certain organic polymers are also used as binders and result in much harder layers than those obtained with plaster of paris.

Preparation of Layers

Thin layers are usually prepared by layering a film of sorbent and water as a slurry on a clean plate and allowing it to dry. The thickness of the slurry should vary with the sorbent used and with the requirements of the chromatographic operation. Slurries that are too thick or too thin should be avoided. A consistency of pea soup is usually best. Most manufacturers recommend optimum water/sorbent ratios for individual products. The slurry is usually made by shaking the sorbent with water in a flask or by mixing in a blender.

Microscope slides can be coated simply by immersing the slide into the slurry, withdrawing it carefully, and allowing it to dry in a horizontal position. Larger plates can be layered by using commerically available apparatus or by employing the technique. In the latter method the tape is placed along the margins of the two opposite sides of the plate. As many layers of tape as necessary to obtain the desired thickness of sorbent are used. The slurry is poured out along one untaped edge of the plate and is spread out with a glass stirring rod held across the plate with both ends resting on the tape. After the slurry has been spread out evenly it can be allowed to dry for about

30 min. Following removal of the tapes the sorbent can be activated by heating at 110°C for 1 hour in an oven.

Prepared plates are available from commercial sources in a wide variety of sorbents and supporting plates. Layers are usually of excellent quality.

For partition chromatography thin layers can be impregnated with a number of different liquids. When water is chosen, the impregnation can be combined with drying the slurry by allowing it first to dry at room temperature followed by heating at 105°C for 10 min. Polar and nonpolar liquids can be impregnated into the layer by dipping the layers into the impregnating liquid after the layer has been dried to remove excess water. Usually a solution of the liquid in some volatile solvent is employed, such as 20% formamide in acetone. The impregnated layers are then removed carefully from the solution and are set in a hood to dry. No heating is used.

Substances called phosphors are sometimes included with the slurry of sorbent. These usually emit visible light when irradiated with ultraviolet light. Many solutes, particularly those that are aromatic or contain conjugated double bonds, will quench the emission of visible light. The presence of these solutes is indicated under ultraviolet irradiation by dark spots in a bright field.

Solvents

Solvents for TLC may be chosen on the basis of the guidelines discussed in the section on chromatographic theory and in the section on column chromatography. In partition systems the relative solubility of the solutes in the two liquid phases must be considered. Typical solvents systems for separating polar solutes include (a) isopropanol/ammonia/water (9 : 1 : 2); (b) butanol/acetic acid/water (4 : 1 : 5); and (c) phenol/water (4 : 1).

Solvents for TLC employing adsorbents may be chosen by running a preliminary series of tests with samples on microscope slides or small strips of film. The chromatogram is developed by different solvents in small scaled bottles. The best solvent systems may be further modified to yield even better results in a second series of tests. Then the best system can be chosen for more sophisticated separations. Optimum resolution is obtained near the center of the chromatogram.

Remember to consult the literature for help in choosing solvent systems.

Application of Samples

The first step in applying a sample to a TLC plate for one dimensional chromatography is to draw a very light pencil line with a straight edge 2 cm from one margin of the plate. Light pencil hash marks are next made at convenient intervals along the straight line (usually 1.5 to 3 cm between marks).

The samples should be applied at the hash marks with a fine capillary tube or a micropipet. Best results are obtained with 1- to 10-microliter (ml) samples and the spots on the layer should be as narrow in diameter as possible. The best procedure is to touch a very small drop to the sorbent and wait for it to dry before superimposing the next small drop. If the sample is applied in a volatile solvent, drying usually takes place very quickly. When less volatile solvents are used drying can be accelerated by blowing a stream of warm air across the plate from a hair dryer. In any event the TLC plate must be absolutely free of solvent before development is begun as its presence may adversely affect the separation.

For two-dimensional chromatography one sample is applied to one corner of the TLC plate 2 cm from each of the nearest two margins.

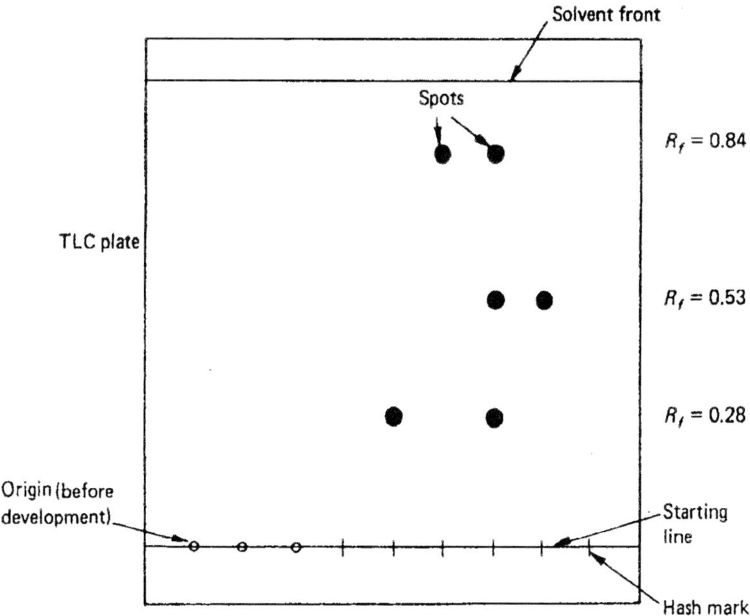

Fig. 14.15. A hypothetical thin-layer plate showing the origin, solvent front, and spots visualized after development.

Development

Thin-layer chromatograms can be developed in a wide variety of chambers. Microscope slide plates can be accommodated in a small jar saturated with the solvent. Larger TLC plates can be developed in chambers. The larger chambers require a period of time for the solvent to saturate the atmosphere of the chamber. This period may be shortened by lining the walls of the chamber with filter paper moistened with the solvent. Sandwich plates, which are frequently used as TLC chambers, have the advantage of not requiring pre-equilibration with solvent because their void space is so small. However, one must be sure that the edges of these are absolutely tight. Evaporation of solvent from around the edges adversely affect the flow of solvent along the margins of the TLC plate. The chamber should be placed in an area free of sharp temperature fluctuations.

When the chamber has been equilibrated, the solvent is adjusted to a depth of about 1cm and the TLC plate is placed in the chamber

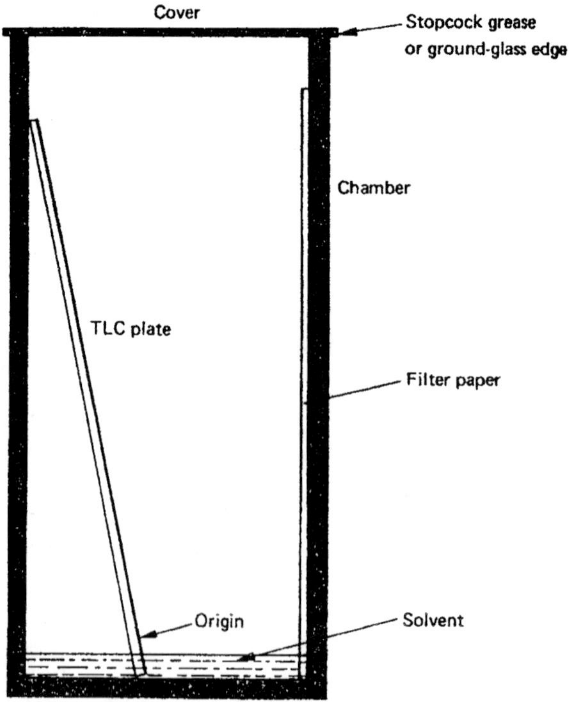

Fig. 14.16. Diagram of a thin-layer chromatography plate in a development chamber showing the location of the solvent and filter paper soaked with solvent to saturate the atmosphere of the chamber.

with the end nearest the samples standing level in the solvent. The cover of the chamber is set in place and development allowed to proceed. The progress of the chromatography can be observed by following the movement of the solvent front visually. Development can be terminated when the solvent front is within about 2 cm of the upper margin of the plate. One must be certain that the level of solvent in the bottom of the chamber does not fall below the sorbent on the plate. On completion of the development the chromatogram is removed from the chamber, the solvent front is delineated with a soft pencil, and the plate is dried in a hood.

With two-dimensional chromatograms the TLC plate is developed in one direction by the first solvent and dried; then, after rotating the plate 90° in a clockwise direction, it is developed by the second solvent.

Detection

Chromophoric solutes of intense colour present no problems in visualization on TLC plates. Other solutes can be detected by their fluorescence under uv light or by their radioactivity. Most solutes, however, must be visualized by their reaction with some reagent to produce a coloured complex or by their quenching of light produced by the fluorescence of phosphors that had been included with the sorbent. Reaction to produce a coloured complex has the more general applicability. The reaction can be accomplished either by spraying a suitable reagent onto the chromatogram or by placing the plate in a chamber that is saturated with the vapour of a reagent. Solutes will appear as spots on the surface of the chromatogram.

Spray reagents are considered to be universal or specific. Universal reagents are those which produce a coloured spot with any organic compound. The most common of these are H_2SO_4 and I_2. The H_2SO_4 is used as a spray of concentrated acid either alone or containing 5% HNO_3. After heating the plate for a few minutes at 100 to 110°C, the organic compounds appear as black spots. The I_2 is most commonly used as a vapour in a closed glass chamber. A few crystals of I_2 are placed in the chamber and allowed to vapourize. The TLC plate is then exposed to the I_2 vapour in the chamber until the organic compounds appear as brown spots.

Specific reagents usually react with some particular classes or types of compounds. A few of the specific spray reagents and their uses are shown in the Table 14.7. This list is by no means exhaustive and the student should consult the references under Thin-Layer Chromatography for a description of additional reagents.

Table 14.7. Specific Spray Reagents for Thin-layer Chromatography.

Reagent	*Preparartion*	*Procedure*	*Use*	*Colour*
Aniline phthalate	Dissolve 0.93 g aniline and 1.66g o-phthalic acid in 100 ml water saturated n-butanol	Spray soln. and heat for 10 min. at 105°C.	Reducing sugars	Various colours
Anisaldehyde in H_2SO_4	Mix 0.5 ml anisaldehyde and 0.5 ml conc. H_2SO_4 in 9 ml 95% ethanol	spray soln and heat for 10 min at 10—100°C	Carbohydrates and steroids	Shades of blue
2,4-dinitrophenyl-hydrazine	Dissolve 0.1 g 2, 4,-di-nitrophenyl hydrazine in 100 ml 95% ethanol; add 1 ml conc. HCl	spray soln. and let dry at room temp.	Aldehydes and ketones	Yellow to red
Ninhydrin	Disolve 0.3 g ninhydrin in 100 ml n-butanol; add 3 ml glacial acetic acid	Spray soln. and heat for 10 min at 110°C	Amino acids, amino sugars, aminophos-phatides	Shades of violet except proline and hydroxy-proline, yellow.
Rhodamine B	Dissolve 0.5 Rhodamine B in 100 ml 95% ethanol	Spray soln. and dry at room tem.	Lipids	Dark violet spot on pink background in Uv*light
Antimony chloride in acetic acid	Mix equal weights of antimony trichloride and glacial acetic acid	Spray soln. and heat for 5 min. at 95°C	Steroids, lipids light.	Various colours in uv

uv*-ultraviolet,

Various sprayers are available commercially for TLC spray reagents or they may be constructed in the laboratory. Most reference books on TLC contain instructions for their fabrication. In addition, pressurized cans of the more common spray reagents are available commercially.

To spray a TLC plate, first place it nearly vertically in a hood in front of some heavy wrapping paper or corrugated paper. The sprayer should be held 8 to 12 in. from the plate. Spraying should be done in sweeping motions across the plate. The plate must ***never*** be soaked since this will cause the spray to run on the plate and possibly move the solute.

When the solutes on a TLC plate have been visualized, the best procedure is to outline the spots with a soft pencil. This is absolutely necessary with those solutes detected using ultraviolet light and is advisable with others since some spots have a tendency to fade and disappear with time. A record of the chromatogram can be made either by photography or by using a suitable photocopier. Colour slides or prints are recommended to preserve a record of coloured spots. Otherwise photocopying is most convenient. If equipment for photorecording is not available, a tracing of the chromatogram may be made for the laboratory notebook.

The location of the spots on a TLC plate is defined by the R_f value. The center or density of the spot is used for purposes of measurement. This presents little difficulty with circular spots of symmetrical distribution. However, in the case of spots with no symmetry, such as those exhibiting tailing, a little common sense is necessary to locate the center of density.

The R_f values of some common amino acids in several different solvent systems are shown in Table 14.8.

Errors

Some of the errors encountered in column chromatography occur also in TLC. Overloading, tailing, and changes in temperature are three of the problems common to both techniques, and their solutions are the same. However, in TLC, in addition to overloading of particular solutes, a physical overloading may also occur. This usually happens when the solutes to be separated constitute a small percentage of the total sample and the bulk of the contaminants are practically insoluble in the solvent system. The effect is to encase the solutes in the contaminant, from which they are only slowly extracted by the solvents; the solutes, therefore, will be spread out over a long portion of

Table 14.8. R_f values of common amino acids in various solvent systems

	R_f			
Amino acid	*1*	*2*	*3*	*4*
Alanine	0.47	0.27	0.29	0.40
β-Alanine	0.33	0.27	0.30	0.29
Aspartic acid	0.55	0.21	0.06	0.06
Arginine HCl	0.04	0.08	0.19	0.07
Cysteic acid	0.69	0.14	0.04	0.21
Cystine	0.39	0.16	0.12	0.22
Glutamic acid	0.63	0.27	0.10	0.15
Glycine	0.43	0.22	0.24	0.34
Hystidine HCl	0.33	0.06	0.32	0.42
Hydroxyproline	0.44	0.20	0.38	0.31
Isoleucine	0.60	0.46	0.49	0.58
Leucine	0.61	0.47	0.48	0.58
Lysine HCl	0.03	0.05	0.09	0.11
Methionine	0.59	0.40	0.49	0.60
Phenylalanine	0.63	0.49	0.55	0.60
Proline	0.35	0.19	0.50	0.30
Serine	0.48	0.22	0.20	0.31
Threonine	0.50	0.25	0.26	0.40
Tryptophan	0.65	0.56	0.63	0.58
Tryosine	0.65	0.47	0.47	0.56
Valine	0.55	0.35	0.40	0.51

chromatogram. When such results are obtained, the sample must be purified before chromatography is attempted.

Another error occassionally encountered in TLC, particularly in partition systems, results from a failure to control the amount of water in the liquid stationary phase of the TLC plate. The source of the error rests in the use of solvent systems that are saturated with water. Chromatograms operated under these conditions never have good reproducibility. The best solution to the problem is to use solvent systems which are unsaturated with respect to water.

Lastly, one must be sure to attain equilibrium of the atmosphere in the chromatographic chamber before beginning development. Failure to do so results in a serious lack of reproducibility.

SPECIAL TECHNIQUES

Preparative TLC

TLC can be used as a preparative technique, and quantities of solute up to 100g have been separated successfully under optimum conditions. Usually, however, quantities less than 2 g are chromatographed.

In preparative TLC the sample to be chromatographed is applied in a streak by drawing a capillary or a pipet containing the sample along a thin line 2.5 cm from the lower edge of the plate. The sample is developed in a direction perpendicular to the streak. The solutes must be visualized by some nondestructive means, following which the adsorbent containing the solute is scrapped off the glass plate. The solute is then eluted from the adsorbent.

Good sorbent layers for preparative TLC are difficult to prepare since they must be thicker than 1 mm. Unless the use of a large quantity of these TLC plates is anticipated, it seems advisable to purchase commercially prepared thick layers. These products generally are of high quality.

The spotting of the sample for preparative TLC is the most critical step, since volumes upto 2 ml must be applied without disturbing the layer. Many devices have been developed to accomplish this task. They are described in the literature and in commercial catalogs.

Development of plates for preparative TLC is similar to that for normal TLC except that the best solvent system should be ascertained by preliminary TLC on thin-layers. Following development, the chromatograms are dried in a hood with blowing air or with a current of nitrogen.

Visualization of preparative TLC plates is best accomplished by means of ultraviolet light on phosphor-containing layers or by exposure to vapours of I_2. In addition solutes may be detected by spraying a suitable reagent along one edge of the chromatogram (and sacrificing that portion) or by pressing a tape across the suspected bands and then spraying the tape with an appropriate reagent. The tape can then be used as a key to locate the bands on the plate.

Bands of solute can be scraped off the glass plate along with the adsorbent onto aluminium foil or powder paper. The solute is then eluted from the adsorbent with a suitable solvent. This last step can best be carried out by placing the scrapings in a sintered glass funnel or in a cone of filter paper in a glass funnel and washing the scrapings several times with the solvent.

Quantitative TLC

TLC plates can be assayed quantitatively, but the best precision one can expect has an error of 5 to 10%. Application of samples and development require particular care and attention to detail. A preliminary development of the plates before spotting the samples is often recommended to purify the adsorbent.

Assay of solute quantities can be performed on the thin-layer or following elution. Assays in place include the measurement of spot size and the measurement of spot density. To measure density, a photoelectric densitometer is used to scan the spots of solute and measure their density. Appropriate standards are measured also and the unknowns compared to these. Sorbent layers of uniform thickness are absolutely necessary. Assay of spot size is simple and does not require any instrumentation, although its results are not usually as precise. It is based on the observation that the square root of the area of a spot is directly proportional to the logarithm of the weight of the solute as in equation:

$$\sqrt{A} = m \log_{10} W + c$$

where A is the area of the spot, W is the weight of the solute, and m and c are proportionality constants for the substance being measured as determined from a calibration with a pure sample of the compound.

In addition to the spot-size and density methods, the quantity of solute can be measured in place by measuring radioactivity or by using a dilution method. In the latter method a series of dilutions of a standard solution of the compound to be assayed are chromatographed and the limit of detection on TLC is determined. A series of dilutions of the unknown are chromatographed and the solution at which the solute is just detectable is determined. The quantity of solute in the unknown is then calculated from the limit of detection and the dilution factor.

Quantitative TLC can also be carried out by eluting the solute from the TLC plate with an appropriate solvent and then determining the concentration spectrophotometrically. The recommended procedure is to chromatograph the unknown with a series of standard concentrations, visualize the spots, delineate the spots into rectangles of identical size, scrape the adsorbent within each rectangle off the plate with a razor blade onto a sheet of aluminium foil, transfer the adsorbent containing the solute to a sintered glass funnel, and wash the adsorbent with an appropriate solvent into a calibrated container.

After bringing the extracts to a known volume with solvent, their spectral adsorbance can be determined in a spectrophotometer. The value for the unknown is then calculated on the basis of the adsorbances of the standards and appropriate blanks.

Gas Chromatography

Gas chromatography is a form of chromatography which employs an inert gas as the mobile phase and, generally, a liquid as the stationary phase. The general principles when using the liquid stationary phase are the same as those which apply to partition chromatography; in this form the technique is often referred to as gas-liquid chromatography (GLC).

Gas chromatography may be used to separate any compounds that can be vapourized without decomposition. This restriction is the limiting factor in choosing gas chromatography as an analytical tool, although nonvolatile compounds can sometimes be converted into highly volatile derivatives. Gas chromatography has two advantages over other forms of chromatography : separations can be made in a much shorter period of time, and the technique can be made relatively precise quantitatively.

In GLC a column (either metal or glass) is placed with some inert packing which is coated with a layer of some nonvolatile hydrocarbon such as paraffin oil. The sample is introduced with the inert gas used as the mobile phase and is heated to volatilize its components. The vapourised components of the sample will dissolve in the liquid stationary phase but, because they are volatile, will equilibrate with the gaseous mobile phase, each with a definite

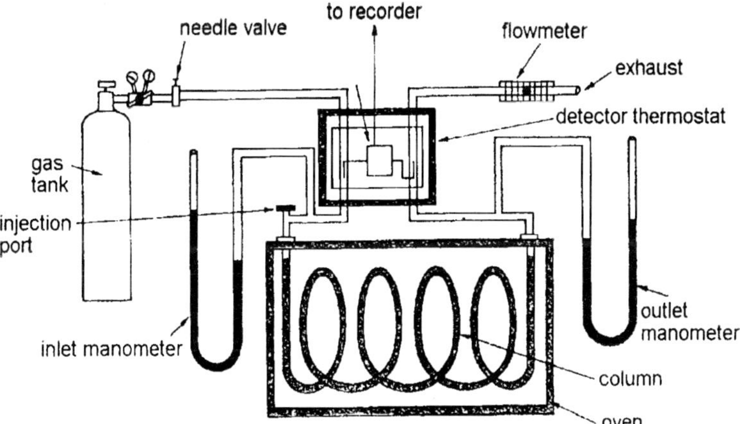

Fig. 14.17. Diagram of a simplified gas chromatography apparatus.

distribution coefficient. Thus, the conditions for a chromatographic separation are satisfied and the compounds will pass through the column at a rate determined by their distribution coefficients. Upon reaching the end of the column the compounds are detected, usually by means of thermal conductivity with a detector connected to a suitable strip-chart recorder. Thermal conductivity is convenient because most organic vapours have much lower thermal conductivities than carrier gases such as helium.

As mentioned earlier samples can be developed by gas chromatography much faster than those methods that employ a liquid mobile phase. The rapidity of development is related to the differences between diffusion rates of compounds in a gas and in a liquid. Diffusion rates are approximately 10,000 times greater in air than in water. Since the rate of development of a chromatogram is limited by the rate of the diffusion equilibrium between the stationary and mobile phases, the higher diffusion rates of the gas allows gas chromatograms to be developed much faster than those with a liquid mobile phase. Development times of gas chromatograms are as much as 1000 times faster than those with liquids.

Column and Stationary Phases

The fact that gas chromatography must be performed at elevated and controlled temperatures requires that the chromatographic column be placed in an oven. Thus, the column must often be coiled so that it can be accommodated even though a straight column would be preferable. Columns may be constructed of plastic, aluminium, copper, glass, or stainless steel. Stainless steel is most often used although glass is required for certain separations.

Columns are of two types; packed and open-tubular. Packed columns, as their name implies, are packed with a solid support for the liquid stationary phase. *Open tubular columns* are usually prepared by coating the wall of the column with the liquid stationary phase. The internal diameters of columns vary greatly according to their use. Preparative columns may be as large as 4 in, whereas some open-tubular columns are as narrow as 0.01 in.

The length of a column is governed by two main factors : the degree of separation desired and the pressure drop from the inlet end of the column to the outlet end. Generally, the longer to column the better will be the separation. The pressure drop, however, is also greater as the length of the column increases although it is also affected by the particle size of the support. When the pressure drop is large, the column is more difficult to operate. Thus, a compromise in column length must occasionally be made. Most packed columns are 4 to 15 ft in length.

Stationary Phases

Diatomaceous earth is usually used as support for packing a GLC column. Other materials such as crushed Teflon and fine glass beads have also been employed successfully. In open-tubular columns the surface of the tubing provides the support for the liquid stationary phase. Such columns are much longer than packed columns in order to obtain the necessary resolution of solutes.

Stationary phases must be chosen on the basis of the nature of the sample to be separated. Most of the references on gas chromatography, as well as commercial catalogs of GLC supplies, provide information to help make a proper choice. For compounds of biochemical interest silicone grease, paraffin oil, or polyethyleneglycols (Carbowaxes) are useful.

Preparation of Columns

Several methods are available for applying the stationary phase to the support. Each involves first, dissolving an amount of the stationary phase in a volatile solvent ; second, mixing the solution with the solid support; and third, evaporating the volatile solvent. The stationary phase usually comprises 5 to 20% by weight of the solid support.

The column is packed by slowly filling a suitable piece of tubing with the coated support after plugging the other end with glass wool. The tubing must be tapped often during the filling to pack the column evenly. When the column has been packed, the open end is plugged and the tubing is coiled to fit inside the oven.

Open-tubular columns are prepared by using an inert gas to force a solution of the stationary phase in some volatile solvent through the tubing. As the solution is forced through some of the solvent evaporates, leaving a film of the stationary phase on the wall of the column. Such a procedure requires the use of a specific amount of solution of known concentration.

All new columns must be *conditioned* before use. This is necessiated by the fact that new columns will bleed some stationary phase which will foul the detector. Conditioning is done by running carrier gas through the column, with the detector disconnected, at a temperature about 20°C higher than that at which the column will be operated. Conditioning normally takes about 10 to 12 hours.

Mobile Phase and Operation

Carrier Gas

The gas which is used as the mobile phase must be inert with respect to the components of the chromatographic system as well as to the components of the sample. Helium is the most commonly used carrier although chromatographs that have flame-ionization or electron-capture detectors require nitrogen or argon.

Temperature

The temperature of operation is one of the most important aspects of a gas chromatographic column. Since the R_f values of compounds rise with temperature, it becomes a useful variable. The selection of temperature is largely a matter of experimentation, although the preliminary choice can be made by starting with a temperature a few degrees below the boiling point of the presumed major component of the sample. A suitable compromise of temperature must often be made since columns at high temperatures operate more efficiently but separate less effectively. A satisfactory alternative can be made by *temperature programming*, the GLC equivalent of gradient elution. In temperature programming a predetermined pattern of raising the temperature during the development is established. The net result is to separate readily the more volatile solutes at the lower temperatures and then later develop the less volatile solutes at higher temperatures.

Detectors

Detectors for gas chromatography are of two types—integral and differential. The integral type indicates a sum of a property related to the mass of each compound developed by the column. The *differential type* indicates the instantaneous concentrations and is the type almost always encountered in commercial instruments. The detectors described below are of the latter type. The two most common detectors are the thermal-conductivity cell and the flame-ionization detector. The sensitivity of these detectors is about 2 mg and 10^{-5} mg, respectively.

The thermal conductivity cell consists of a double arrangement of fine platinum or alloy wires, maintained at a constant temperature, in the axis of a cylindrical channel. One channel is used as a reference channel with carrier gas only passed through it, whereas the other channel is the measuring channel used for the effluent of the column. The thermal conductivity cell is based on the fact that the temperature of the electrically heated wire depends not only on the electrical energy supplied but also on the thermal energy lost by radiation, conduction, and convection. The nature of the gas surrounding the wire will affect the loss of thermal energy by the wire. Thus, every variation in the composition of the gas changes the temperature of the wire and consequently its resistance. In practice, the column effluent containing the solute and the carrier gas has a lower conductivity than the carrier gas alone. Thus, the wire will heat when in contact with the solute, and the resistance of the wire will increase in comparision to that of the reference channel. The increase will result in a decrease in the flow of a current across the wire which then unbalances a Wheatstone bridge. The current required to balance the Wheatstone bridge is then measured by a recorder.

Another type of detector, with considerably better sensitivity than the thermal conductivity cell, is the flame-ionization detector. This detector employs a flame produced by burning hydrogen mixed with air. The effluent from the column is mixed into the hydrogen-air mixture before the gas is burned and will produce carbon dioxide as the result of the combustion of any organic compounds it may contain. The carbon dioxide molecules are then ionized and the ion concentration is measured by converting the changes in current into changes of potential. These changes can be highly amplified to produce a great increase in sensitivity. Since the response to the detector is linear, it is excellent for quantitative assays.

Operation

A constant flow of carrier gas is required for satisfactory operation of the gas chromatography column. Needle valves are available for this purpose and can be inserted into the line between the column and the diaphragm valve of the gas tank. The rate of flow can be monitored by using a bubble flowmeter on the exit port of the apparatus.

Selection of the best flow rate depends on the diameter of the column and to some extent on its individual idiosyncracies. Selection of a proper flow rate can be aided by chromatographing two closely boiling liquids such as benzene and carbon tetrachloride and adjusting the flow rate to produce good resolution. If the flow rate is too fast, resolution will be poor; if it is too slow, considerable tailing will result.

As discussed earlier the choice of a temperature for the operation of a gas chromatograph is largely a matter of experience. If the boiling point of the major component of the sample is known choose a temperature a few degrees lower. Many commercial gas chromatographs come equipped with a temperature programmer or can be modified with a temperature-programming accessory. This is a very desirable feature for a chromatograph. Usually the increase in temperature is linear and can be made at some chosen rate.

The sample is injected into the chromatograph through the injection port by means of a microsyringe. The injection port is covered by a silicone rubber septum which is punctured by the needle of the syringe. In order to vapourize the sample immediately the temperature of the injection port should be about 10°C higher than that of the column.

The amount of sample injected depends on the column and on the type of detector being used. With thermal-conductivity detectors the initial injection should be about 10 microliters (ml) of a 1% solution of the sample. When a flame-ionization detector is used then 1 ml of a similar solution should suffice. The injection of the sample should be done as quickly as possible since a slow injection will result in smearing the solutes through the column. On puncturing the rubber septum the plunger of the microsyringe must be held firmly to keep it from being pushed out by the gas pressure of the system.

Chromatograms are usually recorded by means of a stripchart recorder which is connected to an amplifier that increases the strength of the signal produced by the detector. Most recorders for gas chromatography have a 1 millivolt (mV) fullscale deflection. In most commercial instruments the signal may be controlled by an attenuator which can lower the signal in steps if it is too large to fit entirely on the chart paper. This is especially important where the highly sensitive flameionization detector is being used.

Identification of Components

Solutes in a sample assayed by gas chromatography may be identified by means of several different methods. The two most commonly used are the addition of the suspected compound to the sample and the use of retention times.

Retention time is analogous to the R_f value of other forms of chromatography. It is the time that is required for a compound to emerge from a column as compared to the time required by a standard air peak. An example of a recording of a chromatogram is illustrated in Figure. Although each compound should have the same retention time in a column at a given temperature and gas flow rate, in reality the time will vary somewhat. This difference is caused by the variability of other solutes in the sample which will affect the retention time of their neighbours and by the "age" of the column.

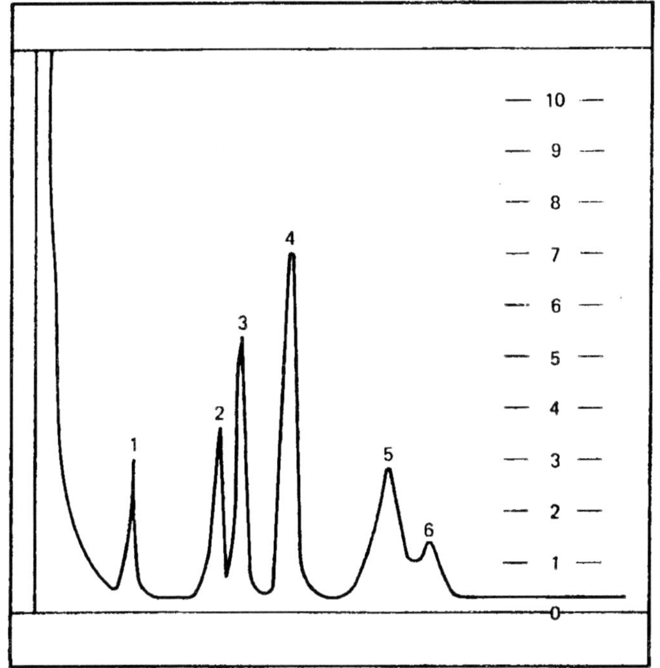

Fig. 14.18. Recorder trace of fatty acid methyl esters separated from a mixture by gas chromatography.

Compounds in a homologous series can be identified by plotting their retention times logarithmically as shown in Figure, where the values obtained for the methyl esters of fatty acids illustrated in Figure are plotted as a function of chain length.

Identification of a compound can be aided by adding a small amount of a known compound to the sample and chromatographing the mixture. If the unknown is the same as the known, then the height of the peak should increase and no additional peak or shoulder should be evident. If the unknown is different from the standard, then an additional peak or shoulder will appear on the recorder trace.

Quantitative Assays

The concentration of a compound separated by gas chromatography is usually determined by measuring the area of its peak. This area may be measured by (*a*) integration, (*b*) planimetry, (*c*) triangulation, and (*d*) cutting and weighing.

The most primitive and least precise of these is cutting and weighing which involves cutting out the peak from the recorder sheet with scissors

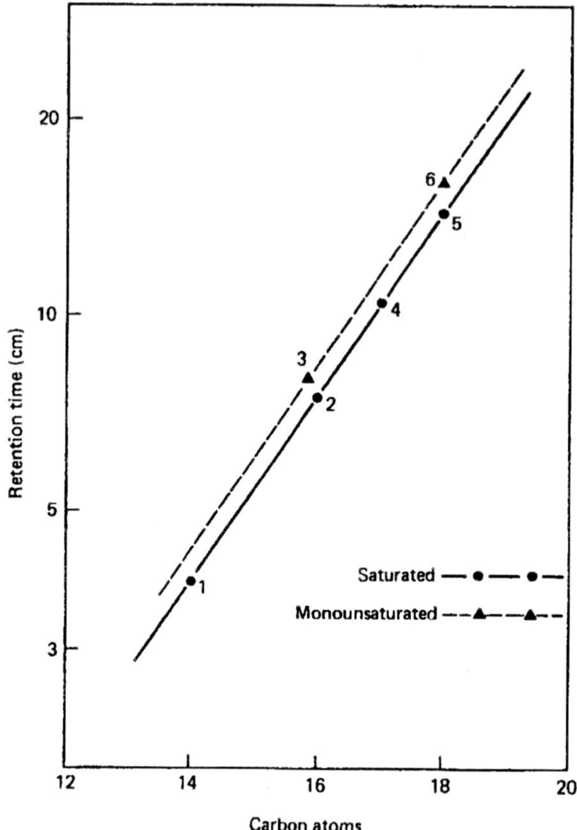

Fig. 14.19. A graph illustrating the relationship between the number of carbon atoms in homologues of fatty acids and the logrithm of their retention time in a gas chromatograph.

and weighing the paper on an analytical balance. The weight of the unknown can be calculated by comparision to the weight of a standard. This method is affected by the accuracy of the cutting and by variations in the thickness and moisture content of the chart paper.

Direct integration is the most precise method of measurement. This may be accomplished by mechanical or electronic means. The electronic method is the more precise (and the more expensive) of the two. These integrators are available from commercial sources and their operation is described by the manufacturers.

One fairly simple method of integration can be accomplished with a ruler, pencil, paper, and a calculator. This method is based on the observation that, for some compounds, the total area of a peak is a function of the product of the peak height and its retention time. A standard of known concentration is included for calibration purposes.

Advantages of Gas chromatography

1. *Speed.* Once the sample is prepared for injection into the gas chromatograph, most analyses require between 5 and 60 min to determine 8-10 compounds. There are a few uncommon procedures which require upto 24 hr for final elution of the last component ; but for the most part, gas chromatography is a rapid technique after the lengthy purification process is completed.
2. *Wide temperature range.* Substances may be measured in a temperature range from -200°C to + 1000°C because freezing and boiling-point problems are eliminated when using gases.
3. *Recovery of sample.* It is possible to insert a device called a splitter into the system so that only a portion of the sample is detected by the detector; the remaining material is vented to the outside, where it may be collected for further analyses.
4. *Sensitivity.* The gas chromatograph is an extremely sensitive analytical tool capable of determining substances in the nanogram range.
5. *Small sample sizes required.* Less than 1μl of material may be injected into the gas chromatograph.
6. *Automation.* It is possible to obtain gas chromatographs with automatic setting devices so that the recording will be accomplished automatically and the analyst need not be in attendance at the instrument for the entire run.

Disadvantages of Gas Chromatography

1. Lengthy preparation time for samples.
2. Contamination of detectors in the case of impure samples.
3. Time-consuming column preparation.

Index